£30

Reproduction and

Development of

Marine Invertebrates

Edited by

W. Herbert Wilson, Jr.
Department of Biology, Colby College,
Waterville, Maine

Stephen A. Stricker
Department of Biology, University of
New Mexico, Albuquerque, New Mexico

George L. Shinn
Division of Science, Northeast Missouri
State University, Kirksville, Missouri

The Johns Hopkins University Press
Baltimore and London

© 1994 The Johns Hopkins University Press
All rights reserved. Published 1994
Printed in the United States of America
on acid-free paper
03 02 01 00 99 98 97 96 95 94   5 4 3 2 1

The Johns Hopkins University Press
2715 North Charles Street
Baltimore, Maryland 21218-4319
The Johns Hopkins Press Ltd., London

Library of Congress Cataloging-in-Publication Data

Reproduction and development of marine
    invertebrates / edited by W. Herbert Wilson, Jr.,
    Stephen A. Stricker, George L. Shinn.
        p.   cm.
    Papers from a symposium held at Friday Harbor
Laboratories of the University of Washington,
June 9–11, 1992.
    Includes index.
    ISBN 0-8018-4777-X (hc : acid-free paper)
    1. Marine invertebrates—Reproduction—
Congresses.   2. Marine invertebrates—Larvae—
Congresses.   3. Marine invertebrates—
Development—Congresses.   I. Wilson, W.
Herbert.   II. Stricker, Stephen A.   III. Shinn, George
Loren.
QL364.15.R43   1994
592'.016—dc20   93-41979

A catalog record for this book is available from the
British Library.

Christopher G. Reed
1951–1990

The symposium documented in this book was held on June 9–11, 1992, in honor of Christopher G. Reed, who made fundamental contributions to the understanding of the development of marine invertebrates. Chris received his B.S. from the University of Washington, where he remained to earn his Ph.D. in 1980 under Richard Cloney with significant guidance from Robert Fernald. His dissertation demonstrated the importance of cilia in morphogenetic movements during the settlement and metamorphosis of bryozoan larvae. Chris left Seattle for a two-year postdoctoral fellowship at Harvard University with Robert Woollacott, and in 1982 he joined the Department of Biological Sciences at Dartmouth College, where he received tenure in 1989. In his all too brief life, Chris published over thirty papers on various aspects of the development of bryozoans, brachiopods, nemerteans, and polychaetes. The transmission and scanning electron microscopes were major tools in Chris's research. His fastidious attention to detail produced micrographs unsurpassed in quality and definition. In addition to being a prolific researcher, Chris was a consummate teacher. He was twice honored with Distinguished Teaching Awards at Dartmouth College. Chris cared deeply about his teaching and was successful in bringing a sense of discovery to his students during his lectures.

Chris was an extremely likable fellow whose zeal for practical jokes is legendary. His early death from cancer has not only deprived biology of an extraordinarily talented researcher and teacher, it has left a hole in the lives of those who survive him—his wife, his daughter, and his numerous relatives and friends. It is a fitting tribute to Chris that so many of his colleagues gathered together so eagerly to hold this symposium. We dedicate this volume to the memory of Christopher G. Reed.

Contents

Introduction

This symposium on the reproduction and development of marine invertebrates was designed to promote an exchange of ideas concerning current research in the field and to provide some focus on new directions to be taken. All levels of biological organization were represented, from organelles to communities of organisms. This broad range of interests forced the participants to expand their perspectives. In essence, the symposium allowed us to assess the state of current studies, take a brief look back, and peer into the future to identify avenues of promising research.

The proceedings of this symposium have been organized into four parts. The first part, composed solely of a chapter by Keith Benson, provides a historical perspective, identifying the conceptual underpinnings of twentieth-century invertebrate zoology and embryology.

The second part, Gametogenesis, Fertilization, and Early Development, contains chapters organized by developmental sequence. Chapters on oogenesis are followed by chapters on spermatogenesis. Two chapters deal with current problems in fertilization. One of these, the chapter by Richard Miller, investigates the influence of the hydrodynamic regime in facilitating fertilization. Finally, there are two chapters concerned with aspects of embryonic development.

The third part of the volume, Larval Morphology and Evolution, includes a set of chapters that deal with invertebrate larvae. The wide-ranging chapters in this part are arranged phylogenetically to encompass annelids, priapulans, molluscs, bryozoans, and echinoderms.

The final part, Larval Dispersal and Reproductive Ecology, is concerned with larval or adult populations of marine invertebrates. Two contributions (Koehl and Powell; Brumbaugh et al.) address the role of hydrodynamics in innovative ways to aid in the understanding of larval dispersal and settlement.

We believe that this volume provides a record of current directions of research on the reproduction and development of marine invertebrates. We hope each reader will be better able to see gaps and anticipate profitable new research directions. The broad scope of this volume, from spermatogenesis and oogenesis to larval and adult dispersal, forces us to recognize that it is unwise to try to limit an inquiry of any biological phenomenon to a single level of biological organization.

Part I

History of Invertebrate
Reproductive Biology

1 Germ Layers, *Gastraea-theorie,* and Cell Lineage: The Emergence of Invertebrate Embryology and Invertebrate Zoology from Nineteenth-Century Morphology

Keith R. Benson

ABSTRACT The modern fields of invertebrate embryology and invertebrate zoology trace their development to major changes within the nineteenth-century tradition in morphology, a tradition that was primarily German in origin. In the United States, morphology gave rise to a new field of investigation, embryology. Within this area of study, American biologists reformulated the existing German tradition and ultimately recast it into two twentieth-century fields of study, invertebrate embryology and invertebrate zoology. Of critical importance for the American development was the pedagogical role of studying the embryology and zoology of marine invertebrates.

Introduction

Much of the classic literature from the American embryological community at the end of the nineteenth century refers to studies conducted in Europe from the 1860s and 1870s, primarily by German biologists. These studies stressed the importance of the germ layers and universal gastraea forms for clarifying evolutionary relationships, especially among the invertebrates. Within this same classic literature, the American biologists moved quickly beyond the European tradition and, at the same time, made pioneering contributions to what was initially known as embryological morphology and then later branched into invertebrate embryology and invertebrate zoology. While these American developments in embryology gave a new direction to research in development, they were also conducted against the backdrop of important changes within the new biology community in this country. In this essay, I will discuss the European roots for American morphology, the reformulation of that tradition within the United States, and the gradual emergence of invertebrate embryology and invertebrate zoology as the tradition became used for both research and teaching purposes.[1]

The European Beginnings

The subdisciplines that characterize modern biology, for example, ecology, genetics, and general physiology, are the results of developments in the biological sciences during the early twentieth century. In contrast, the birth of invertebrate zoology dates from an earlier period and was associated with the reorganization of the eighteenth century's most important museum, the Jardin du Roi in Paris during the French Revolution. The new arrangement assigned the established botanist Jean Baptiste P. A. de Monet de Lamarck to the cabinet of "insects and worms" in the new

Keith R. Benson, Department of Medical History and Ethics, University of Washington, Seattle, WA 98195

Muséum d'Histoire Naturelle, and Lamarck renamed his department *animaux sans vertebras,* or "invertebrates." The effect of his efforts, despite Lamarck's putative allegiance to his own version of a unified scale in nature (*marche de la nature*), was to break the traditional and long-honored Aristotelian *scala naturae.* Lamarck and his peers systematically and programmatically removed the invertebrate forms from a classification built around vertebrates.

There was no real crisis associated with this arrangement in the nineteenth century, especially after Lamarck's brilliant and powerful colleague the Baron Georges Cuvier provided the nineteenth century with his type concept of the natural world, based on four discrete, permanent, and unrelated types. Cuvier's favorite creatures, and the ones he was in charge of at the Muséum d'Histoire Naturelle, were grouped among Vertebrata. The rest of the animal world, with the exception of organisms lumped into the confused *mélange* known as Infusoria, was divided among the three other types, Radiata, Mollusca, and Articulata, the first two representing most of the specimens in Lamarck's cabinet and the last looked over by the final member of the museum's famous triumvirate, Geoffroy Saint-Hilaire. The three types under Lamarck's and Geoffroy Saint-Hilaire's jurisdiction, all animals without a vertebral column, were frequently lumped together as "invertebrate" organisms.

Naturalists who studied Cuvier's four *embranchements* in the first half of the nineteenth century referred to themselves as morphologists and pursued investigations of the similarities and differences of various animal forms. Some of these morphologists, particularly several Germans working with the new and more powerful achromatic microscopes that were developed in the late 1820s and 1830s, soon directed their interests to the developmental stages of animals, the study of which became known as embryology, thus adding another facet to morphology. Christian Pander and Karl Ernst von Baer supplied the empirical evidence in the 1820s for the ubiquity of two germ layers in the embryonic development of all higher animals (Metazoans), the ectoderm and the endoderm; von Baer later pointed to the probable existence of a third germ layer, the mesoderm. During the 1830s, their colleagues Matthias Schleiden and Theodor Schwann illustrated, through microscopical examinations of the development of plants and animals, the elemental nature of cells in all living tissue; and in 1858 Rudolf Virchow completed the formulation of modern cell theory when he claimed, again using microscopical information, "*omnis cellula e cellula*" or, all cells come from previously existing cells. One year before Darwin's famous book, therefore, morphology was presented with an intriguing method to investigate questions of form: the uniform processes of cell and germ-layer formation provided a uniform developmental path behind the diversity in the natural world.

Darwin's book, *On the Origin of Species* (1859), while often credited with revolutionizing the life sciences with respect to the understanding of diversity, may have served an equally important role in the nineteenth century by providing the initial theoretical foundation to argue for the unity of life through his doctrine of unity of descent, now observed in the similarity of embryonic stages. Using this idea, morphologists and naturalists could easily link the former radiates, molluscs, and articulates into related groups based on the new arguments for the natural system. But a troubling gap in the natural system remained between these animals and the exalted vertebrates. Ernst Haeckel suggested a possible bridge in 1866 when he claimed that embryological information indicated all higher Metazoans descended from a gastrulalike ancestral form. He observed this common embryological stage in representatives of all taxa he examined, thereby providing the initial argument for his *gastraea-theorie* of universal descent.[2] Then, in 1871 the Prussian morphologist Alexander Kowalewsky described the larval stages of the primitive chordate *Amphioxus* and found them to be remarkably similar to the larval stages of the tunicates, heretofore a systematically nomadic group of invertebrates. With his interpretation of the linkage between *Amphioxus* and the tunicates justified by the similarity of their developmental stages, the taxonomic gap between invertebrates and vertebrates was considered bridged, and the animal world once again enjoyed unity.[3]

Through most of the 1870s, Ernst Haeckel's suggestive and empirically based embryological interpretation of Darwinian evolutionary theory, stressing the monophyletic nature of the Metazoa, directed biological investigations to examine embryology for information concerning the natural system. Some morphologists began to refer to themselves as embryological morphologists and embraced the new developmental strategies of Haeckel because this approach seemed to hold the key that could unlock nature's phylogenetic secrets. Morphologists theorized they could recognize the laws of development, determine the homologies of organs, and deduce the ancestral history of the entire Metazoa by studying the early developmental stages of organisms. This orientation characterized Francis M. Balfour's famous and much-read text *A Treatise on Comparative Embryology* (1885), and it was the same theme repeated in Ernst Korschelt's and Karl Heider's well-known book, which appeared in English translation as *Textbook of the Embryology of Invertebrates* (1895).

The American Development of Morphology

But almost as soon as these morphologists migrated to the ocean's shores to pursue their research in embryology, for this was where the greatest variety of organisms with rich embryonic histories was found, the hegemony of Haeckelian morphology was attacked, shaken, and then toppled. By the end of the 1870s, careful examinations of early cell events in closely related organisms suggested that there was no universal germ layer homology throughout the Metazoa.[4] Much of this work was a product of the nineteenth century's second wave of microscopical improvements. This involved technical improvements of the microtome and related refinements in microtechnique, including the discovery of new stains, new fixatives, and new embedding materials that contributed to impressive observations from microscopical laboratories at the end of the 1860s and the beginning of the 1870s. With a new microscopical armamentarium, morphologists were able to conduct increasingly sophisticated examinations of early developmental events, including fertilization. New research laboratories located at the seaside, particularly in Naples, Plymouth, and Woods Hole, aided in the institutionalization of these research activities and brought morphologists together to share in their research problems. Additionally, criticisms of historical explanations in morphology, suggested in the 1870s by Wilhelm His, who argued that a description of stages in development did not constitute an explanation for development, led many biologists to reframe embryological questions within a more causal framework, often in sympathy with their colleagues in physiology. Many of these critics were young American graduate students or recent graduates of the new graduate programs that emerged in the United States in the 1880s. These individuals sought to replace the speculative elements of the older tradition in general morphology with new and more specific research programs emphasizing laboratory methods. Finally, both the rediscovery of Mendel's laws in 1900 and the growth of physiological ecology in the early twentieth century created an excitement in biology for experimental studies, thus continuing the impetus to sweep aside the descriptive and speculative Haeckelian program.

Nowhere were this reformulation and redefinition of morphology more clearly evident than in the United States, a country lacking a professional tradition in biology until the last two decades of the nineteenth century. Interestingly, the move to a more professional community in biology can be largely credited to two American morphologists, William Keith Brooks and Charles Otis

Whitman, neither of whom overtly rejected the older morphological tradition, but both of whom pointed their students, who were to represent almost the entire first generation of young American biologists, toward the new developments in biology. And at the core of their work was the redefinition of morphology into invertebrate zoology and embryology.[5]

When the Johns Hopkins University was established in 1876, its first morphologist, William Keith Brooks, found himself facing the difficult task of establishing a laboratory tradition in morphology in a country that lacked any well-established pedagogical tools. To address these needs he wrote America's first book of invertebrate zoology, *Handbook of Invertebrate Zoology*, published in 1882. Brooks' new book lacked any theoretical claims for biology; but after all, it was intended to serve as a laboratory manual for students and teachers. "This book is a handbook, not a text-book, and the entire absence of generalization and comparison is not due to indifference to the generalizations of modern philosophical morphology, but rather to a wish to aid beginners to study them . . . I have, therefore, attempted to show the student how to acquire a knowledge of the facts for himself, in order to remove this burden from lecturers and text-books."[6] Brooks then called the beginner's attention to structural features of typical organisms that illustrated the "types" of marine organisms. Using this book, Brooks and his students spent their summers at a variety of locations along the Chesapeake, the middle Atlantic, and the Caribbean investigating invertebrate zoology and embryology.

Likewise, when C. O. Whitman returned to his native country after receiving his Ph.D. from Rudolf Leuckart at Leipzig in 1879, he confronted problems similar to those of Brooks. Among his first tasks, and representing one of Whitman's most important contributions to American biology, was the publication of *Methods of Research in Microscopical Anatomy and Embryology* in 1885, again a handbook without an overt theoretical commitment but with a clear purpose to expose American students to the latest microscopical technique. Then, after he became director of the Marine Biological Laboratory (MBL) in Woods Hole in 1888, Whitman worked to establish a new direction of teaching at marine stations by offering advanced instruction in addition to the more common introductory courses in marine botany and zoology. He finally succeeded with these plans in 1893, teaching this country's first course in comparative embryology at Woods Hole; invertebrate zoology followed in 1894.

Not surprisingly, therefore, the students of Brooks at Johns Hopkins and those who either attended the MBL or were students in Whitman's department at the University of Chicago never adopted Haeckel's approach in morphology nor did they refer to themselves as morphologists. Instead, many of these young investigators began to examine fertilization and the events surrounding early development, asking questions of form *and* function from a more causal perspective. This was certainly the intent of those involved in cell-lineage work at the MBL, an approach in embryology first suggested by Whitman in his German dissertation, then adopted from Whitman by Wilson in his graduate research, and finally perfected by Conklin at the MBL in his classic 1897 study of the cell lineage of *Crepidula*. Together, these studies provided the empirical information, especially in terms of comparative invertebrate embryology, for Wilson's magisterial *The Cell in Development and Inheritance,* first published in 1896. This book, along with Wilson's classic embryological studies and the contributions of many others, led to a new field of research in embryology, called cytology.[7] The cytologists, skeptical of an over-reliance upon germ-layer explanations, critical of the deterministic nature of cell theory at century's end, and antagonistic against Haeckel's *gastraea-theorie,* called for a reconstruction of embryology based upon careful studies of the entire developmental history of organisms beginning with fertilization.

Of interest, many of these reformations served as topics of the Friday evening lectures given during the summers at Woods Hole.[8]

Teaching about Invertebrates

The same changes fueled a renovation within American biology education. Recognizing the ambiguous nature of the word *morphology,* biologists began to subdivide the study of form, and soon courses began to appear in the curriculum under the heading of invertebrate morphology, invertebrate zoology, or invertebrate embryology. Cornell was the first institution to teach invertebrate zoology, listed officially as general zoology of invertebrates in 1883, using Brooks' introductory text. Michigan followed in 1884 with its own course in invertebrate zoology, this time emphasizing the importance of the course for instruction in systematics. Pennsylvania and Princeton added a course in invertebrate morphology in 1889, again based on Brooks' text, although the course at Pennsylvania represented a restructured general biology course that had become essentially an invertebrate zoology course beginning in 1885. Berkeley adopted a course in invertebrate zoology in 1891 as part of its introduction to the "laboratory method" of zoology, taught and organized by its recently arrived faculty member William Emerson Ritter. Whitman initiated his new department at Chicago in 1892 with a course in embryology of higher invertebrates, followed by comparative anatomy of invertebrates in 1897, and finally invertebrate zoology in 1902. The University of Washington offered its first course in invertebrate zoology in 1894, Harvard started a course in anatomy and development of vertebrates and invertebrates in 1895, and Yale began its first course in invertebrate zoology in 1907.[9]

While many of these classes were initially organized following the model in Brooks' book on the invertebrates, most eventually adopted the text of his student, J. Playfair McMurrich, *A Textbook of Invertebrate Morphology* (1894).[10] McMurrich retained the "type"-based thinking he learned under Brooks at Johns Hopkins. However, unlike his mentor and unlike Whitman, he provided a more theoretical claim for the importance of studying invertebrates when he stated that the "type" illustrated the affinities peculiar to all the organisms represented by the type form. Structural characteristics are the bricks and stones, McMurrich claimed, and the typological affinities represent the mortar to hold the group together. His book listed twelve "types," all corresponding to present-day phyla.[11]

Similar curricular changes characterized the new laboratories organized along the nation's shorelines. The MBL offered its first course in invertebrate zoology, intended for more advanced students in biology, the same year McMurrich's book was published. One year later, in 1895, McMurrich joined the instructional staff at the MBL, and the course was organized specifically after the chapter headings in his book. The course was offered again in 1896, 1897, and 1899, always attracting the most students of all the courses offered at the MBL. Then, as Whitman sought to move the MBL toward more advanced courses (its longtime patron, the Woman's Education Association, withdrew its support of the MBL at the end of the century as a result), invertebrate zoology was shifted into the zoology and embryology courses.

As a result of these developments at Woods Hole, marine zoology essentially became invertebrate zoology, although the course often retained its older name at other stations. This represented, after all, the orientation of both Brooks and Whitman; the teaching of invertebrate zoology and invertebrate embryology allowed instructors to expose students to all the essential conceptual and methodological foundations of modern biology as it was presented at marine laboratories. Therefore, in the discussions surrounding the elimination from the MBL of all introductory courses for teachers in 1907, the result was a complete revision of the zoology course and its subsequent reformulation as invertebrate zoology. Gilman Drew, another product of Brooks at Johns Hopkins, wrote a new laboratory guide for the course, *A Laboratory Manual of Invertebrate Zoology* (1907), and it soon became the most widely used manual for instruction in invertebrate zoology in the United States. Finally, in 1909 the zoology course was renamed invertebrate zoology. This marks the beginning of the American tradition of teaching invertebrate zoology as the introductory class for students in marine biology.

These curricular changes in universities and at the MBL reveal the actual roots of invertebrate zoology as a biological subdiscipline. The emphasis was upon teaching students about characteristic types of invertebrates, at the same time teaching the students how to observe and learn from the organism. But in the first decade of the twentieth century the essential nature of the "type" changed; it became expanded and diluted as biologists used the same term to refer to characteristic *types* of cleavage patterns, *types* of embryos, *types* of mesoderm development, etc. Furthermore, Drew's manual overtly referred to the "type method of laboratory study," cautioning against too great a reliance upon types such that one would not consider unique variations for adaptation, the first time invertebrate zoology moved explicitly away from its typological tradition of the nineteenth century and adopted a clear evolutionary perspective, now emphasizing adaptation.

Not surprisingly, these same developments occurred at other marine laboratories. The three other major teaching institutions in the marine zone at the fin de siècle, the laboratory of the San Diego Marine Biological Association, Hopkins Marine Station, and Puget Sound Biological Station, also offered courses in marine zoology, again emphasizing invertebrates. After all, instructors on the West Coast realized that these organisms provided wonderful examples of the "types" of Metazoans, usually twelve, and the study of these same organisms provided an opportunity for the instructors to introduce students to laboratory techniques, the major need within the American biological community.

Invertebrate Embryology and Invertebrate Zoology in the Early Twentieth Century

While invertebrate zoology became the central focus for courses in general zoology by the early twentieth century, research biologists examined invertebrates in a slightly different light. One of the major foci for the American biology community, as is clear from both McMurrich's text in invertebrate zoology and Wilson's important book, concerned questions of development, or "inheritance and variation."[12] These biologists adopted their research methods from cytology, the late nineteenth-century formulation for embryology. Oriented toward descriptive studies of cell structure, careful tracings of cell lineage, and comparisons of developmental patterns and larval types, the new embryologists sought information about questions of organismic form from the study of the cell and the developing organism. Overtly eschewing the German interpretation of causal and mechanical explanations for embryology based on the germ layer, cell theory, or *gastraea-theorie,* these Americans were both more critical and more cautious in their approach to the problem. Perhaps too suspicious of the older hypothetical work, especially the studies that included the construction of speculative phylogenies, comparative embryologists sought to base their work only upon empirical information from careful laboratory investigations. As a result, when Mendel's work was rediscovered in 1900, when Wilson discovered a potential connection between chromosome structure and the sex of insects in several studies with his students between 1902 and 1905, and when Morgan developed the chromosome theory from his work on white-eyed *Drosophila* in 1910, many embryologists refused to place their complete faith in explanations that emphasized the role of the nuclear material for inheritance and variation. For example, during their entire careers Wilson and Conklin expressed a high degree of caution for explanations of inheritance that addressed only the role of the chromosomes, especially because they believed all cellular processes were involved, including the intractable cytoplasm. Wilson expressed his reserve in 1923, claiming that:

> genetics, it is true, has of late made remarkable advances toward the study of the nuclear organization but leaves unsolved the problem of the "organism as a whole" . . . We begin to see more clearly that the whole cell-system may be involved in the production of every character. How, then, are hereditary traits woven together in a typical order of space and time?[13]

Wilson's rhetoric closely matched that of Conklin, expressed a decade before.

> I hold no brief for this doctrine [chromosome theory] and have repeatedly urged (1893, 1899, 1905, 1908, etc.) its too narrow outlook on the activities of the cell as a whole. It seems to me incredible that this most general of all cell functions [inheritance], which includes differentiation, metabolism and reproduction, should be the property of only a single cell constituent—the chromosomes.[14]

Both Wilson and Conklin held to the belief that, in order to explain inheritance, chromosome theory had to connect with the important contributions from embryology.

By the early twentieth century, therefore, morphology based upon the germ-layer theory and *gastraea-theorie* had given way to two new and important traditions in American biology. One tradition, invertebrate zoology, traces its roots to pedagogical reforms in this country and stressed both the complexity of any attempt to formulate hypothetical ancestral claims for invertebrate taxa and the growing understanding of the importance of adaptations in the complicated evolutionary history of organisms. The second tradition, invertebrate embryology, traces its roots from the emergence of an American research tradition in biology, most clearly illustrated by the reformulation of cytology, as in the embryological work of Wilson, Conklin, and others. In rejecting the older ideas that the Americans inherited from their European predecessors, ideas which they considered to be outmoded, they developed their own approaches to the study of invertebrates and to investigations of the questions surrounding inheritance and variation.

Contemporary invertebrate zoology in the United States finds its disciplinary roots in the work of Ralph Buchsbaum, *Animals without Backbones* (1938), and the monumental contribution of Libbie Henrietta Hyman, *The Invertebrates* (beginning in 1940), both representing the first comprehensive treatments of invertebrates in America since the beginning of the twentieth century. Here, for the first time the type method was completely replaced by an emphasis on the range of morphological variations among invertebrates, placing new emphasis on the morphology, embryology, physiology, and biology of invertebrates. Once again, students using these texts were encouraged to learn and observe from nature. This time, however, they were not instructed to examine typological affinities but were encouraged to look for adaptational variations. As Hyman stated in her work, "the 'type' method of presentation has been eschewed since one of the major purposes of this treatise is to give an extensive account of the range of morphological variation to be found within each group."[15] Nevertheless, the pedagogical value of invertebrate zoology is clear on every page of Buchsbaum and Hyman.

Cytology, on the other hand, did not fare as well from 1920 until after World War II. Largely passed over by the impressive and productive work in genetics, embryological research declined as the first-generation of American embryologists *cum* cytologists retired and the hoped-for connection with genetics remained elusive. In addition, skepticism concerning the reality of many subcellular elements, such as microtubules, led many others to distrust the findings of the embryologists, especially when compared to the successes in genetics and the growing prestige of the new field of molecular biology. However, impressive developments in electron microscopy after World War II confirmed many of the suggestions of earlier cytologists, and numerous cytological observations contained in Eric Davidson's *Gene Activity in Early Development* (1968) have once again pointed to the importance of this approach. It is also clear, from an examination of Davidson's bibliography, just how indebted the present embryological community is to the work of its early-twentieth-century predecessors. Gary Reverberi makes this point explicitly in the introduction to *Experimental Embryology of Marine and Fresh-Water Invertebrates* (1971).[16]

> The recognition of what has been done before us is our essential condition for being able to carry out what we wish to do in the future. This book aims to be a bridge between the past and the future. One of its hopes is to try and convince molecular embryologists that the world does not begin with them, and that their work is not meaningful if it does not follow and carry further the work of the classical embryologist. The chapters contain a large number of references and the reader is continuously advised to go to the "origins." There is no doubt that one's scientific background is much enlarged by dipping into the original papers as some of them are the real classics which demonstrate the foundations on which embryology is built.

Both invertebrate zoology and invertebrate embryology have made and will continue to make important contributions to the understanding of basic biological problems. At the same time, it is important to emphasize their roots in an older tradition in biology, morphology, a tradition that was crucially important in serving as the foundation to the development of biology in the United States; additionally, it is important to expose ourselves to the work and conclusions of those invertebrate zoologists and embryologists who have laid "the foundations" on which our present understanding is dependent.

Notes

1. My initial exposure to this subject dates from the summer of 1979, when I walked into Robert Fernald's laboratory at Friday Harbor and was astounded to see many of the classic monographs of invertebrate zoology and embryology on the reserve shelf, including works by E. B. Wilson, E. G. Conklin, and C. O. Whitman. My astonishment was not that Fernald knew of these studies, for I had already known of his historical interests, but I was impressed that he required his students to become acquainted with this literature while they studied more contemporary publications. In later conversations, he told me how important it was to transfer his appreciation for the work of the first-generation of American embryologists because he traced his own lineage, through J. Franklin Daniel and Richard Eakin at University of California at Berkeley, to William Keith Brooks, the first American-trained Ph.D. in zoology and the mentor of many of this country's first embryologists (including Brooks' son-in-law, Daniel).

I relate this anecdote because Christopher Reed considered himself to be one of Fernald's students and a part of the same lineage (Reed's dissertation adviser was Fernald's colleague and student, Richard Cloney). During that same summer of 1979, I cemented my relationship with Chris Reed, especially after we discovered we shared many interests, among them the history of biology. Of note, this chapter represents material from a seminar I originally presented to the Biology Department at Dartmouth College in 1983 at Reed's invitation and revised for an American Society of Zoologists symposium on the teaching of invertebrate zoology in 1989. I am presenting it to the memory of one of the finest young embryologists of the twentieth century.

2. Haeckel, E. 1866. Generelle Morphologie des Organismen. Jena.

3. Kowalewsky, A. 1871. Weitere Studien über die Entwicklung der einfachen Ascidien, Arch. Micro. Anat. 7. For a treatment of the impact of Kowalewsky's research, see Beeson, R. J. 1978. Bridging the gap: The problem of vertebrate ancestry, 1859–1875. Ph.D. dissertation, Oregon State University, Corvallis.

Kowalewsky's ideas were popularized in English by Haeckel, E. 1874. The gastraea-theory, the phylogenetic classification of the animal kingdom and the homology of the germlamellae. Q. J. Micro. Sci. 14: 142–165 and 223–247.

4. Brooks, W. K. 1879. The development of the digestive tract in molluscs. Proc. Bost. Soc. Nat. Hist. 20: 325. For a more complete treatment of Brooks' position vis-à-vis Haeckel's work, see Benson, K. R. 1981. Problems of individual development: Descriptive embryological morphology in America at the turn of the century. J. Hist. Bio. 14: 115–128.

5. For several treatments of the development of American biology, see Rainger, R., K. R. Benson, and J. Maienschein, eds. 1988. The American Development of Biology. University of Pennsylvania Press, Philadelphia. An excellent treatment of the continuing changes in American biology may be found in Maienschein, J. 1991. Transforming Traditions in American Biology, 1880–1915. Johns Hopkins University Press, Baltimore.

6. Brooks, W. K. 1882. Handbook of Invertebrate Zoology for Laboratories and Seaside Work. S. E. Cassino, Boston. For more information on Brooks as a teacher see Benson, K. R. 1987. H. Newell Martin, W. K. Brooks, and the reformation of American biology. Am. Zool. 27: 759–771.

7. Wilson, E. B. 1896. The Cell in Development and Heredity. The Macmillan Company, New York. This book was the first American text to draw international interest, especially among German embryologists. For information on subsequent developments in cytology, see Maienschein, J. 1991. Cytology in 1924: Expansion and collaboration. *In* K. R. Benson, J. Maienschein, and R. Rainger, eds. The Expansion of American Biology. Rutgers University Press, New Brunswick, N.J.

8. For a selection of these fascinating and enlightening lectures, see Maienschein, J., ed. 1986. Defining Biology: Lectures from the 1890s. Harvard University Press, Cambridge, Mass.

9. These details are from a survey of the college catalogs from 1880–1920 of University of California at Berkeley, Cornell, Yale, Michigan, Harvard, Chicago, Pennsylvania, the Johns Hopkins University, Princeton, and the University of Washington. I was aided in this work by Jacquie Ettinger.

10. McMurrich's book is the first actual invertebrate zoology text because, as Brooks pointed out in the introductory remarks to his own book, his was intended to serve as a handbook.

11. For details on McMurrich's life see, Benson, K. R. In press. McMurrich, James Playfair (16 October 1859–9 February 1939), American National Biography, Oxford University Press, New York.

12. The title of Wilson's book reads *Development and Inheritance,* but both terms were used interchangeably, and somewhat uncritically, with the terms *variation* and *heredity* at the end of the century.

13. Wilson, E. B. 1923. The Physical Basis of Life. Yale University Press, New Haven, Conn.

14. Conklin, E. G. 1912. Experimental studies on nuclear and cell division in the eggs of *Crepidula*, J. Natl. Acad. Sci. 15: 551.

15. Hyman, L. H. 1940. The Invertebrates: Protozoa through Ctenophora. McGraw-Hill, New York, p. vi.

16. Reverberi, G. 1971. Experimental Embryology and Marine and Fresh-Water Invertebrates. American Elsevier Publishing, New York.

Part II

Gametogenesis,

Fertilization,

and Early Development

2 Oocytic Nutrition in the Lower Metazoa: The Scyphozoa

Kevin J. Eckelbarger

ABSTRACT Recent comparative studies of the ovaries of representative species from four scyphozoan orders have shown that they are far more complex than previously reported. Coronate species have a relatively simple ovary in which developing oocytes grow without the aid of accessory cells. However, in the ovaries of semaeostomes and rhizostomes, oocytes develop while closely associated with accessory cells called trophocytes. The stauromedusan ovary differs from that of other jellyfish because oocytes develop within a relatively complex follicle while surrounded by follicle cells. This finding suggests that scyphozoans are among the most primitive extant metazoans with ovarian accessory cells. Differences in ovarian morphology within the Scyphozoa reflect current views on the phylogenetic relationship between the four orders and lend additional support to the view that the Scyphozoa are more closely related to the Anthozoa than to the Hydrozoa. Studies of the mechanisms of yolk synthesis in scyphozoan oocytes indicate that many species utilize heterosynthetic methods commonly observed in higher metazoans but generally believed to be absent from primitive metazoans such as the Scyphozoa.

Introduction

When considering invertebrate life histories, marine ecologists often focus their attention on the relationship between egg size, energy content (maternal investment), fecundity, and larval developmental mode (planktotrophy versus lecithotrophy). This interest in invertebrate eggs and larvae has resulted in the creation of a new discipline of larval ecology that explores the factors influencing the distribution and abundance of species (Young, 1990). Surprisingly, the ovary and the process of oogenesis have been largely ignored in discussions of the evolution of invertebrate life histories. This is unfortunate because variations observed in patterns of egg organization and yolk accumulation appear to be directly related to interspecific differences in embryonic development, subsequent larval biology, and reproductive patterns (Eckelbarger, 1983, in prep.). The ovary plays a pivotal role in egg formation because it is generally involved in mediating or directly contributing to the biosynthesis of yolk materials. Despite general impressions to the contrary, invertebrate ovaries show enormous variation in their complexity and the degree to which they are actively involved in yolk synthesis (see Adiyodi and Adiyodi, 1983; Wourms, 1987). A knowledge of ovarian function offers insight into the evolution of life histories and should be a major focus for investigators attempting to understand variability in invertebrate reproductive patterns.

In order to comprehend the relationship between oocyte organization and life history patterns in invertebrates, it is essential to understand ovarian morphology and oogenesis, particularly during vitellogenesis, when yolk is synthesized for later use during embryogenesis. It is especially enlightening to examine ovarian morphology and oogenesis from a phylogenetic viewpoint in

Kevin J. Eckelbarger, Darling Marine Center, University of Maine, Walpole, ME 04573, and Department of Animal, Veterinary, and Aquatic Sciences, University of Maine, Orono, ME 04469.

order to (1) better understand their evolution within the Metazoa and (2) understand how this diversity relates to observed variations in spawning patterns and nutrient utilization.

This chapter reviews current knowledge regarding ovarian structure and oocyte nutrition in the true jellyfish, the Scyphozoa, primitive diploblastic metazoans with a low grade of tissue organization (Hyman, 1940). New information is presented based on comparative studies of species from all four scyphozoan orders including the free-swimming Semaeostomeae, Rhizostomeae, and Coronatae and the sessile Stauromedusae. Although the ovaries of lower metazoans have been assumed to be simple and to lack mechanisms of yolk synthesis that usually characterize higher metazoans (Boyer, 1972; Anderson, 1974; Gremigni, 1979, 1983; Eckelbarger, 1984; Weglarska, 1987), recent studies involving the Scyphozoa have shown this assumption to be false (Eckelbarger and Larson, 1988, 1992). In fact, jellyfish ovaries show surprising morphological diversity, and the mechanisms of yolk synthesis probably differ as well. Significantly, ovaries of Stauromedusae contain folliclelike accessory cells resembling those commonly seen in the ovaries of many higher invertebrates but which have not been reported from the Cnidaria. These unexpected findings suggest that scyphozoans were among the first metazoans to utilize accessory cells during their reproductive evolution.

Materials and Methods

Ultrastructural studies were conducted on ovarian tissues removed from freshly collected jellyfish. Medusae of *Aurelia aurita* (Order Semaeostomeae) and *Linuche unguiculata* (Order Coronatae) were collected from Nassau Harbor, Bahama Island, in November 1987 and June 1989, respectively, while those of *Stomolophus meleagris* (Order Rhizostomeae) were collected from Fort Pierce Inlet, St. Lucie County, Florida, in October 1988. Adult specimens of *Haliclystus salpinx* (Order Stauromedusae) were collected from Victoria, British Columbia, in August 1990.

For electron microscopy, pieces of the ovary were excised and fixed by immersion for 1 h at room temperature in 2.5 percent glutaraldehyde containing $0.3 M$ Millonig's phosphate buffer and $0.14 M$ sodium chloride. Tissue was then rinsed for 15 min in three changes each of buffer wash ($0.4 M$ Millonig's phosphate mixed 1:1 with $0.6 M$ sodium chloride) and postfixed for 1 h at room temperature in 1 percent osmium tetroxide buffered in $0.1 M$ Millonig's phosphate buffer and $0.38 M$ sodium chloride. Following fixation, tissue was rinsed in distilled water and dehydrated for 2 h in ascending concentrations of ethanol, transferred through two changes of propylene oxide over 10 min, and embedded in Epon. Thick sections were cut on a Porter-Blum MT2-B ultramicrotome with a Histoknife and stained with Richardson's stain. Thin sections were cut with a diamond knife, stained for 10 min each with aqueous saturated uranyl acetate and lead citrate, and examined with a Phillips 201 transmission electron microscope.

Results

Scyphozoans are acraspedote (lacking a velum), tetramerous medusae or medusaelike polypoids in which the floor of the gastrovascular cavity (stomach) bears gonads in the form of bands or sacs that may hang down into the subumbrellar cavity (Hyman, 1940) (Fig. 2.1). Thus, developing gametes are closely associated with potential sources of nutrients from the digestive system. In all but a few scyphozoans, the sexes are separate. The gastrovascular cavity is divided into separate gastric pouches. The floor of each pouch is lined with a layer of gastrodermis that evaginates into the stomach lumen to form a ribbonlike genital fold. This "ovary" consists of a two-layered sandwich of gastrodermal cells filled with a thin layer of mesoglea in which the oocytes arise and

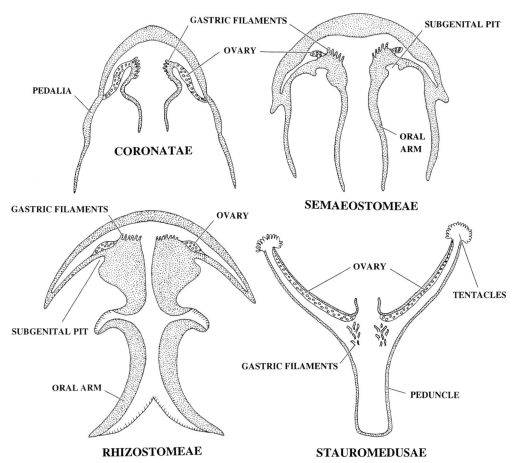

FIGURE 2.1. Diagrammatic transverse sections showing the gross morphology of medusae representative of all four orders of scyphozoans.

differentiate. In small medusae, the genital folds are relatively flat but become more convoluted as the animal matures.

Early Oogenesis

Oocytes arise endodermally from cells within the gastrodermis (germinal epithelium) and gradually migrate into the mesoglea as they grow (Figs. 2.2b–e). No distinct mitotically dividing population of oogonia has been observed. Oocytes are easily identified at the light microscopical level by their round shape, prominent nucleus, single nucleolus, and strongly basophilic ooplasm (Figs. 2.2d and 2.3b). In the rhizostome, *Stomolophus meleagris*, young oocytes are interconnected by intercellular bridges (Fig. 2.2a). Groups of two to four oocytes have been observed together, two of which are connected by a single intercellular bridge. Later germ cells can be distinguished from surrounding gastrodermal cells by the presence of synaptonemal complexes indicting the zygotene/pachytene stage of meiosis I (Eckelbarger and Larson, 1988). As oocytes grow, they become obvious at the light microscopic level because they are larger than surrounding cells and gradually bulge into the mesoglea from the inner surface of the germinal epithelium (Figs. 2.2b–e).

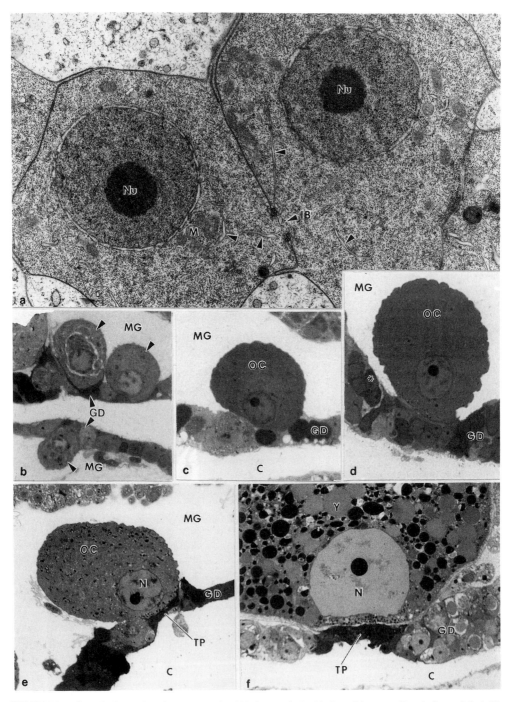

FIGURE 2.2. a. Stomolophus meleagris—two oogonia within the germinal epithelium of the ovary. Note the intercellular bridge (IB) connecting the two cells and the intraooplasmic channels (arrowheads) within the ooplasm (Nu, nucleolus; M, mitochondrion); (b–f) Ovary of Aurelia aurita (light micrographs): b. Previtellogenic oocytes (arrow heads) migrating from the gastrodermis (GD) into the mesoglea (MG); c. Larger oocyte (OC) entering the mesoglea (MG) from the gastrodermis (GD) (C, coelenteron); d. Previtellogenic oocyte (OC) attached to the inner surface of the gastrodermis (GD) (MG, mesoglea; * indicates young oocyte); e. Early vitellogenic oocyte (OC) in close association with trophocyte cells (TP) within the gastrodermis (GD) (C, coelenteron; N, nucleus; MG, mesoglea); f. Late-stage vitellogenic oocyte with prominent nucleus (N) in close association with trophocyte cells (TP) (Y, yolk; GD, gastrodermis).

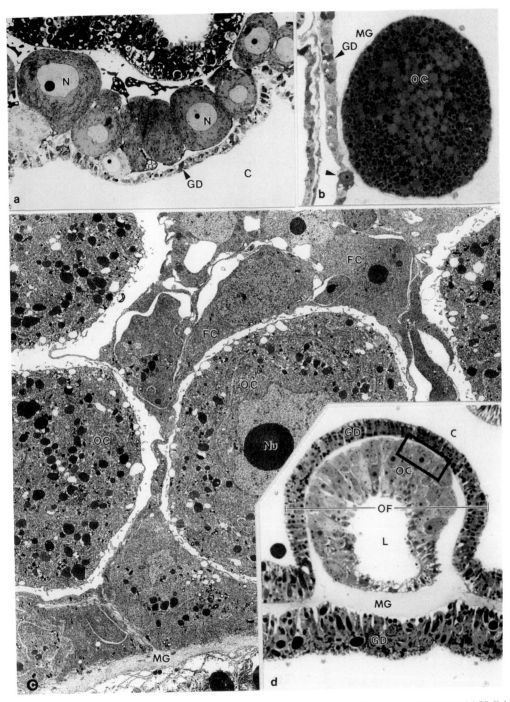

FIGURE 2.3. *a*. Cross-section through ovary of *Linuche unguiculata* showing developing oocytes with large nuclei (N) (light micrograph) (GD, gastrodermis; C, coelenteron); *b*. Ovary of *Linuche unguiculata* showing vitellogenic oocyte (OC) free within the mesoglea (MG) (light micrograph). Note oogonium (arrowhead) still within the gastrodermis (GD); *c*. Electron micrograph of area outlined by rectangle in *d*. Early vitellogenic oocytes (OC) surrounded by a thin layer of follicle cells (FC) (Nu, nucleolus; MG, mesoglea); *d*. Cross-section through the ovarian follicle (OF) of *Haliclystus salpinx* showing young oocytes (OC) developing (light micrograph) (MG, mesoglea; GD, gastrodermis; C, coelenteron). In this specimen, the central lumen (L) contains no differentiated oocytes.

In a recent study of ovarian morphology in six species of scyphozoans from three orders, it was determined that at least two types of ovaries exist (Eckelbarger and Larson, 1992). In members of the Semaeostomeae and Rhizostomeae, oocytes migrate into the mesoglea and become closely associated with a group of unique gastrodermal cells, the trophocytes, which differ morphologically from other gastrodermal cells (Figs. 2.2e,f, 2.4a, and 2.5). This association persists until oogenesis is complete and the eggs are spawned. In members of the Coronatae, the ovary lacks trophocytes, and oocytes enter the mesoglea and differentiate autonomously (Figs. 2.3a,b and 2.5).

The ovaries of two species of sessile jellyfish within the Stauromedusae, *Haliclystus salpinx* and *Lucernaria* sp., are substantially different from those reported for other jellyfish. A series of ovaries form a band running from near the mouth to the cluster of tentacles located on the tip of each arm (Fig. 2.1). Each ovary consists of a spherical follicle formed by the evagination of the endoderm layer of the subumbrella (Fig. 2.3d). Young oocytes develop near the periphery of the follicle while mature eggs enter the central lumen of the follicle upon completion of vitellogenesis. Unlike the semaeostome and rhizostome ovaries, oocytes do not develop within the mesoglea, which forms a thin layer around the outer region of the ovarian follicle. Individual oocytes are surrounded by a thin basal lamina (Figs. 2.4b and 2.5c). The oocytes are isolated from the mesoglea and from each other (Figs. 2.3c,d and 2.5) by squamous follicle cells (Fig. 2.3c), which are sometimes shared between adjacent oocytes. The follicle cells have a prominent nucleus with a single nucleolus, but the cytoplasm is largely devoid of organelles. Throughout oogenesis, individual oocytes remain isolated from each other by the follicle cells. There are no comparable cells in the ovaries of jellyfish belonging to other orders.

Vitellogenesis

During active yolk synthesis, the oocytes of all species examined develop a unique system of surface invaginations that penetrate deeply into the ooplasm, forming a complex system of branching ooplasmic channels enclosing an intracellular space (Eckelbarger and Larson, 1992). The channels are so prominent they are readily visible even at the light microscopic level. These structures become extensive during the vitellogenic phase; however, they appear first in early oocytes when they are present in the germinal epithelium and connected by intercellular bridges (Fig. 2.2a). In transverse section, they appear to be smooth, intracellular membranes resembling endoplasmic reticulum (Fig. 2.4b). A single cisternum of rough endoplasmic reticulum (RER) typically parallels each profile of a channel (Fig. 2.4c). Initially, the channels are occluded, but as the oocytes grow, they dilate to form a complex, branching network of spaces that extend throughout the oocyte from the oolemma to the perinuclear region. The channels disappear as vitellogenesis ends.

The oocytes of all scyphomedusae examined show similarities with respect to the type and abundance of organelles believed to have a proteosynthetic function during vitellogenesis. At least three types of storage products have been observed in oocytes, including glycogen particles, lipid droplets, and membrane-bounded yolk bodies. The Golgi complex and RER appear to coordinate the autosynthetic production of yolk bodies in all species examined. Golgi complexes are common and in close association with a single cisternum of RER (Fig. 2.6c). In semaeostome and rhizostome species, acquisition of yolk precursors appears to occur also through a process of receptor-mediated endocytosis manifested by the presence of numerous coated pits and vesicles and smooth-surfaced tubules along the oocyte surface (Figs. 2.6a,b,d). These endocytotic vesicles and tubules appear to fuse to form yolk bodies. Endocytotic activity during vitellogenesis

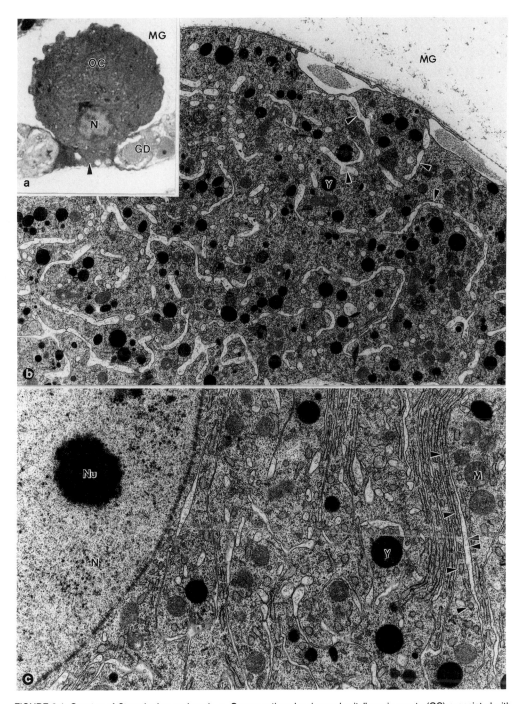

FIGURE 2.4. Oocytes of *Stomolophus meleagris*. *a*. Cross-section showing early vitellogenic oocyte (OC) associated with early trophocyte cells (arrowhead) (light micrograph) (MG, mesoglea; GD, gastrodermis); *b*. Cortical ooplasm of early vitellogenic oocyte shown in *a*. Note intracellular channels (arrowheads) throughout ooplasm (Y, yolk; MG, mesoglea); *c*. Perinuclear ooplasm of later vitellogenic oocyte showing intracellular channels (double arrowheads) and closely associated cisternae of RER (single arrowheads) (Y, yolk; N, nucleus; Nu, nucleolus).

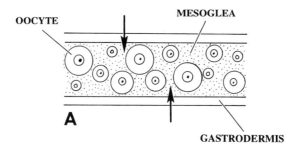

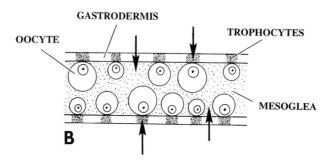

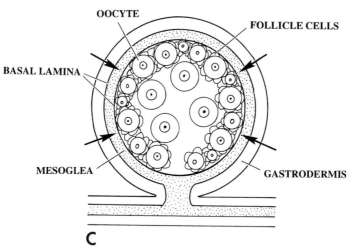

FIGURE 2.5. Diagrammatic representation of jellyfish ovarian morphologies. Arrows indicate probable pathways of yolk precursors from the coelenteron to developing oocytes. *a*. Ovary of a coronate medusa in which oocytes develop within the mesoglea without the aid of accessory cells. *b*. Ovary of semaeostome and rhizostome medusae in which oocytes develop in association with trophocytes. Individual oocytes in *a* and *b* are surrounded by a thin basal lamina (not shown). *c*. Ovary of a stauromedusa in which oocytes develop within a follicle while surrounded by follicle cells.

reaches high levels in the oocytes of *Aurelia aurita* and *Linuche unguiculata*, while the oocytes of *Stomolophus meleagris* show much lower levels. Endocytotic activity in vitellogenic oocytes of *Haliclystus salpinx* was rarely observed. In species in which developing oocytes are in contact with trophocytes, yolklike inclusions appear in the ooplasm immediately adjacent to the zone of trophocyte contact (Eckelbarger and Larson, 1988).

Figure 2.5 summarizes the general structure of the ovaries of four orders of scyphozoans and

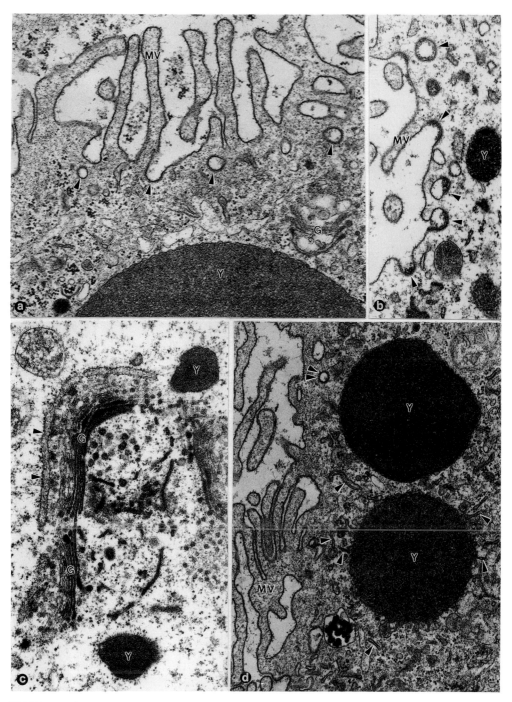

FIGURE 2.6. *a*. Cortical ooplasm of vitellogenic oocyte from *Aurelia aurita* showing endocytotic pits and vesicles (arrowheads) along the cell surface (Y, yolk; MV, microvilli; G, Golgi complex); *b*. Endocytotic pits and vesicles (arrowheads) forming along the surface of a vitellogenic oocyte of *Linuche unguiculata* (Y, yolk; MV, microvilli); *c*. Two adjacent Golgi complexes (G) in association with RER (arrowheads) and nascent yolk bodies (Y) in an oocyte of *Linuche unguiculata*. *d*. Two yolk bodies (Y) in close association with smooth-surfaced tubules (arrowheads) in an oocyte of *Aurelia aurita*. Note endocytotic vesicle (double arrowheads) and folded microvilli (MV).

the relationship, if any, that the differentiating oocytes have with accessory cells. In the coronate ovary (Fig. 2.5a), oocytes develop within the mesoglea and apparently receive nutrients from the coelenteron by way of diffusion or active transport through the gastrodermis and from the by-products of intracellular digestion within the gastrodermal cells. In semaeostome and rhizostome ovaries (Fig. 2.5b), oocytes develop in association with trophocytes and appear to receive nutrients from the coelenteron by way of diffusion or active transport through the gastrodermis, from the by-products of intracellular digestion within the gastrodermal cells and by way of the trophocytes. In the stauromedusan ovary (Fig. 2.5c), oocytes lie outside the mesoglea and apparently receive nutrients from the coelenteron by way of diffusion or active transport through the gastrodermis and from the by-products of intracellular digestion within these cells. Nutrients must pass through the mesoglea, the layer of follicle cells, and the basal lamina that surrounds each oocyte.

Discussion

Ovarian morphology in the Scyphozoa is diverse, and there is a higher level of complexity than one might expect considering the general morphological simplicity of cnidarians. Early light microscopic descriptions of ovarian morphology (Widersten, 1965) did not reveal the degree of complexity now evident from ultrastructural studies (Eckelbarger and Larson, 1988, 1992). The gross morphology of the ovaries ranges from simple mesogleal-filled sacs containing developing oocytes in semaeostome, rhizostome, and coronate species to relatively complex ovarian follicles in the Stauromedusae. In the ovaries of coronate species, oocytes reside in the mesoglea during vitellogenesis and grow without the aid of accessory cells. In semaeostome and rhizostome species, the oocytes apparently receive nutritional support from modified gastrodermal cells (trophocytes), which resemble cells of the trophonema in anthozoans (Fautin and Mariscal, 1991). Experimental studies have shown that the anthozoan trophonema plays a nutritive role during vitellogenesis by channeling nutrients from the coelenteron to the oocyte (Larkman and Carter, 1982), and a similar role has been hypothesized for the trophocytes of semaeostome and rhizostome jellyfish (Eckelbarger and Larson, 1988, 1992). In Stauromedusae, trophocytes are absent, but follicle cells are closely associated with developing oocytes throughout vitellogenesis. This is the first observation in a cnidarian of accessory cells that do not remain a part of the gastrodermal epithelium. This morphological diversity suggests functional differences in the process of vitellogenesis. In addition, ultrastructural examination of mature yolk bodies of all scyphomedusae has shown unexplained interspecific differences (Eckelbarger and Larson, 1992). Whether or not these variations represent differences in vitellogenic mechanisms remains unclear.

Widersten (1965) conducted a comprehensive study of scyphozoan ovaries and described the gastrodermal cells associated with the developing oocytes as "nurse cells." Unfortunately, the cells do not fit the modern definition of nurse cells, currently defined as sibling germ cells that remain in cytoplasmic contact with oocytes by way of intercellular bridges (Anderson, 1974). Since Widersten's study, several reviews have continued to describe these cells as nurse cells (Campbell, 1974; Lesh-Laurie and Suchy, 1991), and in other ultrastructural studies of scyphozoan ovaries, these cells were collectively termed the *paraovular body* (Rotinni-Sandrini et al., 1983; Avian and Rotinni-Sandrini, 1991). Recent studies of oogenesis in semaeostomes and rhizostomes have determined that these cells are not classical nurse cells and should be termed *trophocytes,* a more generic term that reflects the uncertainty of their origins (Eckelbarger and Larson, 1988, 1992). Currently there is no evidence that cnidarians possess true ovarian nurse cells, and thus use of the term should be avoided.

The presence of folliclelike cells in association with developing oocytes makes the Cnidaria among the most primitive phylum in which these accessory cells have been described. Folliclelike cells have been reported from sponges, but most species appear to lack permanent, well-defined ovaries observed in other phyla, and ultrastructural studies of the accessory cells have been limited (Fell, 1983; Harrison and de Vos, 1991). Follicle cells have been reported from many invertebrate phyla that possess well-defined ovaries (see Adiyodi and Adiyodi, 1983), but until now, turbellarians were the most primitive metazoans in which they had been documented (Rieger et al., 1991). Numerous functions have been attributed to invertebrate follicle cells including the synthesis or regulation of heterosynthetic yolk precursors, the recycling of metabolites from moribund eggs, the production of egg envelopes, and the production of hormones and other regulatory molecules (Wourms, 1987). Based on limited observations in the Stauromedusae, their role is unknown. However, because they lack proteosynthetic organelles, they presumably do not have a nutritive function. They segregate developing oocytes from surrounding cells and thus may control the movement of molecules into the microenvironment surrounding the oocyte. Further studies will be required to ascertain their specific function.

It is noteworthy that the stauromedusan ovary is more highly organized than that of other scyphozoan orders. In addition to an ovarian follicle and the presence of accessory cells, oocytes are segregated within the ovary by developmental stage, that is, vitellogenic oocytes reside in the outer region of the follicle in close association with follicle cells, while mature eggs are located in the interior. In contrast, in the ovaries of other jellyfish orders, oocytes of all sizes and stages of development are randomly distributed throughout the mesoglea. Another significant feature of the stauromedusan ovary is the fact that developing oocytes are segregated from the mesoglea. Mesoglea is present in the periphery of each ovarian follicle, but individual oocytes are separated from it by surrounding follicle cells. In other scyphozoan ovaries, oocytes are located in the mesoglea and are separated from it by a thin basal lamina. In the ovaries of semaeostomes, rhizostomes, and coronates, the mesoglea appears to play the role of a circulatory system because the oocytes are actively incorporating metabolites from the surrounding environment via endocytotic processes. The ultimate source of nutrients in all jellyfish is the coelenteron where prey digestion occurs. For oocytes to receive digestive by-products, molecules must pass first through the ovarian epithelium (gastrodermis) and enter the mesoglea within which they diffuse. However, in the stauromedusan ovary, metabolites must pass through the gastrodermis, the mesoglea, the follicle cells, and finally the basal lamina surrounding each oocyte. It is assumed also that some yolk precursors originate from the gastrodermal cells themselves, as by-products of intracellular digestion.

The origin of germ cells in scyphozoans has been attributed to interstitial cells within the gastrodermis of the germinal epithelium (Hyman, 1940; Widersten, 1965; Campbell, 1974). To my knowledge, the present study is the first to document cytoplasmic continuity between young oocytes in any cnidarian. While not a dramatic finding, it adds to our fragile knowledge of early germ cell development in this primitive phylum. Some ultrastructural documentation of early germ cell development has been published for anthozoans (Larkman, 1983), hydrozoan medusae (Kessel, 1968), and *Hydra* (Aizenshtadt,, 1974; Noda and Kanai, 1977, 1980; Honegger et al., 1989), but for only one scyphozoan, *Aurelia aurita* (Eckelbarger and Larson, 1988). The present study indicates that oogonia divide mitotically in the germinal epithelium but fail to complete cytokinesis initially. The intercellular bridges presumably disappear before individual oocytes migrate into the mesoglea where vitellogenesis begins.

The presence of intracellular channels in growing jellyfish oocytes is a feature unique to the Cnidaria, for nothing comparable has been described in the eggs of metazoans belonging to any other phylum. Kessel (1968) was the first to note this peculiar feature in the oocytes of an unidentified trachyline hydromedusa. He hypothesized that they represented infoldings of the oolemma, an hypothesis that has received recent support (Eckelbarger and Larson, 1988, 1992). In the present study, even the oogonia show signs of developing a few infoldings within their ooplasm, the earliest stage in which these structures have been observed in any cnidarian oocyte. The function of these structures remains unknown. However, the infoldings significantly increase the surface area of the oocyte and might play a role in the passage of materials both into and out of the oocytes during development. It is also noteworthy that single cisternae of RER typically lie parallel to the infoldings, although the significance of this apparent association is problematic.

In the semaeostome, rhizostome, and coronate species examined, vitellogenesis appears to occur through a combination of autosynthetic and heterosythetic pathways. In the semaeostomes *Aurelia aurita* and *Diplumularis antarctica*, and the rhizostomes *Stomolophus meleagris* and *Cassiopea xamachama*, yolk bodies arise from three separate sources: (1) from the uptake of precursors in the region of oocyte-trophocyte contact through receptor-mediated endocytosis (heterosynthetic), (2) from the endocytotic uptake of precursors along the oocyte surface having direct contact with the surrounding mesoglea (heterosynthetic), and (3) from the combined synthetic efforts of the Golgi complex and RER (autosynthetic). In the coronate *Linuche unguiculata*, yolk precursors appear to be restricted to the latter two sources, since coronates lack trophocytes (Eckelbarger and Larson, 1988, 1992). In the Stauromedusae *Haliclystus salpinx*, yolk synthesis appears to occur primarily through the autosynthetic activity of the Golgi complex and RER because endocytotic activity during vitellogenesis was rarely observed. The degree of endocytotic activity in growing oocytes appears to be correlated with the length of oogenesis. The oocytes of *Aurelia aurita* and *Linuche unguiculata* demonstrate high levels of endocytosis, and both produce eggs at a high rate—in excess of 100 eggs per day (Kremer et al., 1990; Larson, unpubl. obs.). In contrast, the oocytes of *Haliclystus auricula* show almost no signs of endocytotic activity during vitellogenesis, and oocyte development appears to occur slowly between annual summer breeding seasons (Berrill, 1963). The correlation between rapid egg production and heterosynthetic incorporation of yolk precursors has been demonstrated in polychaetous annelids (Eckelbarger, 1983) and other invertebrate phyla (Eckelbarger, in prep.) and indicates the importance of ovarian function when attempting to comprehend the evolution of life history patterns.

Some features of the reproductive biology of cnidarians can be useful in deciphering phylogenetic relationships between the Hydrozoa, Anthozoa, and Scyphozoa. In the Hydrozoa, germ cells have an ectodermal origin, and developing oocytes use a primitive method of nutrition involving the uptake of nutrients from somatic or other germ line cells involving phagocytic processes (Campbell, 1974; Spracklin, 1984; Thomas and Edwards, 1991) similar to that reported in sponges (Harrison and de Vos, 1991). In the Anthozoa and Scyphozoa, the endodermal origin of germ cells is one reason for suggesting a close phylogenetic relationship between the two classes (Thiel, 1966; Barnes, 1980). Until recently, ovarian morphology was not used as a character in cnidarian phylogenetic speculations. However, recent studies have revealed striking similarities between the relatively complex trophonema-oocyte relationship in the anthozoan ovary and the simpler trophocyte-oocyte relationship in semaeostome and rhizostome jellyfish (Eckelbarger and Larson, 1992). This supports the theory that anthozoans and scyphozoans are more closely related to each other than either is to the Hydrozoa (Hyman, 1940). Furthermore, the existence of trophocyte-oocyte complexes in the ovaries of both the Semaeostomeae and Rhizostomeae supports the hypothesis that these orders are the most closely related within the

Scyphozoa and that the Coronatae and Stauromedusae are nearer the cnidarian stem form (Hyman, 1940; Thiel, 1966). There is, in fact, considerable evidence that the Rhizostomeae are derived from the Semaeostomeae (Uchida, 1960; Thiel, 1966). These findings support the efficacy of using features of the reproductive biology of organisms to decipher invertebrate phylogenetic relationships.

Acknowledgments

I wish to thank Pamela Blades-Eckelbarger for assistance with tissue fixation, ultramicrotomy, and darkroom work during the preparation of this manuscript. I am also grateful to Seth Tyler and Kelly Edwards for use of the Electron Microscopy Center in the Department of Zoology at the University of Maine, Orono. Partial support was provided by National Science Foundation grant OCE-877922. This chapter is dedicated to the memory of Chris Reed, a friend and superb invertebrate morphologist who set the standards for the field. This represents contribution number 258 of the Darling Marine Center.

Literature Cited

Adiyodi, R. G., and K. G. Adiyodi. 1983. Reproductive Biology of Invertebrates, Vol. 1. Oogenesis, Oviposition, and Oosorption. John Wiley & Sons, New York.

Aizenshtadt, T. B. 1974. Investigation of oogenesis in hydra. Communication I. Ultrastructure of interstitial cells at early stages of their transformation into oocytes. Ontogenez 5: 13–20.

Anderson, E. 1974. Comparative aspects of the ultrastructure of the female gamete. *In* G. H., Bourne, J. F. Danielli, and K. W. Jean, eds. Aspects of Nuclear Structure and Function. Academic Press, New York.

Avian, M., and L. Rottini-Sandrini. 1991. Oocyte development in four species of scyphomedusa in the northern Adriatic Sea. Hydrobiologia 216/217: 189–195.

Barnes, R. D. 1980. Invertebrate Zoology. W. B. Saunders, New York.

Berrill, M. 1963. Comparative functional morphology of the Stauromedusae. Can. J. Zool. 41: 1249–1262.

Boyer, C. B. 1972. Ultrastructural studies of differentiation in the oocyte of the polyclad turbellarian, *Prosteceraeus floridanus*. J. Morphol. 136: 273–296.

Campbell, R. D. 1974. Cnidaria. *In* A. C. Giese and J. S. Pearse, eds. Reproduction of Marine Invertebrates, Vol. 1. Acoelomate and Pseudocoelomate Metazoans. Academic Press, New York, pp. 133–199.

Eckelbarger, K. J. 1983. Evolutionary radiation in polychaete ovaries and vitellogenic mechanisms: Their possible role in life history patterns. Can. J. Zool. 61: 487–504.

———. 1984. Comparative aspects of oogenesis in polychaetes. Fortschr. Zool. 29: 123–148.

Eckelbarger, K. J., and R. L. Larson. 1988. Ovarian morphology and oogenesis in *Aurelia aurita* (Scyphozoa: Semaeostomae): Ultrastructural evidence of heterosynthetic yolk formation in a primitive metazoan. Mar. Biol. 100: 103–115.

———. 1992. Ultrastructure of the ovary and oogenesis in the jellyfish *Linuche unguiculata* and *Stomolophus meleagris*, with a review of ovarian structure in the Scyphozoa. Mar. Biol. 114: 633–643.

Fautin, D. G., and R. N. Mariscal. 1991. Cnidaria: Anthozoa. *In* F. W. Harrison and J. A. Westfall, eds. Microscopic Anatomy of Invertebrates, Vol. 2. Placozoa, Porifera, Cnidaria and Ctenophora. Wiley-Liss, New York, pp. 267–358.

Fell, P. 1983. Porifera. *In* K. G. Adiyodi and R. G. Adiyodi, eds. Reproductive Biology of Invertebrates, Vol. 1. Oogenesis, Oviposition, and Oosorption. John Wiley & Sons, New York, pp. 1–29.

Gremigni, V. 1979. An ultrastructural approach to planarian taxonomy. Syst. Zool. 28: 345–355.

———. 1983. Platyhelminthes-Turbellaria. *In* K. G. Adiyodi and R. G. Adiyodi, eds. Reproductive Biology of Invertebrates. John Wiley & Sons, New York, pp. 67–107.

Harrison, F. W., and L. de Vos. 1991. Porifera. *In* F. W. Harrison and J. A. Westfall, eds. Microscopic

Anatomy of Invertebrates, Vol. 2. Placozoa, Porifera, Cnidaria and Ctenophora. Wiley-Liss, New York, pp. 29–89.

Honegger, T. G., D. Zurrer, and P. Tardent. 1989. Oogenesis in *Hydra carnea*: A new model based on light and electron microscopic analyses of oocyte and nurse cell differentiation. Tissue & Cell 21: 381–393.

Hyman, L. H. 1940. The Invertebrates: Protozoa through Ctenophora, Vol. 1. McGraw-Hill, New York.

Kessel, R. G. 1968. Electron microscope studies on developing oocytes of a coelenterate medusa with special references to vitellogenesis. J. Morphol. 126: 211–248.

Kremer, P., J. Costello, and M. Canino. 1990. Significance of photosynthetic endosymbionts to the carbon budget of the scyphomedusa *Linuche unguiculata*. Limnol. Oceanogr. 35: 609–624.

Larkman, A. U. 1983. An ultrastructural study of oocyte growth within the endoderm and entry into the mesoglea in *Actinia fragacea* (Cnidaria, Anthozoa). J. Morphol. 178: 155–177.

Larkman, A. U., and M. A. Carter. 1982. Preliminary ultrastructural and autoradiographic evidence that the trophonema of the sea anemone *Actinia fragacea* has a nutritive function. Int. J. Invert. Reprod. Dev. 4: 375–379.

Lesh-Laurie, G. E., and P. E. Suchy. 1991. Cnidaria: Scyphozoa and Cubozoa. *In* R. W. Harrison and J. A. Westfall, eds. Microscopic Anatomy of Invertebrates, Vol. 2. Placozoa, Porifera, Cnidaria, and Ctenophora. Wiley-Liss, New York, pp. 185–266.

Noda, K., and C. Kanai. 1977. An ultrastructural observation on *Pelmatohydra robusta* at sexual and asexual stages, with special reference to "Germinal Plasm." J. Ultrastruct. Res. 61: 284–294.

———. 1980. An ultrastructural observation on the embryogenesis of *Pelmatohydra robusta*, with special reference to "germinal dense bodies." *In* P. Tardent and R. Tardent, eds. Developmental and Cellular Biology of Coelenterates. Elsevier–North Holland, New York, pp. 133–138.

Rieger, R. M., S. Tyler, J. P. S. Smith III, and G. E. Rieger. 1991. Platyhelminthes: Turbellaria. *In* F. W. Harrison and B. J. Bogitsh, eds. Microscopic Anatomy of Invertebrates, Vol. 3. Platyhelminthes and Nemertina. Wiley-Liss, New York, pp. 7–140.

Rotinni-Sandrini, L., M. Avian, V. Axiak, and A. Malej. 1983. The breeding period of *Pelagia noctiluca* (Scyphozoa, Semaeostomeae) in the Adriatic and Central Mediterranean Sea. Nova Thalassia 6: 65–75.

Spracklin, B. W. 1984. Oogenesis in *Tubularia larynx* and *Tubularia indivisa* (Hydrozoa, Athecata). Ph.D. dissertation, University of New Hampshire, Durham.

Thiel, H. 1966. The evolution of Scyphozoa, a review. *In* W. J. Rees, ed. The Cnidaria and Their Evolution. Academic Press, New York, pp. 77–117.

Thomas, M. B., and N. C. Edwards. 1991. Cnidaria: Hydrozoa. *In* F. W., Harrison and J. A. Westfall, eds. Microscopic Anatomy of Invertebrates, Vol. 2. Placozoa, Porifera, Cnidaria and Ctenophora. Wiley-Liss, New York, pp. 91–183.

Uchida, T. 1960. Metamorphosis from viewpoint of animal phylogeny. Bull. Biol. Stn. Asamushi 10: 181–188.

Widersten, B. 1965. Genital organs and fertilization in some Scyphozoa. Zool. Bidr. Upps. 37: 45–58.

Weglarska, B. 1987. Yolk formation in *Isohypsibius* (Eutardigrada). Zoomorph. 107: 287–292.

Wourms, J. P. 1987. Oogenesis. *In* A. C. Giese and J. S. Pearse, eds. Reproduction of Marine Invertebrates, Vol. 9. General Aspects: Seeking Unity in Diversity. Blackwell Scientific Publications, Pacific Grove, Calif. pp. 50–157.

Young, C. M. 1990. Larval ecology of marine invertebrates: A sesquicentennial history. Ophelia 32: 1–48.

3 Nucleolar Disassembly during Starfish Oocyte Maturation

Stephen A. Stricker, Kenneth L. Conwell II, Dana J. Rashid,

and Angela M. Welford

ABSTRACT In this study, the timing and patterns of nucleolar disassembly are examined in hormone-treated starfish oocytes undergoing meiotic maturation. Based on serial fixations and video microscopic investigations of living oocytes, nucleolar breakdown typically occurs 1–2 min following the onset of germinal vesicle breakdown. Prior to nucleolar disassembly, an apparently novel form of nucleolar extrusion occurs, wherein the contents of nucleolar vacuoles are deposited into the nucleoplasm. Within 5–10 min after the completion of the nucleolar extrusions, nucleoli begin to disassemble via either a vesicular pattern or a more gradual process that does not involve the formation of discrete vesicles. Electron microscopic examinations of prophase-arrested specimens indicate that the nucleolus typically consists of: (1) a more or less centrally located, electron-lucent region that presumably represents the fibrillar center of the nucleolus; and (2) an electron-dense peripheral region that apparently corresponds to the nucleolar granular component. A conspicuous dense fibrillar component such as found in other nucleoli is characteristically lacking, but prior to the peak of the breeding season nucleoli may also display a distinct polar cap that comprises reticulated nucleonemas of unknown significance. In Western blotting studies and immunofluorescence analyses of nucleoli, a 35-kD antigen corresponding to the conserved nucleolar protein fibrillarin occurs in a discrete patch that may represent the fibrillar center of the nucleolus. In addition to data presented from ultrastructural and immunolocalization studies, aspects of the compositional complexity and dynamic nature of nucleoli are discussed with reference to the growing literature that is available on the shuttling of proteins between the nucleolus and cytoplasm.

Introduction

The nucleolus represents a vital and complex nuclear component in which ribosomal genes are transcribed and the products of this transcription are combined with specific proteins to form preribosomal particles (Jordan and Cullis, 1982; Goessens, 1984; Scheer and Benavente, 1990). Associated with the well-developed biosynthetic activities of the nucleolus is a high degree of structural compartmentalization. When viewed by electron microscopy, nucleoli generally display three regions: (1) one to several lightly staining *fibrillar centers* that contain proteins and the chromosomal nucleolar organizing region(s) with transcriptionally active DNA; (2) a *dense fibrillar component* in which the nascent preribosomal particles that were transcribed in the fibrillar center accumulate; and (3) the peripherally located *granular component* that consists of maturing preribosomal particles (Goessens, 1984; Fischer et al., 1991).

Stephen A. Stricker, Kenneth L. Conwell II, Dana J. Rashid, and Angela M. Welford; Department of Biology, University of New Mexico, Albuquerque, NM 87131.

During the cell cycle, nucleoli undergo extensive reorganizations, typically disassembling at the end of prophase and reforming in telophase (Busch and Smetana, 1970; Smetana and Busch, 1974). Such cataclysmic changes in nucleolar morphology must be accomplished correctly so that the interphase nucleolus attains the proper ultrastructural configuration needed for ribosome production. Although excellent accounts are available on stages of nucleologenesis in various cell types (Busch and Smetana, 1970; Stahl, 1982), the patterns of nucleolar breakdown prior to cell division are not as well documented, partly because it can be difficult to control the onset of nucleolar disintegration in some cell lines.

In this study, the nucleoli of prophase-arrested starfish oocytes are synchronously triggered to undergo breakdown by treatment with maturation-inducing hormone, in order to analyze the patterns of nucleolar disassembly during oocyte maturation. Nucleolar disassembly is examined in living and serially fixed specimens by means of time-lapse video microscopy, electron microscopy (EM), confocal microscopy, and immunolocalization techniques. Based on these studies, the precise timing of nucleolar disassembly relative to germinal vesicle breakdown (GVBD) is determined, and the morphological changes involved in nucleolar disassembly are analyzed. In addition, an apparently novel extrusion of nucleolar vacuoles into the nucleoplasm is described, and the heterogeneous distribution of a conserved nucleolar protein, fibrillarin, is documented.

Materials and Methods

Oocyte Preparations

Adult specimens of *Pisaster ochraceus* and *Astropecten* sp. were purchased from Marinus Inc. (Long Beach, California) and maintained at 13°C in aerated aquaria containing bicarbonate-buffered artificial sea water (Instant Ocean). Nucleolar disassembly in both species followed similar patterns, although the process of oocyte maturation occurred more rapidly in *Astropecten* sp. Unless specified otherwise, observations presented in this chapter pertain to *P. ochraceus* examined during the spring and summer months of 1991 and 1992.

For all studies, prophase-arrested primary oocytes were obtained from excised ovaries as described by Stricker and Schatten (1989). Following five washes in Tris-buffered calcium-free sea water (Schroeder and Stricker, 1983), the follicle-free oocytes were returned to filtered sea water and subsequently kept in a refrigerated incubator set at $15 \pm 1°C$ for up to 8 h before use. All reagents and drugs were purchased from Sigma Chemical Co. (St. Louis, Missouri), unless otherwise noted.

Nucleolar Dynamics and Video Microscopy

To quantify the timing of nucleolar breakdown relative to germinal vesicle breakdown, two types of studies were conducted. In one case, small batches of washed oocytes were maintained in an incubator as described above and subsequently treated with a $5 \times 10^{-6} M$ solution of the maturation-inducing hormone 1-methyladenine to trigger the resumption of meiosis. At various times following the addition of hormone, 80 μl of maturing oocytes were rapidly immersed in a Triton-containing, glycerol-based extraction buffer (Balczon and Schatten, 1983) to which 10 percent methanol was added. The extracted cells were then mounted on slides, and the onsets of GVBD and nucleolar disassembly were scored in one-hundred specimens per time point using a Zeiss Axiovert 10 inverted microscope equipped with a Zeiss Plan-Neofluor 20× objective and differential interference contrast (DIC) optics. In this study, the convention of Stricker and Schatten (1989) was adopted wherein the onset of GVBD was defined as the stage when the nuclear envelope began to vesiculate. This in turn was designated as the time point when the nuclear outline became more folded and the nucleoplasm of Triton-extracted specimens changed from a heterogeneous, threadlike configuration to a more homogeneous and flocculent appearance (Stricker and Schatten, 1989). Similarly, the onset of nucleolar disassembly was counted as the time point when the border of the nucleolus became more irregular.

For the studies described above, seven different females were used, and several replicate fixation series were carried out. Most of the observed variation occurred between individuals, rather than across replicates from the same animal even when the replicates were assayed at varying times following the initial removal of the ovaries.

As an alternative method of monitoring nucleolar dynamics, individual oocytes were viewed by time-lapse video microscopy using the Axiovert 10 microscope and a thermoelectric cooling stage (Cloney et al., 1970). The cooling stage was set at 15°C and periodically checked with a thermocouple probe to ensure that the region near the oocytes remained at 15°C. In these cases, follicle-free oocytes in filtered sea water were either: (1) treated with hormone as described above and rapidly placed in a small drop on a slide before being slightly compressed with a coverslip supported by clay feet; or (2) attached by means of protamine sulfate to a coverslip that had been glued over a hole in the bottom of a petri dish (Stricker et al., 1992a); the attached cells were then covered with 10 ml of filtered sea water and treated with hormone while being imaged by video microscopy. Time-lapse recordings of the compressed slide preparations were carried out using a Zeiss Plan-Neofluor 100×, 1.3 NA oil immersion objective and DIC optics, whereas the noncompressed, attached cells in the petri dish were typically observed with a Zeiss Plan-Neofluor 40×, 0.75 NA dry objective and DIC optics. Since the oocytes mounted on slides were viewed under oil immersion and were thus "thermally coupled" to the microscope, the actual temperatures on the slide approached ambient temperatures rather than the 15°C set on the cooling stage. Thus, only those specimens viewed in a dish by the 40× objective could be reliably used to obtain data on nucleolar dynamics at 15°C. To compare the results of runs conducted at varying temperatures, the timing of nucleolar breakdown relative to nuclear envelope disassembly was normalized by assigning the onset of nuclear envelope breakdown in each run as 1 (Stricker and Schatten, 1991). Thus, the beginning of nucleolar breakdown was reported as a ratio of nucleolar breakdown onset time to nuclear envelope breakdown onset time, wherein ratios >1 represented runs in which nucleolar disassembly occurred after nuclear envelope breakdown began, and ratios <1 signified cases where nucleolar breakdown took place before the nuclear envelope disassembled.

In all time-lapse video microscopy runs, the video signal was captured using a Panasonic BD-400 CCD (charge-coupled device) camera that was attached to the trinocular head of the microscope. The signal was then processed with the aid of the Image-1 image processing system (Universal Imaging Corp., Westchester, Pennsylvania). For routine recordings, the unenhanced, live signal was subjected to background subtraction, a sixteen-frame jumping average, and contrast stretching, before being recorded to videotape using a Panasonic AG-6750 S-VHS time-lapse recorder set at a 36:1 time compression. Individual frames were then freeze framed and photographed from the video monitor using Kodak T-Max 100 film.

Electron Microscopy

Prophase-arrested and maturing oocytes were initially fixed on ice in a modified ruthenium red–sodium cacodylate solution of glutaraldehyde (Stricker et al., 1992b), before being postfixed in bicarbonate-buffered osmium tetroxide (Stricker and Reed, 1981). The samples were dehydrated in ethanol, embedded in LX-112 epoxy resin (Ladd Research Inc.), and cut at 60–70 nm on a diamond knife using a Sorvall MT-5000 ultramicrotome. Thin sections stained in uranyl acetate and lead citrate were examined at 80 kV using a Zeiss EM-109 transmission electron microscope.

Immunofluorescence and Confocal Microscopy

For immunofluorescence localization of nucleolar antigens, prophase-arrested oocytes were washed in calcium-free sea water and stripped of their surrounding vitelline envelope by treatment in a 1 percent sea water solution of thioglycolic acid at pH 10 (Maruyama et al., 1986). The denuded specimens were then: (1) immersed in an extraction buffer containing glycerol and Triton X-100 (Balczon and Schatten, 1983); (2) fixed in ice-cold methanol; and (3) examined by indirect immunofluorescence as described by Stricker and Schatten (1989). Primary antibodies included a commercially available mouse monoclonal antibody that recognizes undetermined components of nucleoli from various human cell types (Chemicon International, antibody MAB1277) and human polyclonal sera from patients with autoimmune diseases (courtesy of Dr. G. Maul, Wistar Institute). A major component of the polyclonal sera was antifibrillarin antibodies directed against the 30–38-kD fibrillarin protein(s) found in the nucleoli of various species (Reimer et al., 1987; Aris and Blobel, 1988; Henriquez et al., 1990). Appropriate secondary antibodies conjugated to FITC (fluorescein isothiocyanate) were used to visualize the localization of the nucleolar antigens, and stained preparations were mounted in a glycerol-based medium that contained p-phenylenediamine or Dabco (1,4-diazabicyclo[2.2.2]octane) (Aldrich Chemical Co.) to retard photobleaching.

For confocal microscopy, the immunofluorescence preparations were examined on a Bio-Rad MRC-600 laser scanning confocal microscope equipped with a Nikon Planapochromatic 60×, 1.4 NA oil immersion objective (Stricker et al., 1992a). The 488-nm excitation of the 25-mW argon-ion laser was attenuated with an ND 1 neutral density filter, and individual optical sections were recorded to the hard disk of the host computer (Stricker, 1994). In most cases, a correlative, nonconfocal brightfield image was also obtained with each confocal image. Hardcopy output was achieved by photographing the video monitor using Kodak T-Max 100 film.

Nucleolar Isolations and Immunoblotting

For mass isolations of nucleoli, prophase-arrested oocytes were obtained from one to several freshly excised ovaries, and their vitelline envelopes were removed by digestion in thioglycolic acid as described above. The denuded oocytes were then incubated for approximately 1 h at 15°C in an imidazole-based, Triton-containing extraction buffer (Bershadsky et al., 1978) with 2.5 μl/ml PMSF (phenylmethylsulfonyl fluoride) as a protease inhibitor. The incubation was continued until the cytoplasm was completely digested and a relatively clean preparation of germinal vesicles was obtained, judging from light microscopic examinations. All subsequent steps were conducted at 4°C to retard protein degradation.

After collection of the isolated germinal vesicles by settlement, the thioglycolate supernatant was decanted, and the nuclear pellet was homogenized through an 18-gauge needle. The homogenate was resuspended in 0.25 M sucrose/imidazole buffer and spun through an 0.88 M sucrose/imidazole-buffered cushion at 3000 g for 20 min. Pellets of nucleoli were collected and then spun again through a 0.88 M sucrose cushion at 1000 g for 3 min. Light microscopic observations revealed relatively clean nucleolar preparations following these steps. To visualize nucleic acids in isolated nucleoli, pellets were stained with 10 μg/ml of DAPI (4′,6-diamidino-2-phenylindole) dissolved in Bershadsky's buffer.

For immunoblotting studies, the nucleolar samples were solubilized in Laemmli sample buffer, boiled, and run on 10 percent SDS PAGE gels. The gels were silver stained (Morrissey, 1981) or transferred to nitrocellulose for Western blotting using the primary antibodies listed above. The blots were then reacted with alkaline phosphatase-conjugated secondary antibodies, according to standard protocols.

Results

Nucleoli of Prophase-arrested Oocytes

At the animal pole of the fully grown prophase-arrested oocyte is a hypertrophied nucleus referred to as the "germinal vesicle" (GV). Each germinal vesicle possesses a single large nucleolus. The diameter of the nucleolus averages 18.9 ± 2.9 μm ($N = 50$) in *Pisaster ochraceus* and 15.5 ± 1.0 μm ($N = 20$) in *Astropecten* sp. In living, noncompressed oocytes, the nucleolus tends to occur in an eccentric position that is randomly situated relative to the animal pole. As in the oocytes of some insects (Macgregor, 1982), the GV of starfish oocytes also contains a single small ($\sim$4-μm) spherical body of unknown composition and function, but there are no supernumerary, full-sized nucleoli, such as observed in amphibian oocytes (Duryee, 1950).

Throughout most of the breeding season, which for our laboratory-maintained specimens extended from about January to July in the case of *P. ochraceus* versus October to May for *Astropecten* sp., the nucleolus has a more or less circular outline and a slight indentation at one end. The indented end of the nucleolus is typically attached to a clump of filamentous material, which may represent the nucleolar-associated heterochromatin that has been noted for the nucleoli of many other cell types (Goessens, 1984). Based on three-dimensional reconstructions of serial sections obtained by confocal microscopy (data not shown), the nucleolus is not perfectly spherical. Instead, it tends to be somewhat flattened perpendicular to the animal-vegetal axis of the oocyte.

Internally, the nucleolus exhibits a distinct substructure when viewed by DIC optics (Fig. 3.1a) or by fluorescence microscopy in the case of DAPI-stained preparations (Fig. 3.1c). The most conspicuous elements of the nucleolus are numerous craterlike spaces of irregular size and shape that are commonly referred to as nucleolar "vacuoles" even though they are not surrounded by a membrane (Jordan, 1984; Kato et al., 1990). Such vacuoles are discernible in living specimens by light microscopy and often contain flocculent material. In sectioned specimens viewed by EM, the content of most vacuoles seems to be similar to that occurring in the surrounding nucleoplasm. Thus, the vacuoles may be contiguous with the nucleoplasm and could represent invaginations such as found in many other nucleoli (Goessens, 1984; Jordan, 1984). In some vacuoles, there is also electron-dense material, ranging from fine particles to a more unusual tubulo-fibrillar form (Figs. 3.1d, e). The composition and significance of this vacuolar material remain unknown.

In addition to the vacuoles, two distinct regions can be discerned in sectioned nucleoli: (1) a single, lightly staining area that presumably corresponds to the fibrillar center, based on its overall appearance and its more or less central position within the nucleolus; and (2) a more dense peripheral region that seems to represent the nucleolar granular component (Figs. 3.1b, h). No delimiting membrane occurs at the periphery of either nucleolar component. Associated with the fibrillar center are numerous patches of dense fibrillar to slightly granular material that may represent parts of a poorly defined dense fibrillar component (Goessens, 1984) or grazing sections through regions of the granular component that are indented in the fibrillar center. In any case, a conspicuous dense fibrillar region, such as observed in most other nucleoli, was not readily evident in the specimens examined in this study.

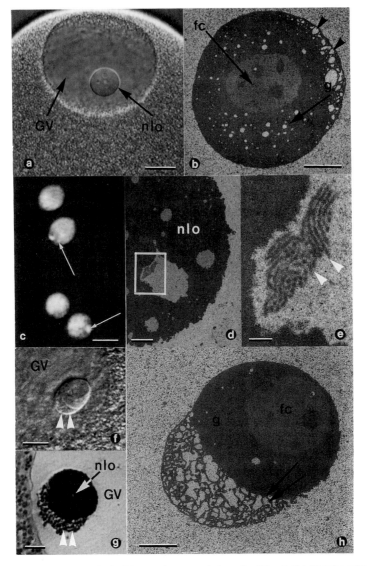

FIGURE 3.1. *a*. Photomicrograph of a living prophase-arrested oocyte of the starfish *Pisaster ochraceus*, showing the large germinal vesicle (GV) and its eccentrically positioned nucleolus (nlo) prior to treatment with maturation-inducing hormone. (scale bar = 20 μm); *b*. Electron micrograph of a nucleolus within the GV of a prophase-arrested *P. ochraceus* oocyte, showing the putative fibrillar center (fc) and granular component (g) with numerous vacuoles (arrowheads) typically present in the nucleolus (scale bar = 5 μm); *c*. Fluorescence photomicrograph of DAPI-stained nucleoli isolated from prophase-arrested *P. ochraceus* oocytes, with arrows marking heterogeneous staining within the nucleoli (scale bar = 20 μm); *d*. Electron micrograph of the periphery of a nucleolus (nlo) from a *P. ochraceus* oocyte, showing vacuoles (the box outlines the region shown at higher magnification in *e*) (scale bar = 1 μm); *e*. Electron micrograph of tubulo-fibrillar structures (arrowheads) located within a nucleolar vacuole in *P. ochraceus* (scale bar = 175 nm); *f*. Photomicrograph of a prophase-arrested *P. ochraceus* oocyte obtained early in the breeding season, showing the germinal vesicle (GV) and nucleolus that possesses a caplike region (double arrowheads) (scale bar = 20 μm); *g*. Photomicrograph of a 1-μm section through an *Astropecten* sp. prophase-arrested oocyte, showing the germinal vesicle (GV) and nucleolus (nlo) with a reticulated caplike region (double arrowheads) (scale bar = 20 μm); *h*. Electron micrograph of the nucleolus in the germinal vesicle of a prophase-arrested oocyte of *P. ochraceus*, showing the putative fibrillar center (fc), granular component (g), and the reticulated nucleonemalike network that constitutes the polar cap of the nucleolus (double arrows) (scale bar = 5 μm).

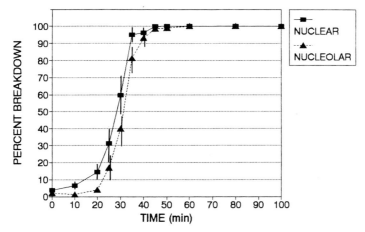

FIGURE 3.2. Graph summarizing the timing of nucleolar breakdown (NUCLEOLAR) relative to germinal vesicle breakdown (NUCLEAR), based on the averages of ten fixation series conducted at 15°C. For these studies, small batches of *P. ochraceus* oocytes that were maintained in an incubator at $15 \pm 1°C$ were treated with 5×10^{-6} *M* 1-methyladenine to trigger the resumption of meiosis and subsequently fixed in a Triton/glycerol extraction buffer at various times. The presence or absence of an intact nuclear envelope and nucleolus was then ascertained in one hundred specimens per time point. Note that the nucleolus tends to break down slightly after the nuclear envelope disassembles. The specimens undergoing breakdown well in advance of 20 min presumably represent "spontaneously maturing" oocytes. Vertical bars indicate standard errors of the mean. TIME = minutes following application of maturation-inducing hormone.

Prior to the peak of the breeding season, there seems to be a transient period (e.g., during March for *P. ochraceus*) when oocytes can contain nucleoli that are distinctly elongated. In addition to the putative fibrillar center and granular components described above, these elongated nucleoli also possess a caplike region that comprises a reticulated network of unknown significance (Figs. 3.1f–h). On an ultrastructural level, such a network resembles the "nucleonemas" that are characteristic of the noncompact stages of nucleoli in other species (Busch and Smetana, 1970; Fawcett, 1981). During this time of year, oocytes obtained from the same ovary may display elongated nucleoli with caps or the more spherical nucleoli lacking a cap. In *Astropecten* sp., some of the elongated nucleoli possess a cap at both of their ends.

Timing of Nucleolar Disassembly in Hormone-treated Oocytes

Following addition of the maturation-inducing hormone 1-methyladenine, prophase-arrested oocytes resume meiotic maturation and undergo germinal vesicle breakdown. Based on serial fixations and extractions of populations of *P. ochraceus* oocytes, GVBD usually begins slightly before the nucleolus starts to disassemble (Fig. 3.2). At 15°C, GVBD has begun in 50 percent of the oocytes by 29.2 ± 2.8 min, whereas one-half of the oocytes have started nucleolar breakdown by 30.6 ± 3.8 min. Similarly, time-lapse studies of individual oocytes monitored at 15°C also indicate that nucleolar disassembly usually begins 1–2 min after the nuclear envelope vesiculates. In a few cases (<15 percent), the nucleolus begins to break down slightly before GVBD. The incidence of such "premature" nucleolar breakdowns is somewhat greater at room temperature and when dealing with *Astropecten* sp. oocytes (data not shown).

Patterns of Nucleolar Disassembly

In time-lapse video sequences conducted at 15°C, no changes can be observed in living nucleoli for the first 5–10 min following treatment with 1-methyladenine, except for the fact that the nucleolus often moves 5–20 μm toward the center of the GV. Such nucleolar movements do not usually occur in control oocytes not treated with hormone. By 15–20 min, many of the nucleolar vacuoles shift positions and sometimes fuse with neighboring vacuoles. In elongated nucleoli with polar caps (Figs. 3.1f–h), the caplike structure gradually becomes incorporated into the rest of the nucleolus (data not shown). Thus, the distinction between the cap and nucleolus proper becomes less evident around this time in oocyte maturation.

At approximately 20–25 min posttreatment, nucleoli invariably undergo one to several vacuolar "extrusions," wherein the nucleolar vacuole rapidly moves toward the edge of the nucleolus and subsequently disappears (Fig. 3.3a). When properly oriented relative to the optical axis, these extrusions can be seen to unload flocculent material that eventually disperses into the nucleoplasm (data not shown). Nucleolar extrusions take place at a relatively defined time that corresponds to about 80 percent of the interval between hormone treatment and the onset of GVBD (Fig. 3.4). Such extrusions occur in oocytes of *P. ochraceus* and *Astropecten* sp. examined at physiological temperatures (i.e., 15°C), as well as at room temperature, and are evident whether or not the oocyte is compressed by an overlying coverslip. No extrusions occur during observations of control specimens (i.e., those oocytes lacking 1-methyladenine treatment and examined by time-lapse microscopy for up to several hours).

Following the onset of GVBD, the nucleolus begins to disassemble by one of two distinct patterns. Both disassembly patterns are evident throughout the breeding season and can be observed in oocytes obtained from the same female, regardless of when the oocytes were actually isolated from the ovary. In approximately one-third of hormone-treated specimens, the nucleolus vesiculates into a morulalike cluster, and within 5–10 min the individual vesicles become scattered in the former nuclear region of the animal pole (Fig. 3.3b). Such vesiculated nucleoli also occur in a few maturing oocytes of *Astropecten* sp. (Fig. 3.3c). In the majority of hormone-treated specimens, however, the nucleolus undergoes a gradual dissolution without forming distinct vesicles (Figs. 3.3d–f). In these cases, a significant portion of the disintegrating nucleolus becomes positioned near the developing meiotic apparatus (Fig. 3.3e). Typically, one to several spherical components of the nucleolus remain visible for an extended period following dissolution of most of the nucleolus (data not shown). By 80 min posttreatment, however, there are no conspicuous nucleolar remnants detectable by conventional brightfield microscopy, except in a few specimens which contain a refractile body that is slightly larger than the chromosomes assembling at the metaphase I plate (Fig. 3.3f).

Because external concentrations of calcium have been shown to affect nucleolar morphology in other systems (Selzer et al., 1991), nucleolar disassembly was also examined in calcium-free sea water. Under these conditions, a morulalike vesiculation occurred in only 1 of 10 time-lapse runs, and nearly all of the nucleoli displayed instead the gradual dissolution pattern lacking distinct vesicles.

Immunolocalizations of Nucleolar Antigens

Currently, we are analyzing nucleolar composition and disassembly patterns by a variety of immunolocalization techniques in order to track various constituents of the nucleolus following nucleolar disassembly. Reported here are the results of studies on the highly conserved nucleolar protein fibrillarin (Ochs et al., 1985) as it occurs in prophase-arrested oocytes. Triton-extracted specimens viewed by confocal microscopy display heterogeneous staining within the nucleolus following treatment with a human autoimmune serum containing antibodies against fibrillarin (Fig. 3.3g). An intensely staining region that may correspond to the fibrillar center of the nucleolus is characteristically evident toward the periphery of the nucleolus. In addition, a weaker

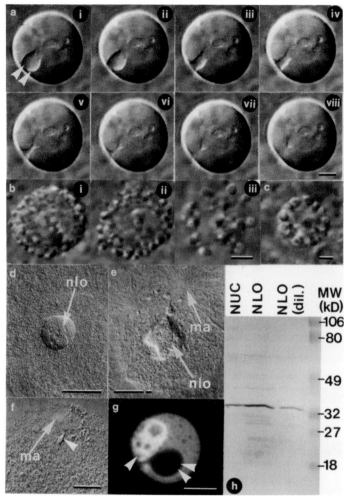

FIGURE 3.3 *a*. Time-lapse video microscopy of a nucleolar extrusion that occurs following addition of maturation-inducing hormone to a *P. ochraceus* oocyte. The double arrowhead marks a nucleolar vacuole that is extruding its contents into the nucleoplasm. In other cases, flocculent material extruded from the vacuole can actually be seen dispersing in the nucleoplasm. Note: this sequence was observed at room temperature, but similar extrusions occur in oocytes maintained at 15°C. (i) 15 min 24 s following addition of hormone; (ii) 15 min 29 s; (iii) 15 min 32 s; (iv) 15 min 34 s; (v) 15 min 36 s; (vi) 15 min 38 s; (vii) 15 min 39 s; (viii) 15 min 41 s. GVBD occurred 19 min after addition of hormone (scale bar = 5 μm). *b*. Time-lapse microscopy of hormone-induced nucleolar disassembly via a vesicular pattern in *P. ochraceus*. This sequence was run at room temperature, but such vesiculated nucleoli also occur at 15°C. GVBD began at 19.5 min after addition of hormone. (i) 21 min 38 s following hormone treatment; (ii) 22 min 49 s; (iii) 26 min 33 s (scale bar = 5 μm). *c*. The nucleolus of a hormone-treated *Astropecten* sp. oocyte, showing a vesicular disassembly pattern (scale bar = 4 μm). *d*. Photomicrograph of a nucleolus (nlo) in a Triton-extracted *P. ochraceus* oocyte at the onset of nucleolar disassembly, showing an early stage in of a gradual, nonvesicular pattern of nucleolar disintegration (compare with a vesicular type of disassembly as shown in *b*(i) (scale bar = 20 μm). *e*. Photomicrograph of a Triton-extracted *P. ochraceus* oocyte, showing a nucleolus (nlo) undergoing a gradual, nonvesicular type of disassembly near the meiotic apparatus (ma) (scale bar = 20 μm). *f*. Photomicrograph of a Triton-extracted *P. ochraceus* oocyte, following nucleolar disassembly. A small refractile body (arrowhead) that may represent a remnant of the nucleolus can be seen near the meiotic apparatus (ma) (scale bar = 20 μm). *g*. Confocal laser scanning photomicrograph of a nucleolus in the germinal vesicle of a Triton-extracted, prophase-arrested oocyte of *P. ochraceus*. The oocyte was stained with a polyclonal serum containing antibodies against fibrillarin. Note the intense staining (arrowhead) in what may represent the fibrillar center of the nucleolus. Unstained vacuoles (double arrowheads) are typically present in these preparations (scale bar = 10 μm). *h*. Western blot of germinal vesicles (NUC) and nucleoli (NLO) isolated from prophase-arrested *P. ochraceus* oocytes and stained with a polyclonal serum containing antibodies against fibrillarin. Note the ~35-kD major band corresponding to fibrillarin, as well as numerous minor bands that are enhanced in the nucleolar preparation. NLO (dil.) represents a sixfold dilution of the nucleoli compared to the amount loaded on lane 2 (NLO).

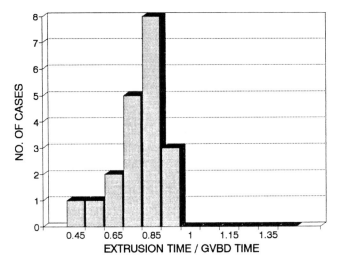

FIGURE 3.4. Graph showing the timing of nucleolar extrusion events relative to germinal vesicle breakdown, based on data obtained from time-lapse microscopy runs using *P. ochraceus* oocytes. Individual oocytes were examined at room temperature using DIC optics and a 100× oil immersion objective. To allow comparisons of data obtained at varying room temperatures, the onset of GVBD in each time-lapse run is normalized as 1, and the timing of nucleolar extrusion is expressed as a ratio of nucleolar extrusion time:GVBD time. Note that extrusions occur prior to GVBD (i.e., ratio <1) and that the average time point for extrusion corresponds to approximately 80 percent of the time interval between hormone application and GVBD.

background signal and numerous unstained vacuoles are scattered throughout the granular component of the nucleolus.

When isolated nucleoli from prophase-arrested oocytes of *P. ochraceus* are tested by Western blotting techniques, the autoimmune serum mainly recognizes an approximately 35-kD protein that corresponds well in molecular weight to fibrillarins isolated from other species. In addition, there are several minor reactive bands of unknown significance (Fig. 3.3h). Changes in the immunofluorescence patterns and immunoblotting characteristics of this fibrillarinlike antigen during germinal vesicle breakdown are currently being investigated.

Discussion

In this study, the timing and patterns of nucleolar disassembly during starfish oocyte maturation are described. Based on serial fixations and time-lapse imaging of living hormone-treated oocytes, the nucleolus characteristically begins to disassemble after the onset of GVBD, which in turn is signaled by the vesiculation of the nuclear envelope. A similar timing of nucleolar breakdown relative to GVBD was noted in a previous study that focused on nuclear lamina dynamics during oocyte maturation in *Pisaster ochraceus* (Stricker and Schatten, 1989). Species-specific variations apparently exist, however, because nucleoli break down prior to GVBD in *Asterias forbesii* (Chambers and Chambers, 1949).

Within 10–15 min following treatment with maturation-inducing hormone, the nucleolus of *P. ochraceus* oocytes often moves a few micrometers toward the center of the nucleus. Such nucleolar translocations also occur in maturing oocytes of *A. forbesi*, where nucleoli "move in germinal vesicles in the direction of gravity" at velocities of up to 0.1 μm/s (Hiramoto, 1976). It is possible that nucleoli simply remain stationary in the relatively viscous GVs of prophase-arrested oocytes but then tend to translocate to a lower position in the GV, following a hormone-induced change in the viscoelastic properties of the nucleoplasm (Hiramoto, 1976). Observations of nucleolar movements made in this study support this view, since nucleoli almost invariably move toward the center of the GV, as would be expected if they were simply rolling toward the lowest point of the spherical nucleus or they were pushed in that direction by compression of the overlying coverslip, as the GV undergoes its microfilament-mediated changes in shape prior to GVBD (Stricker and Schatten, 1991).

By 10–15 min posttreatment with hormone, the vacuoles in the nucleoli move about and on occasion fuse to form larger vacuoles. Shortly thereafter, nucleoli undergo one to several "extrusions," in which the contents of a vacuole are deposited into the nucleoplasm. Such extrusions apparently represent a novel nucleolar process and do not correspond to the nucleolar extrusions noted in other oocytes, in which parts of the nucleolus are transported across the nuclear envelope into the cytoplasm (Duryee, 1950; Kessel and Beams, 1963). As reviewed by Busch and Smetana (1970), controversy exists concerning whether the extrusions of nucleoli into the cytoplasm represent a normal phenomenon or a pathological condition. We believe the intranuclear extrusions of nucleolar vacuoles observed in this study correspond to normal stages in oocyte maturation for the following reasons: (1) they occur at a relatively specific time prior to GVBD; (2) they are not artifactually produced by compression or abnormal temperature; and (3) the process of time-lapse microscopy itself does not lead to nucleolar extrusions, since control oocytes not treated with hormone fail to undergo extrusions during several hours of observation.

Exactly how nucleolar extrusions occur remains unclear, because there is no ultrastructural evidence for a membranous or contractile network within nucleoli that could provide the motive force for extrusions. Alternatively, the eversion of nucleolar contents could be simply due to changes in the physicochemical properties of nucleoli. This hypothesis is based on studies of the "colloidal coacervate" nature of nucleoli obtained from *Xenopus* oocytes, wherein nucleolar vacuoles can be stimulated to evert their contents by treatment with dilute acids (Duryee, 1950). Similarly, changes that occur in the nuclear environment following treatment with maturation-inducing hormone may cause the nucleolar vacuoles of starfish oocytes to undergo extrusion by changing the colloidal properties of the nucleolus, although experimental confirmations of this are lacking. The possible functions of nucleolar extrusions, if any, remain unknown. Currently, we are attempting to define which components of the nucleolus are extruded, so that we can conduct experimental manipulations of the extrusion event and thereby assess the biological significance of this phenomenon.

After the nucleolar extrusions have occurred, the nucleolus begins to disassemble via one of two very different patterns. In a minority of cases, nucleolar breakdown involves the formation of conspicuous vesicles that disperse in the nucleoplasm as it mixes with the surrounding cytoplasm. More commonly, however, nucleoli undergo a gradual dissolution that does not involve vesicle formation. It remains unclear exactly which factors prompt one form of disassembly over the other. Because both types of disassembly are observed throughout the breeding season and can be seen in oocytes obtained from a single female regardless of when the oocytes are isolated from the ovary, the differences are not simply due to obvious seasonal or interspecimen variation.

In experiments utilizing calcium-free sea water, few vesiculated nucleoli are observed, suggesting that a calcium-mediated step may be needed for vesicle formation. However, approximately two-thirds of the nucleolar disassemblies in normal, calcium-containing sea water also do not progress via vesiculation. Hence, the presence of calcium in the external medium is not always sufficient by itself to produce nucleolar vesicles in maturing oocytes. An additional consideration is that the calcium-free sea water used in this study is buffered with Tris, whereas the calcium-containing Instant Ocean solution utilizes bicarbonate to buffer the pH. Thus, the differences observed in disassembly patterns may not necessarily be due to the presence or absence of calcium ions in the medium, but may instead be related to possible changes in pH that occur over the course of the experiment in one solution versus the other.

As is true of nucleoli in other cells (Busch and Smetana, 1970), the ultrastructure of nucleoli in prophase-arrested starfish oocytes is highly complex and typically comprises what appears to be a fibrillar center surrounded by a peripheral granular component. Unlike other nucleoli, however, the specimens observed in this study apparently lack a conspicuous dense fibrillar component, at least during the times of the year that these examinations were carried out. The absence of a recognizable dense fibrillar component could be due to the fact that: (1) preribosomal particles do not accumulate to form a distinct granular component during the stages of oogenesis observed in this study and thus the peripheral region that has been tentatively identified as the granular component actually represents a highly developed dense fibrillar component; (2) the dense fibrillar component blends imperceptibly with the other parts of the nucleolus; (3) a dense fibrillar component is simply lacking throughout the year in these nucleoli; or (4) the oocytes observed in this study were not actively transcribing their ribosomal genes at the time of fixation and hence lacked the dense fibrillar component that is present at other times of the year.

The possibility that nucleolar morphology and composition may vary during the stages of oogenesis and actual months of the breeding season is further supported by the fact that a caplike region with a nucleonemalike composition is transiently observed prior to the peak of the breeding season. The functional significance of these caps is unclear, although it is possible that they may represent mixed fibrillar and granular components that arise during discrete bursts of transcription, as has been noted for nucleoli in other cell types (Smetana and Busch, 1974). However, if the nucleonemalike caps of starfish oocyte nucleoli do in fact represent sites of active transcript accumulation, it remains unclear why they would be positioned such a great distance from the putative fibrillar center (cf. Fig. 3.1h). Thus, further cytochemical analyses and radiotracer studies are needed to monitor rRNA production before the functional significance of the polar caps (or for that matter other components) of the nucleoli can be determined.

Based on electrophoretic studies, over one hundred nucleolar proteins have been detected (Prestayko et al., 1974; Smetana and Busch, 1974). A few of these nucleolar proteins have been well characterized, and information regarding their biological functions has been obtained (e.g., Franke et al., 1981; Schmidt-Zachmann et al., 1984; Lapeyre et al., 1987). For example, the highly conserved protein fibrillarin is a component of the U3 snRNP (small nuclear ribonucleoprotein), which in turn plays a vital role in the initial processing of primary rRNA transcripts (Kass et al., 1990; Fischer et al., 1991).

In this study, we show that a human autoimmune antiserum containing antibodies against fibrillarin crossreacts with an approximately 35-kD band on Western blots of isolated nucleoli from starfish oocytes. In immunofluorescence localizations, the fibrillarinlike antigen tends to localize in a distinct patch within the nucleoli of prophase-arrested oocytes. Whether or not this patch corresponds to the fibrillar center of the nucleolus, as has been shown in other cell types (Aris and Blobel, 1988), needs to be verified by immunoelectron microscopic analyses. The weaker background staining observed throughout the periphery of the nucleolus is probably due to antibodies against nonfibrillarin antigens that show up as minor bands in the immunoblots of this polyclonal serum, since in other species fibrillarin is absent in the granular component of the nucleolus (Fischer et al., 1991).

Recent investigations of a variety of cell types have verified that considerable shuttling of nucleolar proteins takes place between the nuclear and cytoplasmic compartments of the cell (Borer et al., 1989). Export of nucleolar proteins can occur during the translocation of ribosomal particles to the cytoplasm (Mehlin et al., 1992). In the case of trafficking of nucleolar proteins into the nucleus, many nucleolar proteins possess specific amino acid domains, referred to as nucleolar localization signals, that in turn help these proteins to target the nucleolus following their translocation into the nucleus from their cytoplasmic site of synthesis (Peculis and Gall, 1992). Similarly, a 140,000-MW protein which possesses a nuclear localization signal for importing proteins into the nucleus occurs in the nucleolus of Buffalo rat liver cells (Meier and Blobel, 1990). Since the nuclear localization signal of this protein presumably acts on the cytoplasmic side the nuclear envelope, it is postulated that the nucleolar-bound pool of the 140,000-MW protein serves as a reservoir and that normally this protein is shuttled to the cytoplasm where it can aid protein translocation into the nucleus (Meier and Blobel, 1990). Such biochemical and molecular analyses of changing nucleolar composition coupled with the morphological evidence of nucleolar extrusion events and varying disassembly patterns presented in this study underscore the fact that the nucleolus is a highly complex and dynamic structure that undergoes considerable modifications during the cell cycle and development.

Acknowledgments

We are indebted to the following people for their able assistance with various aspects of this project: M. Bellis, V. Centonze, R. Christner, C. Gillas, L. Hertel, and C. Morris. Dr. G. Maul of the Wistar Institute kindly supplied the antifibrillarin sera. Confocal microscopic observations were conducted at the Integrated Microscopy Resource, Madison, Wisconsin (NIH Biomedical Technology Resource RR570). Parts of this study were supported by a grant from the University of New Mexico Research Allocation Committee. The senior author fondly dedicates this chapter to the memory of Christopher G. Reed, the scientist and friend who is sorely missed.

Literature Cited

Aris, J., and G. Blobel. 1988. Identification and characterization of a yeast nucleolar protein that is similar to a rat liver nucleolar protein. J. Cell Biol. 107: 17–31.

Balczon, R., and G. Schatten. 1983. Microtubule containing detergent extracted cytoskeletons in sea urchin eggs from fertilization through cell division: Antitubulin immunofluorescence microscopy. Cell Motil. 3: 213–226.

Bershadsky, A. D., V. I. Gelfand, T. M. Svitkina, and I. S. Tint. 1978. Microtubules in mouse embryo fibroblasts extracted with Triton X-100. Cell Biol. Int. Rep. 2: 425–432.

Borer, R. A., C. F. Lehner, H. M. Eppenberger, and E. A. Nigg. 1989. Major nucleolar proteins shuttle between nucleus and cytoplasm. Cell 56: 379–390.

Busch, H., and K. Smetana. 1970. The Nucleolus. Academic Press, New York.

Chambers, R., and E. L. Chambers. 1949. Nuclear and cytoplasmic interrelations in the fertilization of the *Asterias* egg. Biol. Bull. 96: 270–282.

Cloney, R. A., J. Schaadt, and J. V. Durden. 1970. Thermoelectric cooling stage for the compound microscope. Acta Zool. (Stockholm) 51: 95–98.

Duryee, W. R. 1950. Chromosomal physiology in relation to nuclear structure. Ann. N.Y. Acad. Sci. 50: 920–942.

Fawcett, D. W. 1981. The Cell. W. B. Saunders Co., Philadelphia.

Fischer, D., D. Weisenberger, and U. Scheer. 1991. Assigning functions to nucleolar structures. Chromosoma 101: 133–140.

Franke, W. W., J. A. Kleinschmidt, H. Spring, G. Krohne, C. Grund, M. F. Trendelenburg, M. Stoehr, and U. Scheer. 1981. A nucleolar skeleton of protein filaments demonstrated in amplified nucleoli of *Xenopus laevis*. J. Cell Biol. 90: 289–299.

Goessens, G. 1984. Nucleolar structure. Int. Rev. Cytol. 87: 107–158.

Henriquez, R., G. Blobel, and J. P. Aris. 1990. Isolation and sequencing of Nop1-A yeast gene encoding a nucleolar protein homologous to a human autoimmune antigen. J. Biol. Chem. 265: 2209–2215.

Hiramoto, Y. 1976. Mechanical properties of starfish oocytes. Dev. Growth Differ. 18: 205–209.

Jordan, E. G. 1984. Nucleolar nomenclature. J. Cell Sci. 67: 217–220.

Jordan, E.G., and C.A. Cullis. 1982. The Nucleolus. Cambridge University Press, Cambridge.

Kass, S., J. A. Steitz, and B. Sollner-Webb. 1990. The U3 small nucleolar ribonucleoprotein functions in the first step of the preribosomal RNA processing. Cell 60: 897–908.

Kato, K. H., S. Washitani-Nemoto, A. Hino, and S.-I. Nemoto 1990. Ultrastructural studies on the behavior of centrioles during meiosis of starfish oocytes. Dev. Growth Differ. 32: 41–49.

Kessel, R. G., and H. W. Beams. 1963. Nucleolar extrusion in oocytes of *Thyone briareus*. Exp. Cell Res. 32: 612–615.

Lapeyre, B., H. Bourbon, and F. Amalric. 1987. Nucleolin, the major nucleoprotein of growing eukaryotic cells: An unusual protein structure revealed by the nucleotide sequence. Proc. Natl. Acad. Sci. U.S.A. 84: 1472–1476.

Macgregor, H. C. 1982. Ways of amplifying ribosomal genes. *In* E. G. Jordan and C. A. Cullis, eds. The Nucleolus. Cambridge University Press, Cambridge, pp. 129–152.

Maruyama, Y. K., K. Yamamoto, I. Mita-Miyazawa, T. Kominami, and S.-I. Nemoto. 1986. Manipulative methods for analyzing embryogenesis. *In* T. E. Schroeder, ed. Methods in Cell Biology. Academic Press, Orlando, Fl., pp. 325–344.

Mehlin, H., B. Daneholt, and U. Skoglund. 1992. Translocation of a specific premessenger ribonucleoprotein particle through the nuclear pore studied with electron microscope tomography. Cell 69: 605–613.

Meier, U. T., and G. Blobel 1990. A nuclear localization signal binding protein in the nucleolus. J. Cell Biol. 111: 2235–2245.

Morrissey, J. H. 1981. Silver stain for proteins in polyacrylamide gels: A modified procedure with enhanced uniform sensitivity. Anal. Biochem. 117: 307–310.

Ochs, R. L., M. A. Lischwe, W. H. Spohn, and H. Busch. 1985. Fibrillarin: A new protein of the nucleolus identified by autoimmune sera. Biol. Cell 54: 123–134.

Peculis, B. A., and J. G. Gall. 1992. Localization of the nucleolar protein NO38 in amphibian oocytes. J. Cell Biol. 116: 1–14.

Prestayko, A. W., G. R. Komp, D. J. Schmoll, and H. Busch. 1974. Comparison of proteins of ribosomal subunits and nucleolar preribosomal particles from Novikoff hepatoma ascites cells by two-dimensional polyacrylamide gel electrophoresis. Biochemistry 13: 1945–1951.

Reimer, G., K. M. Pollard, C. A. Penning, R. L. Ochs, M. A. Lischwe, H. Busch, and E. M. Tan. 1987. Monoclonal autoantibody from a (New Zealand black × New Zealand white) F1 mouse and some human scleroderma sera target an Mr 34,000 nucleolar protein of the U3 RNP particle. Arthritis Rheum. 30: 793–800.

Scheer, U., and R. Benavente. 1990. Functional and dynamic aspects of the mammalian nucleolus. Bioessays 12: 14–21.

Schmidt-Zachmann, M. S., B. Hugle, U. Scheer, and W. W. Franke. 1984. Identification and localization of a novel nucleolar protein of high molecular weight by a monoclonal antibody. Exp. Cell Res. 153: 327–346.

Schroeder, T. E., and S. A. Stricker. 1983. Morphological changes during maturation of starfish oocytes: Surface ultrastructure and cortical actin. Dev. Biol. 98: 373–384.

Selzer, P. M., P. Webster, and M. Duszeko. 1991. Influence of Ca 2+ depletion on cytoskeleton and nucleolus morphology in *Trypanosoma brucei*. Eur. J. Cell Biol. 56: 104–112.

Smetana, K., and H. Busch. 1974. The nucleolus and nucleolar DNA. *In* H. Busch, ed. The Cell Nucleus. Academic Press, New York, pp. 73–147.

Stahl, A. 1982. The nucleolus and nucleolar chromosomes. *In* E. G. Jordan and C. A. Cullis, eds. The Nucleolus. Cambridge University Press, Cambridge, pp. 1–24.

Stricker, S. A. 1994. Confocal microscopy of living eggs and embryos. *In* J. K. Stevens, L. R. Mills, and J. E. Trogadis, eds. Three Dimensional Confocal Microscopy. Academic Press, San Diego. (in press).

Stricker, S. A., and C. G. Reed. 1981. Larval morphology of the nemertean *Carcinonemertes epialti* (Nemertea: Hoplonemertea). J. Morph. 169: 61–70.

Stricker, S. A., and G. Schatten. 1989. Nuclear envelope disassembly and nuclear lamina depolymerization during germinal vesicle breakdown in starfish. Dev. Biol. 135: 87–98.

———. 1991. The cytoskeleton and nuclear disassembly during germinal vesicle breakdown in starfish oocytes. Dev. Growth Differ. 33: 163–171.

Stricker, S. A., V. E. Centonze, S. W. Paddock, and G. Schatten. 1992a. Confocal microscopy of fertilization-induced calcium dynamics in sea urchin eggs. Dev. Biol. 149: 370–380.

Stricker, S. A., A. M. Welford, and C. A. Morris. 1992b. Somatic cell-oocyte interactions during oogenesis in the acoel flatworm *Childia groenlandica*. Invert. Reprod. Dev. 21: 57–77.

4 Review of Dynamic Changes in the Cytoskeleton during Early Stages of Starfish Oocyte Maturation

Joann J. Otto

ABSTRACT The synchronous maturation of starfish oocytes can be induced in vitro by treatment with the hormone 1-methyladenine (1-MA). This makes it possible to study events that occur at particular phases of the meiotic cell cycle. This chapter reviews changes that occur in the cytoskeleton during the meiotic cell cycle. Within 1 min of 1-MA treatment, actin begins to polymerize to form microfilaments that form the structural core of spikes that protrude from the cell surface. These spikes are transient structures and disappear within 15–20 min after 1-MA addition. Their formation and disassembly are accompanied by the redistribution of two actin-associated proteins, fascin and a homologue of fodrin. Immature oocytes contain an extensive array of cortical microtubules. These microtubules disappear shortly after maturation has begun and do not reappear until it is complete. A network of fibers that stain with antibodies against the intermediate filament protein, cytokeratin, is present in immature oocytes. Similar to the cortical microtubule array, this network disassembles shortly after 1-MA addition; however, it does not appear to reassemble when maturation is completed.

Introduction

The cytoskeleton of animal cells, in conjunction with cell adhesion mechanisms, determines the shape and motility of cells. The cytoskeleton also functions in the anchorage of organelles and intracellular transport, particularly of vesicles. Microfilaments, microtubules, and intermediate filaments compose the cytoskeleton of animal cells. These cytoskeletal filament systems are present in arrays which differ among cell types. In many somatic cells, each filamentous array undergoes changes in its organization during the progress of the cell through the cell cycle. This is perhaps most dramatically illustrated by the microtubule-based cytoskeleton. During interphase in most somatic cells, a cytoplasmic microtubule array is present which extends throughout the cell. At the beginning of the mitotic phase, the interphase microtubules disassemble, and tubulin, the protein subunit of microtubules, polymerizes to form the mitotic spindle which separates the chromosomes. Although the reorganization of the cytoskeleton in somatic cells was well documented by the late 1970s, similar studies in gametes and germ line cells did not begin until the early 1980s. The starfish oocyte provides an excellent system to study since the cell cycle can be triggered to resume in vitro. In this chapter, after introducing key aspects of starfish oocyte maturation, I review studies on the reorganization of microfilaments, microtubules, and an intermediate filament protein-containing network during starfish oocyte maturation.

Joann J. Otto, Department of Biological Sciences, Purdue University, West Lafayette, IN 47907–1392.

Starfish Oocyte Maturation

Fully developed starfish oocytes arrested in prophase of meiosis I are stored in the ovary. Upon stimulation by the hormone 1-methyladenine (1-MA), meiosis resumes (reviewed by Meijer and Guerrier, 1984). In vivo, 1-MA is released by follicle cells surrounding the oocyte. In vitro, isolated oocytes treated with 1-MA undergo meiotic maturation. This ability to induce meiotic maturation in vitro has allowed various effects of 1-MA action to be studied.

To induce meiosis, 1-MA must be present at or above a threshold concentration for a certain length of time, called the hormone-dependent period. The hormone-dependent period varies with species and temperature, but generally is on the order of 20–40 min (reviewed by Meijer and Guerrier, 1984). The hormone treatment need not be continuous. If the periods of hormone exposure are discontinuous, they are additive; thus, the cell must have a mechanism for "remembering" the previous length of hormone exposure. When the hormone-dependent period ends, the oocyte undergoes germinal vesicle (nuclear envelope) breakdown, the first obvious sign of reentering the cell cycle.

The receptor for 1-MA and the second messenger system or systems for hormone action remain unknown. Both the phosphatidylinositol and the adenylate cyclase pathways may be involved in controlling oocyte maturation, but fluctuations in at least some of the individual messengers in these pathways alone are insufficient to stimulate oocyte maturation. Alterations in the activity of protein kinase C (Kishimoto et al., 1985) or the levels of inositol trisphosphate (Chiba et al., 1990), calcium (Witchell and Steinhardt, 1990), or cAMP (Meijer et al., 1989) are not sufficient for reinitiation of meiosis.

Presumably, the second messengers in the 1-MA pathway ultimately lead to the activation of maturation promoting factor (MPF), which has been shown to be a universal mediator of entry into the mitotic and meiotic (M) phases of the cell cycle (reviewed by Nurse, 1990). MPF is composed of two subunits: cyclin and homologues of the yeast cdc2 gene which encodes a serine/threonine protein kinase called $p34^{cdc2}$. Recently, Ookata et al. (1992) demonstrated that active MPF relocates from the cytoplasm into the germinal vesicle immediately prior to germinal vesicle breakdown in starfish oocytes. At this time MPF phosphorylates histone H1 (Standart et al., 1987) and presumably other substrates that are known to be phosphorylated by MPF in vitro (reviewed by Nurse, 1990; Yamashiro and Matsumura, 1991). In *Asterina pectinifera*, the one species in which the relative timing of these events has been examined in detail, the hormone-dependent period lasts 10–12 min (Kishimoto et al., 1985), MPF activation (as measured by H1 kinase activity) begins at about 7 min and is maximal at 20 min, and germinal vesicle breakdown begins at 18 min (Ookata et al., 1992). Thus, the activation of MPF begins as the hormone-dependent period is nearing completion.

Reorganization of the Cytoskeleton during the Cell Cycle

In most somatic animal cells that have been examined, the cytoplasmic cytoskeletal systems reorganize as the cells enter mitosis (for review, see Alberts et al., 1989). Cytoplasmic arrays of microtubules and microfilaments disassemble, and intermediate filaments usually aggregate. Microtubules then reassemble to form the mitotic spindle. Early in anaphase, the contractile ring, a microfilament-based structure, assembles and constricts the cell during cytokinesis. These reorganizations are thought to be mediated by the kinase activity of MPF; indeed, elements of each filament system have been identified as substrates of MPF in vitro (reviewed by Nurse, 1990; Yamashiro and Matsumura, 1991).

Although these changes in the cytoplasmic cytoskeleton immediately prior to mitosis have been well studied in somatic cells (for review, see Alberts et al., 1989), they have not been well documented in oocytes during meiosis. In this chapter, I review several dynamic changes that occur in the cytoskeleton of oocytes of the starfish, *Pisaster ochraceus*, during the hormone-dependent period. Presumably, these changes are either caused by, or involved in, the second messenger signaling that occurs during this time. In this species, the hormone-dependent period is about 20 min long, and germinal vesicle breakdown occurs at about 50 min at 12–15°C. The results reviewed here were primarily obtained in collaboration with Dr. Thomas Schroeder (Friday Harbor Laboratories) and are mainly derived from our studies of cortical preparations of oocytes. These preparations are made by attaching oocytes or eggs to polylysine-coated cover-slips, shearing them with a buffered solution to retain the cortex of the cell (the approximately 1-μm-thick layer of cytoplasm immediately beneath the plasma membrane), fixing the cortices, and then staining them with various antibodies against cytoskeletal proteins. The molecules that are localized in the cortex must be attached firmly enough to survive the shearing process. The dynamic changes that occur during early maturation involve all three cytoskeletal networks known in animal cells, namely, microfilaments, microtubules, and intermediate filaments.

Microfilaments

The actin-based cytoskeleton in the cortex of starfish oocytes rapidly reorganizes upon exposure to 1-MA. Monomeric or oligomeric actin, which is tightly bound to the cortex in unstimulated oocytes, begins to polymerize within 1 min of hormone addition (Otto and Schroeder, 1984a). Within 5–10 min the polymerized actin is organized into bundles that form the cores of spikes protruding from the surface of the oocyte (Schroeder, 1981; Schroeder and Stricker, 1983; Otto and Schroeder, 1984a). These spikes are transient structures and begin to recede about 10–15 min after the start of 1-MA treatment. After the spikes disappear, very little actin remains bound to the cortex.

Fascin, an actin-cross-linking protein, and an additional 220-kD protein, which is probably a homologue of fodrin, are localized within the spikes. Fascin presumably contributes to the bundling of actin in the spikes, and the 220-kD protein likely mediates the binding of the plasma membrane to the actin core of the spike (Otto and Schroeder, 1984a). Both these proteins exhibit dynamic associations with the cortex as visualized by immunofluorescence microscopy of cortical preparations. Fascin appears to be present in the cortex only in association with spikes while the 220-kD protein is localized in the cortex before spike formation in a diffuse pattern. Some of the 220-kD protein appears to remain after the spikes have disassembled. Before 1-MA addition, the 220-kD protein may function to bind oligomers of actin to the membrane just as spectrin, the fodrin homologue in vertebrates, does in erythrocytes.

Myosin II appears to be tightly and uniformly associated with the cortex throughout the period of spike formation and disassembly, but it is not incorporated into the spikes (Otto and Schroeder, 1984a). Correspondingly, the spikes are not seen to move or exhibit contractility (Schroeder, 1981). The observation that myosin II is associated with the plasma membrane in the apparent absence of filamentous (F-) actin is unusual; however, we find that myosin II in the contractile ring of dividing sea urchin eggs also may not require F-actin to associate with the plasma membrane (Schroeder and Otto, 1988).

In addition to the qualitative changes in the cortical actin cytoskeleton, I have quantified the amount of F-actin throughout oocyte maturation with an assay based on the specific binding of phalloidin to F-actin. Total F-actin declines approximately 25 percent between the time of 1-MA addition and germinal vesicle breakdown (Otto, unpublished). This decline in global F-actin parallels the decline in oocyte stiffness which occurs in starfish oocytes during maturation (Shoji et al., 1978; Nemoto et al., 1980) and may contribute to this loss of stiffness.

Microtubules

An extensive array of long criss-crossing microtubules is present in the cortex of unstimulated oocytes (Otto and Schroeder, 1984b). The microtubules in this array are sometimes longer than 60 μm and are present in at least the first 3 μm of cytoplasm beneath the plasma membrane; we do not yet know if the microtubules extend throughout the cytoplasm. The centrosome is localized at the animal pole quite near the cortex, and some of the microtubules emanate from it. However, the observation that short cortical microtubules, which are not connected to the centrosome, are present after nocodozole or colchicine treatment suggests that not all the microtubules are associated with the centrosome.

Within 20 min after 1-MA treatment (the approximate length of the hormone-dependent period), cortical microtubules begin to disassemble (Schroeder and Otto, 1984). Cortical microtubules remain absent until meiosis is complete. Somewhat later, they reassemble but are not as dense as before oocyte maturation. Although cortical microtubules are not present during most of meiosis, the meiotic spindles that contain microtubules are often isolated with the cortices, suggesting that the spindles may be attached to the cortex. Intriguingly, a portion of the MPF present in starfish oocytes appears to be associated with the meiotic spindles whereas prior to meiotic reinitiation it is primarily cytoplasmic (Ookata et al., 1992). This relocation of MPF correlates with the time of the disassembly of the cortical microtubules. It would be interesting to know if the cytoplasmic MPF is associated with cortical or cytoplasmic microtubules and is retained in the cytoplasm by this interaction.

Intermediate Filaments

Antibodies against cytokeratin, a subunit of one class of intermediate filaments, stains an extensive fibrous network in the cortex of immature oocytes (Schroeder and Otto, 1991). The network is a loose anastomosis of fibers about 0.05 to 0.3 nm in diameter and resembles the mesh pattern of a hairnet or "snood" (Schroeder and Otto, 1991). The animal pole per se does not exhibit anticytokeratin staining, but linear elements of stained fibers radiate from around the centrosome. Snood fibers often exhibit periodic antibody staining and have a 0.75-nm repeat substructure in electron micrographs (Schroeder and Otto, 1991); we do not yet know which element of the structure visualized by electron microscopy is staining with anticytokeratin. From electron micrographs, we also know that the snood is localized in the cortex of the oocyte; it does not appear to extend below the plasma membrane more than about 1 nm into the cytoplasm. The substructure of the snood is unusual because the filaments composing the snood fibers measure only 5 nm in diameter versus the typical 10 nm for intermediate filaments. The filaments may be protofilaments or a precursor pool for a keratin network which may arise later in development.

When oocytes are treated with 1-MA and the anticytokeratin staining pattern is examined as a function of time during oocyte maturation, the snood appears to disassemble in a polarized fashion from the vegetal pole to the animal pole of the oocyte. This disassembly begins near the end of the hormone-dependent period and is complete by the time of germinal vesicle breakdown. The snood does not reappear later during oocyte maturation. Similar observations have been made in amphibian oocytes except that the disassembly occurs from the animal to vegetal pole (Dent and Klymkowsky, 1989).

Perspectives

First and most obvious, it will be useful to understand the functions of cytoskeletal arrays in oocytes. The cytoskeletal elements may represent pools of stored subunits to be used later in development, or they may function in maintaining the organization of oocytes stored in the ovary or play a role in maturation events. These problems are extremely difficult to address experimentally because drugs that inhibit microtubule and actin function also disrupt mitosis (and meiosis) and cytokinesis, which are defining and necessary events in oocyte maturation and throughout subsequent development. There are also no drugs that specifically disrupt events mediated by intermediate filaments. The most useful approach will probably be to identify proteins that associate with only the cytoplasmic arrays of microfilaments, microtubules, and intermediate filaments and to design probes such as antibodies or antisense RNA to disrupt the function or synthesis of these proteins. Recently, the antisense RNA approach has shown that cytokeratin arrays occurring in *Xenopus* oocytes are required for proper epithelial structure and for morphogenesis during gastrulation (Torpey et al., 1992).

Second, once the second messenger system for 1-MA action is clarified, it will be interesting to determine if the messengers directly modulate the assembly states of cytoskeletal arrays. It is also conceivable that an investigation of the molecules regulating disassembly of the cytoskeleton will lead to the identification of the regulatory path that activates MPF.

Acknowledgments

I thank R. Heil-Chapdelaine, T. Schroeder, and J. Wulfkuhle for their helpful comments on the manuscript. The work described here was supported in part by National Science Foundation grants PCM-8020984 and MCB-9012165.

Literature Cited

Alberts, B., D. Bray, J. Lewis, M. Raff, K. Roberts, and J. Watson (eds.). 1989. Molecular Biology of the Cell. Garland Publishing, Inc., New York.

Chiba, K., R. T. Kado, and L. A. Jaffe. 1990. Development of calcium release mechanisms during starfish oocyte maturation. Dev. Biol. 140: 300–306.

Dent, J. A., and M. W. Klymkowsky. 1989. Whole-mount analyses of cytoskeletal reorganization and function during oogenesis and early embryogenesis in *Xenopus*. *In* H. Schatten and G. Schatten, eds. The Cell Biology of Fertilization. Academic Press, Orlando, Fl., pp. 63–103.

Kishimoto, T., M. Yoshikuni, H. Ikadai, and H. Kanatani. 1985. Inhibition of starfish oocyte maturation by tumor-promoting phorbol esters. Dev. Growth Differ. 27: 233–242.

Meijer, L., W. Dostmann, H. G. Genieser, E. Butt, and B. Jastorff. 1989. Starfish oocyte maturation: Evidence for a cyclic AMP-dependent pathway. Dev. Biol. 133: 58–66.

Meijer, L., and P. Guerrier. 1984. Maturation and fertilization in starfish oocytes. Int. Rev. Cytol. 86: 129–199.

Nemoto, S.-I., M. Yoneda, and I. Uemura. 1980. Marked decrease in the rigidity of starfish oocytes induced by 1-methyladenine. Dev. Growth Differ. 22: 315–325.

Nurse, P. 1990. Universal control mechanism regulating onset of M phase. Nature 344: 503–508.

Ookata, K., S-I. Hisanaga, T. Okano, K. Tachibana, and T. Kishimoto. 1992. Relocation and distinct subcellular localization of p34^{cdc2}-cyclin B complex at meiosis reinitiation in starfish oocytes. EMBO J. 11: 1763–1772.

Otto, J. J., and T. E. Schroeder. 1984a. Assembly-disassembly of actin bundles in starfish oocytes: An analysis of actin-associated proteins in the isolated cortex. Dev. Biol. 101: 263–273.

Otto, J. J. and T. E. Schroeder. 1984b. Microtubule arrays in the cortex and near the germinal vesicle of immature starfish oocytes. Dev. Biol. 101: 274–281.

Schroeder, T. E. 1981. Microfilament-mediated surface change in starfish oocytes in response to 1-methyl-adenine: Implications for identifying the pathways and receptors for maturation-inducing hormones. J. Cell Biol. 90: 362–371.

Schroeder, T. E., and J. J. Otto. 1984. Cyclic assembly-disassembly of cortical microtubules during maturation and early development of starfish oocytes. Dev. Biol. 103: 493–503.

———. 1988. Immunofluorescent analysis of actin and myosin in isolated contractile rings of sea urchin eggs. Zool. Sci. 5: 713–725.

———. 1991. Snoods: A periodic network containing cytokeratin in the cortex of starfish oocytes. Dev. Biol. 144: 240–247.

Schroeder, T. E., and S. A. Stricker. 1983. Morphological changes during maturation of starfish oocytes: Surface topography and cortical actin. Dev. Biol. 98: 373–384.

Shoji, Y., Y. Hamaguchi, and Y. Hiramoto. 1978. Mechanical properties of the endoplasm in starfish oocytes. Exp. Cell Res. 117: 79–87.

Standart, N., J. Minshull, J. Pines, and T. Hunt. 1987. Cyclin synthesis, modification and destruction during meiotic maturation of the starfish oocyte. Dev. Biol. 124: 248–258.

Torpey, N., C. C. Wylie, and J. Heaysman. 1992. Function of maternal cytokeratin in *Xenopus* development. Nature 357: 413–415.

Witchell, H. J., and R. A. Steinhardt. 1990. 1-Methyladenine can consistently induce a Fura-detectable transient calcium increase which is neither necessary nor sufficient for maturation in oocytes of the starfish *Asterina miniata*. Dev. Biol. 141: 393–398.

Yamashiro, S., and F. Matsumura. 1991. Mitosis specific phosphorylation of caldesmon: Possible molecular mechanism of cell rounding during mitosis. Bioessays 13: 563–568.

5 Membrane Ultrastructure in Early *Strongylocentrotus purpuratus* Embryos: Improved Resolution Using High-Pressure Freezing and Freeze Substitution

Kent McDonald

ABSTRACT One- and two-celled embryos of *Strongylocentrotus purpuratus* were prepared for conventional transmission electron microscopy (EM) by high-pressure freezing followed by freeze substitution in osmium/acetone and embedding in epoxy resin. This method of specimen preparation for EM provides excellent preservation of all cytoplasmic components as revealed by smooth, continuous membrane profiles in the nuclear envelope, mitochondria, rough and smooth endoplasmic reticulum, Golgi apparatus, vesicles, and other membrane-bound organelles. Cytoskeletal elements, such as the microtubules of the mitotic apparatus, are well preserved, and there is little, if any, extraction of the ground cytoplasm. These low-temperature fixation methods are recommended for future EM studies of these usually difficult-to-fix cells.

Introduction

Free calcium within cells regulates a number of fundamental cellular processes which in turn influence development. The structural component of cells responsible for regulating free calcium is the endomembrane system, notably the endoplasmic reticulum (ER). The model for calcium regulation in cells is the sarcoplasmic reticulum of striated muscle, and it is likely that some membranous component of nonmuscle cells functions in an analogous fashion. To study the effects of calcium on development in nonmuscle cells, researchers have focused on the embryos of a variety of organisms, including sea urchins (Vacquier, 1975; Kiehart, 1981; Inoue and Yoshioka, 1982; Swann and Whitaker, 1986; Oberdorf et al., 1986; Payan et al., 1986; Steinhardt and Alderton, 1988). Using calcium-specific probes such as aequorin (Eisen and Reynolds, 1985), the Fura dyes (Poenie et al., 1986; Hafner et al., 1988), or antibodies against calcium-binding proteins (Henson et al., 1989, 1990; Petzelt and Hafner, 1986) and new technologies such as video enhanced, laser confocal light microscopy (White et al., 1987), these investigators have been able to build up an impressive body of data on the changes in free intracellular calcium following fertilization and during initial mitoses in these organisms. The evidence from both light (Henson et al., 1989; Terasaki and Jaffe, 1991; Terasaki and Sardet, 1991; Sardet et al., 1992) and electron microscopy (Poenie and Epel, 1987; Henson et al., 1989, 1990) support the idea that the ER component of developing embryos is key in regulating free calcium. While light microscope

Kent McDonald, Laboratory for Three Dimensional Fine Structure, Department of Molecular, Cellular and Developmental Biology, University of Colorado, Boulder, CO 80309–0347.

observations are invaluable for giving an overview of the redistribution of ER during embryo development, they lack the resolving power (in embryos) to show the detailed spatial relationship between ER and other cellular components such as microfilaments and microtubules. Electron microscopy can achieve the needed resolution, but most embryos are notoriously hard to fix well for this kind of observation.

Harris (1975, 1986) has explored the problems associated with fixation of sea urchin (mostly *Strongylocentrotus purpuratus*) embryos and has concluded that the usual combination of glutaraldehyde followed by osmium tetroxide does not work well. The best fixation seems to be 1 percent OsO_4 in 0.4 M acetate buffer at pH 6 (Harris, 1975). Unfortunately, embryos fixed by this method still show a lot of cytoplasmic extraction (Harris, 1975; Paweletz et al., 1984). The extent of extraction and the likelihood that cellular components are rearranged during EM processing mean that this fixation scheme would not be useful for accurate descriptions of subcellular architecture and especially not for high-resolution studies of ER-cytoskeleton interactions.

One solution to preserving ultrastructure in hard-to-fix cell types is ultrarapid freezing followed by freeze substitution. Mycologists discovered the value of this approach about a decade ago and were able to make significant improvements in the ultrastructure of certain fungi (Howard and Aist, 1979; Heath and Rethoret, 1982; Hoch and Staples, 1983; Heath et al., 1985). Until recently, however, such methods were not available for preservation of embryo ultrastructure because the size of cells which could be routinely well frozen was about 40 μm and more usually around 10 μm. Ding et al. (1991) have recently shown that double propane jet freezing can give reliable freezing to a depth of over 100 μm in plant cells, and this method may also be useful for sea urchin embryos. However, the development of the high-pressure freezer by Balzers Union (Lichtenstein) based on the ideas of Moor and Riehle (1968) has expanded the upper size limit even further. With this technology it is possible to achieve good freezing in samples of up to 600 μm (Moor, 1987; Dahl and Staehelin, 1989). We have been using high-pressure freezing (HPF) in combination with freeze substitution (FS) to study the ultrastructure of *Drosophila* and sea urchin embryos (McDonald and Morphew, 1989, 1993). In this chapter, we report on how HPF-FS can be used to improve the preservation of membrane ultrastructure in early embryos of *Strongylocentrotus purpuratus*.

Materials and Methods

Sea Urchins. *Strongylocentrotus purpuratus* were obtained from Marinus Inc. (Westchester, Calif. 90083), and eggs were collected by injection of 1 M KCl into the coelomic cavity. Fertilization in vitro was accomplished by mixing 100 ml of eggs with about 0.1 ml of sperm. Development was monitored by light microscopy, and when they reached the appropriate stage the embryos were concentrated by hand centrifugation, resuspended in "freezing solution" (see below), concentrated again, and loaded into the specimen holders for the Balzers HPM 010 High Pressure Freezer (Bal-Tec Products, Inc., Middlebury, Conn. 06762). Fertilization envelopes were *not* removed prior to freezing. The micrographs in this chapter were all taken from one- or two-celled stage embryos.

Freezing Solutions. Embryos were resuspended in one of the following mixtures prior to freezing: (1) sea water plus 15 percent dextran (79,000 MW), (2) sea water plus dry baker's yeast made up to the consistency of a "runny" paste, (3) as in mixture 2 except with 10 percent methanol added, (4) 1-hexadecene, or (5) cold water fish gelatin (Sigma). Sea water alone was not used because it forms ice crystals which impede the removal of heat during freezing, and consequently, most of the cells do not freeze well. To facilitate heat transfer, it is customary to use a nonpenetrating cryoprotectant, that is, a substance which will not penetrate into cells or induce rapid osmotic effects, but which will interfere with ice crystal formation in the external medium. Because the embryos are in these solutions for usually less than a minute, we believe that gross alterations of the cytoplasmic architecture are unlikely. Furthermore, we left embryos in these solutions for up to 5 min and then returned them to sea water, and they continued normal development. It is possible that the fertilization envelope provided a barrier to the passage of these compounds into the cells. The external cryoprotectants also serve another function, to fill in the air spaces in the specimen holder. Air spaces present during the pressurization step of high-pressure freezing, will collapse and consequently so will the tissues in the holder.

High-Pressure Freezing. Embryos were loaded into special interlocking specimen holders (Craig et al., 1987), covered with one of the freezing solutions above, and immediately frozen in the Balzers HPM 010. Samples were stored in liquid nitrogen until needed for further processing.

Freeze Substitution. Samples were transferred from liquid nitrogen to 1 percent OsO_4 in anhydrous acetone at $-90°C$ and left for about 3 days in a homemade freeze substitution device (Kiss and McDonald, 1993). They were allowed to warm to $0°C$ over a period of 6–8 h, then rinsed in acetone at $0°C$ three times, allowed to warm to room temperature, and then flat-embedded in Epon-Araldite resin.

Flat Embedding. Glass microscope slides were sprayed with a release agent (MS-122, Miller-Stephenson, Los Angeles) that was allowed to dry, then buffed so the slide was clear again. Embryos in resin were placed between the coated slides, then into a 60°C oven for 48 h to allow the resin to cure. One of the slides in the sandwich was then removed by inserting a razor blade between it and the resin, which caused the slide to pop off. Cells were scanned on a light microscope using phase-contrast optics, and appropriate embryos were cut out and remounted for sectioning.

Electron Microscopy. Thin sections (75 nm) were cut on a Reichert Ultracut E microtome, picked up on slot grids, poststained with 1 percent aqueous uranyl acetate (7 min) and lead citrate (3 min), then viewed and photographed in a Philips CM10 electron microscope operating at 80 kV.

Results

The illustrations that follow are a survey of different membrane structures in the early embryo. A comprehensive documentation of all membrane systems and their dynamics in early embryogenesis is beyond the scope of this chapter. Our goal is simply to show how well high-pressure freezing and freeze substitution can preserve membrane structure.

Figure 5.1 shows a low-magnification view of one-half of a two-celled embryo. One can see the fertilization envelope and hyaline layer surrounding the embryo and the general organization of the cytoplasm. Organelles show no particular distribution pattern except that the yolk platelets tend to be in the periphery of the cell and the area surrounding the nucleus is filled with vesicles. The nuclear envelope is a smooth, continuous circle, and plasma membranes between adjacent cells are smooth and evenly spaced.

Figure 5.2 is a low-magnification view of the mitotic spindle region in a one-celled embryo.

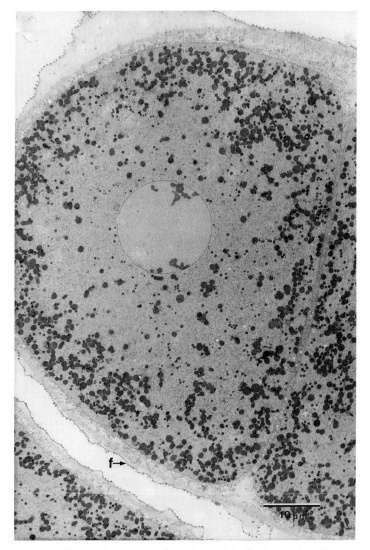

FIGURE 5.1. Low-magnification view of a two-celled embryo with fertilization envelope (f) intact (scale bar = 10 μm).

The yolk platelets are somewhat radially arranged as are sheets of ER in the region outside the spindle. The lamellar nature of the ER in cells at this stage is shown in higher magnification in Figure 5.3. The region just outside the spindle in Figure 5.2 shows membranes that are more vesicular than lamellar. These membranes and the astral spindle microtubules associated with them are shown at higher magnification in Figure 5.4. Within the spindle proper are numerous small vesicles, and occasionally one sees membranous material on the surface of the chromosomes (Fig. 5.5). At the centrosome, there is a dense concentration of microtubules but membranes of diverse size and shape are also present (Fig. 5.6).

The Golgi system in sea urchin embryos is distributed throughout the peripheral cytoplasm, and the stacks of Golgi cisternae show a characteristic size and arrangement. There are typically 6–8 cisternae per Golgi apparatus and many budding or fusing vesicles at the *trans* face (Figs. 5.7 and 5.8). Golgi paired at the *cis* (Fig. 5.8) and *trans* (data not shown) face are common.

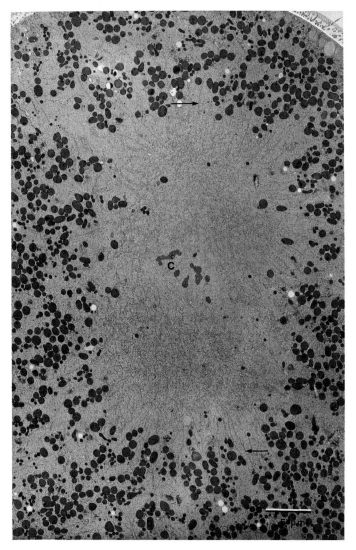

FIGURE 5.2. Low-magnification view of a single-celled embryo in mitosis. The clear central area represents the spindle area where one can see chromosomes (c). At the margins of the asters, lamellar ER can be seen (arrows) (scale bar = 5 μm).

Mitochondria (Fig. 5.9) are also evenly distributed throughout the cytoplasm, and their outer and inner membranes are smooth and continuous.

Figure 5.10 shows the cortex of the dividing cell in Figure 5.2. The most notable feature is the array of acidic vesicles just under the plasma membrane. Between this layer and the yolk granules is a zone of membranes that appear to be elongated tubes.

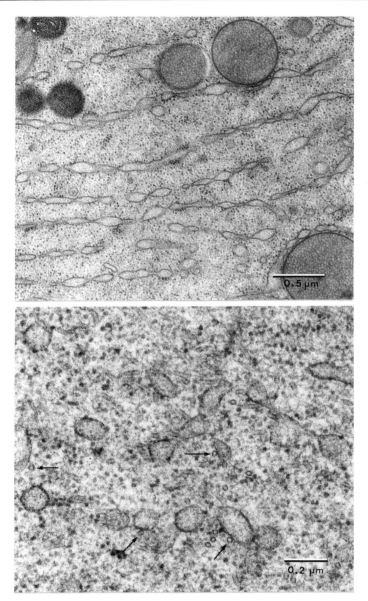

FIGURE 5.3 *(top).* Higher-magnification view of the region denoted by arrows in Figure 5.2 (scale bar = 0.5 μm).
FIGURE 5.4 *(bottom).* Higher-magnification view of the aster region of the spindle. Arrows indicate some of the microtubules that come into close proximity to membranes (scale bar = 0.2 μm).

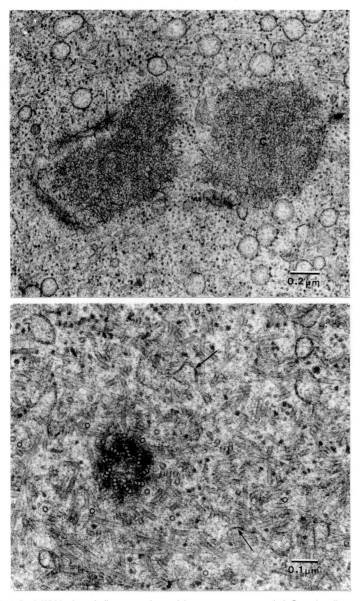

FIGURE 5.5 *(top)*. Within the spindle proper, the vesicles appear more rounded. Occasionally, one sees membrane (arrow) associated with the chromatin (c) (scale bar = 0.2 μm). FIGURE 5.6 *(bottom)*. Cross-section through one of the centrioles at a spindle pole. Membranes (arrows) are intermingled with microtubules in the pole region (scale bar = 0.1 μm).

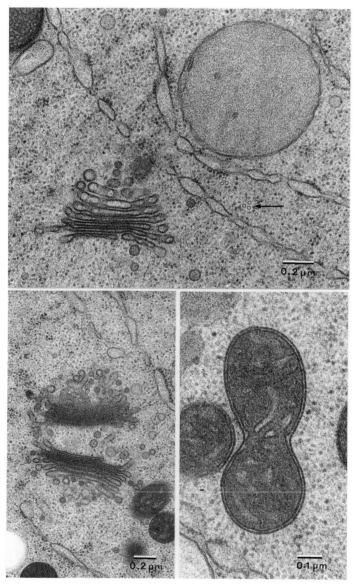

FIGURE 5.7 *(top).* Golgi apparatus in the peripheral cytoplasm of a mitotic cell. Microtubules in cross-section (arrow) are evident (scale bar = 0.2 μm). FIGURE 5.8 *(bottom left).* Two Golgi apparatuses paired at their *cis* face (scale bar = 0.2 μm). FIGURE 5.9 *(bottom right).* High-magnification view of mitochondria (scale bar = 0.1 μm).

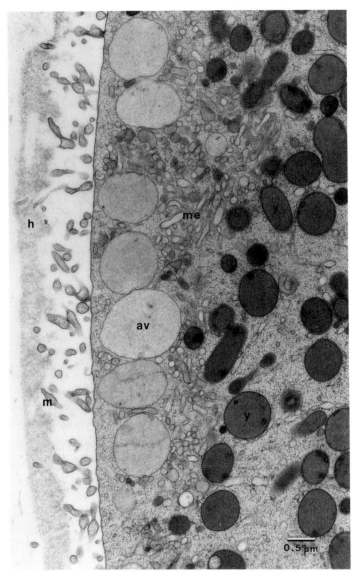

FIGURE 5.10. Cortex of dividing cell shown in Figure 5.2 (scale bar = 0.5 μm).

Discussion

Because living cells cannot be observed directly in the electron microscope, they must necessarily be treated in some way to avoid the destructive effects of high vacuum conditions. All specimen preparation methods will alter the "true" ultrastructure of the cell, but some are better than others and it should be the goal of any EM study to choose the least damaging method. In general, methods that involve fixation by ultrarapid freezing (cryofixation) are better than those which involve cross-linking with chemicals at room temperature (Hayat, 1989). For one thing, cryofixation does not rely on specific chemical cross links to fix cell components. All chemical species are equally immobilized by freezing. Second, ultrarapid freezing is orders of magnitude faster than conventional chemical fixation. While it may take seconds to fix by chemical cross-linking, it takes milliseconds to arrest all subcellular activity by ultrarapid freezing (Gilkey and Staehelin, 1986). If the cells being fixed have walls or other permeability barriers around them, then chemical fixation may take minutes (Hayat, 1981), during which time osmotic shocks and autolytic events may further alter the cell ultrastructure. The layers surrounding the sea urchin embryo (Figs. 5.1 and 5.10) are likely to be responsible for some of the poor fine-structural preservation typically seen in these cells. Researchers studying fungal cell fine structure found that removing the cell walls with enzymes resulted in improved chemical fixation (Peterson and Ris, 1976). In studies of sea urchin embryo ultrastructure, the fertilization envelope is typically removed, but the hyaline layer may still act as a barrier to fixative penetration. Because we do not remove either the fertilization envelope or the hyaline layer, the methods described here offer a way of minimizing the perturbation of the cells physiology prior to fixation. The resuspension in "freezing solutions" for 1 to 2 min is a potential problem, but by using high-molecular-weight compounds that do not act osmotically we believe that the cells are not adversely affected.

Although it has long been obvious that cryofixation is a superior method for preserving structure, it was not until the development of the high-pressure freezer that samples much larger than 100 μm thick (Ding et al., 1991) could be frozen without cryoprotection. Cryofixation by high-pressure freezing is a relatively new technique, although the theoretical basis for the method was published in 1968 (Moor and Riehle, 1968). In 1985, Balzers Union (Lichtenstein) made instruments commercially available (Moor, 1986), and since that time, a number of studies have been published that demonstrate its usefulness as a cryofixation method. Plant tissues have been the most intensively studied (Craig and Staehelin, 1988; Kaeser et al., 1988; Lancelle and Hepler, 1989; Kiss et al., 1990; Hippe et al., 1989; Staehelin et al., 1990), followed by animal tissues (Draznin et al., 1988; Hunziker et al., 1984a,b; Moor et al., 1980; Keene, 1990; McDonald and Morphew, 1989, 1993) and fungi (Welter et al., 1988; Knauf and Mengden, 1988). Studies done before 1985 were carried out in Zurich in Hans Moor's lab using a homemade device.

The principles behind high-pressure freezing are discussed at length by Moor (1987) and Dahl and Staehelin (1989) and need be only briefly summarized here. Water molecules in the solid phase (ice) occupy more volume than the same number of molecules in the liquid phase. High pressure (optimum 2045 bar; Moor, 1987) slows down this expansion of water molecules as they grow into ice crystals. This means there is more time to remove heat from cells and tissues, which in turn means that good freezing can be achieved at greater depths, up to 600 μm by theoretical calculations (Riehle, 1968). The actual depths of well-frozen tissue will vary due to the water content and other chemical characteristics of the cells.

Once frozen, there are several options for further treatment of the cells. They can be cryosectioned and observed at low temperature in an electron microscope fitted with a cold stage (Michel et al., 1991), they can be freeze-fractured and etched (Craig and Staehelin, 1988), and they can be freeze substituted and then either critically point dried or embedded in epoxy resin for normal TEM viewing or methacrylate resins for immunocytochemistry. Most studies include some thin sectioning of epoxy-embedded material because it is the easiest and most familiar way to evaluate the quality of fixation or ice damage. A detailed discussion of why freeze substitution works better than conventional fixation and dehydration is beyond the scope of this chapter; however, the article by Kellenberger (1987) is an excellent summary of the problem.

Comparison of our results with previous studies of sea urchin embryo membranes, especially the work of Harris (1975), shows mainly differences in quality rather than specific details. Changes in the membranes from vesicular at interphase to tubulovesicular at mitosis were observed by Harris (1975), as well as the association of microtubules and vesicles in the mitotic apparatus. Henson et al. (1990) observed the Golgi apparatus in the course of looking at EM immunocytochemistry of calcium-sequestering proteins in oocytes. We find it interesting that the Golgi are more like plant dictyosomes (Staehelin et al., 1990) than the typical animal Golgi apparatus. The cortex ultrastructure of *S. purpuratus* eggs and embryos has been studied by both Sardet (1984) and Chandler (1984), but because they were studying isolated cortices rather than intact cells, it is difficult to compare the structures we see in Figure 5.10 with their results. It is also difficult to know what changes occur in membranes during the preparation of isolated cortices. Sardet (1984) has reported that different populations of components remain in these preps depending on the strength of the shear forces used to make them. By using the methods described in this chapter, one could study the structure in intact cells and the fixation would give millisecond time resolution so that even the membrane rearrangements that occur during chemical fixation (Kiss et al., 1990; Staehelin et al., 1990; Draznin et al., 1988) would be kept to a minimum.

In summary, we conclude that ultrarapid freezing/freeze substitution is superior to conventional chemical fixation for preserving the ultrastructure of *S. purpuratus* embryos. It is recommended for those EM studies requiring high temporal or spatial resolution.

Acknowledgments

I would like to thank Dr. L. Andrew Staehelin for providing unlimited access to his high-pressure freezer, Mary Morphew for excellent technical assistance, and Corey Nislow for preparing the embryos for freezing. This work was supported in part by National Institutes of Health grant RR03744 to K. M. and grant RR00592 which supports the Boulder Laboratory for Three Dimensional Fine Structure, J. R. McIntosh, Principal Investigator.

Literature Cited

Chandler, D. E. 1984. Exocytosis *in vitro*: Ultrastructure of the isolated sea urchin cortex as seen in platinum replicas. Exp. Cell Res. 89: 198–211.

Craig, S., J. C. Gilkey, and L. A. Staehelin. 1987. Improved specimen support cups and auxiliary devices for the Balzers high pressure freezing apparatus. J. Microsc. 148: 103–106.

Craig, S., and L. A. Staehelin. 1988. High pressure freezing of intact plant tissues: Evaluation and characterization of novel features of the endoplasmic reticulum and associated membrane systems. Eur. J. Cell Biol. 46: 80–93.

Dahl, R., and L. A. Staehelin. 1989. High-pressure freezing for the preservation of biological structure: theory and practice. J. Electron. Microsc. Tech. 13: 165–174.

Ding, B., R. Turgeon, and M. V. Parthasarathy. 1991. Routine cryofixation of plant tissue by propane jet freezing for freeze substitution. J. Electron. Microsc. Tech. 19: 107–117.

Draznin, B., R. Dahl, N. Sherman, K. E. Sussman, and L. A. Staehelin. 1988. Exocytosis in normal anterior pituitary cells. J. Clin. Invest. 81: 1042–1050.

Eisen, A., and G. T. Reynolds. 1985. Sources and sinks of the calcium released upon fertilization of single sea urchin eggs. J. Cell Biol. 100: 1522–1527.

Gilkey, J. C., and L. A. Staehelin. 1986. Advances in ultrarapid freezing for the preservation of cellular ultrastructure. J. Electron. Microsc. Tech.3: 177–210.

Hafner, M., C. Petzelt, R. Nobiling, J. B. Pawley, D. Kramp, and G. Schatten. 1988. Wave of free calcium at fertilization in the sea urchin egg visualized with fura-2. Cell Motil. Cytoskel. 9: 271–277.

Harris, P. 1975. The role of membranes in the organization of the mitotic apparatus. Exp. Cell Res. 94: 409–425.

————. 1986. Cytology and Immunocytochemistry. *In* T. E. Schroeder, ed. Methods in Cell Biology, Vol. 27. Academic Press, Orlando, Fl., pp. 243–262.

Hayat, M. A. 1981. Fixation for Electron Microscopy. Academic Press, New York.

————. 1989. Principles and Techniques of Electron Microscopy. CRC Press, Boca Raton, Fl.

Heath, I. B., and K. Rethoret. 1982. Mitosis in the fungus *Zygorhynchus molleri*: Evidence for stage specific enhancement of microtubular preservation by freeze substitution. Eur. J. Cell Biol. 28: 180–189.

Heath, I. B., K. Rethoret, A. L. Arsenault, and F. P. Ottensmeyer. 1985. Improved preservation of the form and contents of wall vesicles and the Golgi apparatus in freeze substituted hyphae of *Saprolegnia*. Protoplasma 128: 81–93.

Henson, J. H., D. A. Begg, S. M. Beaulieu, D. J. Fishkind, E. M. Bonder, M. Terasaki, D. Lebeche, and B. Kaminer. 1989. A calsequestrin-like protein in the endoplasmic reticulum of the sea urchin: Localization and dynamics in the egg and first cell cycle embryo. J. Cell Biol. 109: 149–161.

Henson, J. H., S. M. Beaulieu, B. Kaminer, and D. A. Begg. 1990. Differentiation of a calsequestrin-containing endoplasmic reticulum during sea urchin oogenesis. Dev. Biol. 142: 255–269.

Hippe, S., K. During, and F. Kreuzaler. 1989. In situ localization of a foreign protein in transgenic plants by immunoelectron microscopy following high pressure freezing. Freeze substitution and low temperature embedding. Eur. J. Cell Biol. 50: 230–234.

Hoch, H. C., and R. C. Staples. 1983. Ultrastructural organization of the non-differentiated uredospore germling of *Uromyces phaseoli* variety *typica*. Mycologia 75: 795–824.

Howard, R. J., and J. R. Aist. 1979. Hyphal tip ultrastructure of the fungus *Fusarium*: Improved preservation by freeze-substitution. J. Ultrastruct. Res. 66, 224–234.

Hunziker, E. B., W. Hermann, R. K. Schenk, M. Muller, and H. Moor. 1984a. Cartilage ultrastructure after high pressure freezing, freeze substitution and low temperature embedding. I. Chondrocyte ultrastructure—implications for the theories of mineralization and vascular invasion. J. Cell Biol. 98: 267–276.

————. 1984b. Cartilage ultrastructure after high pressure freezing, freeze substitution and low temperature embedding. II. Intercellular matrix ultrastructure—preservation of proteoglycans in their native state. J. Cell Biol. 98: 277–282.

Inoue, H., and T. Yoshioka. 1982. Comparison of Ca^{2+} uptake characteristics of microsomal fractions isolated from unfertilized and fertilized sea urchin eggs. Exp. Cell Res. 140: 283–288.

Kaeser, W., H.-W. Koyro, and H. Moor. 1988. Cryofixation of plant tissues without pretreatment. J. Microsc. 154: 279–288.

Keene, D. R. 1990. The connective tissue matrix of cartilage as revealed by standard fixation, ruthenium staining, high-pressure freezing, and embedding in water-soluble media: A comparison of methods. Proc. Twelfth Int. Cong. Electron. Microsc. pp. 326–327.

Kellenberger, E. 1987. The response of biological macromolecules and supramolecular structures to the physics of specimen preparation. *In* R. A. Steinbrecht and K. Zierold, eds. Cryotechniques in Biological Electron Microscopy. Springer-Verlag, Berlin, pp. 35–63.

Kiehart, D. 1981. Studies on the *in vivo* sensitivity of spindle microtubules to calcium ions and evidence for a vesicular calcium-sequestering system. J. Cell Biol. 88: 355–365.

Kiss, J. Z., T. H. Giddings, L. A. Staehelin, and F. D. Sack. 1990. Comparison of the ultrastructure of conventionally fixed and high pressure frozen/freeze substituted root tips of *Nicotiana* and *Arabidopsis*. Protoplasma 157: 64–74.

Kiss, J. Z., and K. McDonald. 1993. EM immunocytochemistry following cryofixation and freeze substitution. *In* D. Asai, ed. Antibodies in Cell Biology, Methods in Cell Biology, Vol. 37.

Knauf, G. M., and K. Mendgen. 1988. Secretion systems and membrane-associated structures in rust fungi after high pressure freezing and freeze-fracturing. Biol. Cell 64: 363–370.

Lancelle, S. A., and P. K. Hepler. 1989. Immunogold labeling of actin on sections of freeze-substituted plant cells. Protoplasma 150: 72–74.

McDonald, K., and M. K. Morphew. 1989. Preservation of embryo ultrastructure by high pressure freezing and freeze substitution. Proc. Forty-Seventh Electron. Microsc. Soc. Am. Meeting. San Francisco Press: San Francisco, pp. 994–995.

———. 1993. Improved preservation of ultrastructure in difficult-to-fix organisms by high pressure freezing and freeze substitution. I. *Drosophila melanogaster* and *Strongylocentrotus purpuratus* embryos. Microsc. Res. Tech. (in press).

Michel, M., T. Hillman, and M. Muller. 1991. Cryosectioning of plant material frozen at high pressure. J. Microsc. 163: 3–18.

Moor, H. 1986. Recent progress in high pressure freezing. Proc. Eleventh Int. Cong. Electron. Microsc., Kyoto, 1961–1964.

———. 1987. Theory and practice of high pressure freezing. *In* R. A. Steinbrecht and K. Zierold, eds. Cryotechniques in Biological Electron Microsocopy. Springer-Verlag, Berlin, pp. 175–191.

Moor, H., and U. Riehle. 1968. Snap-freezing under high-pressure: A new fixation technique for freeze-etching. *In* D. S. Bocciarelli, ed. Electron Microscopy 1968, Proc. Fourth Eur. Reg. Conf. Electron. Microsc. Rome, pp. 33–34.

Moor, H., G. Bellin, C. Sandri, and K. Akert. 1980. The influence of high pressure freezing on mammalian nerve tissue. Cell Tissue Res. 209: 201–216.

Oberdorf, J. A., J. F. Head, and B. Kaminer. 1986. Calcium uptake and release by isolated cortices and microsomes from the unfertilized sea urchin, *Strongylocentrotus droebachiensis*. J. Cell Biol. 102: 2205–2210.

Paweletz, N., D. Mazia, and E.-M. Finze. 1984. The centrosome cycle in the mitotic cycle of sea urchin eggs. Exp. Cell Res. 152: 47–65.

Payan, P., J. P. Girard, C. Sardet, M. Whitaker, and J. Zimmerberg. 1986. Uptake and release of calcium by isolated egg cortices of the sea urchin *Paracentrotus lividus*. Biol. Cell 58: 87–90.

Peterson, J. B., and H. Ris. 1976. Electron-microscopic study of the spindle and chromosome movement in the yeast *Saccharomyces cerevisiae*. J. Cell Sci. 22: 219–242.

Petzelt, C., and M. Hafner. 1986. Visualization of the Ca^{++} transport system of the mitotic apparatus of sea urchin eggs with a monoclonal antibody. Proc. Natl. Acad. Sci. 83: 1719–1722.

Poenie, M., J. Alterton, R. Steinhardt, and R. Tsien. 1986. Calcium rises abruptly and briefly throughout the cell at the onset of anaphase. Science 233: 886–889.

Poenie, M., and D. Epel. 1987. Ultrastructural localization of intracellular calcium stores by a new cytochemical method. J. Histochem. Cytochem. 35: 939–956.

Riehle, U. 1968. Schnellgefrieren organischer Preparate fur die Elektronen-Mikroskopie. Chem. Ing. Tech. 40, 5: 213–218.

Sardet, C. 1984. The ultrastructure of the sea urchin egg cortex isolated before and after fertilization. Dev. Biol. 105: 196–210.

Sardet, C., J. Speksnijder, M. Terasaki, and P. Chang. 1992. Polarity of the ascidian egg cortex before fertilization. Development 115: 221–237.

Staehelin, L. A., T. H. Giddings, J. Z. Kiss, and F. D. Sack. 1990. Macromolecular differentiation of Golgi stacks in root tips of *Arabidopsis* and *Nicotiana* seedlings as visualized in high pressure frozen and freeze-substituted samples. Protoplasma 157: 75–91.

Steinhardt, R. A., and J. Alderton. 1988. Intracellular free calcium rise triggers nuclear envelope breakdown in the sea urchin embryo. Nature 332: 364–366.

Swann, K. S., and M. J. Whitaker. 1986. The part played by inositol triphosphate and calcium in the propagation of the fertilization wave in sea urchin eggs. J. Cell Biol. 103: 2333–2342.

Terasaki, M., and L. A. Jaffe. 1991. Organization of the sea urchin egg endoplasmic reticulum and its reorganization at fertilization. J. Cell Biol. 114: 929–940.

Terasaki, M., and C. Sardet. 1991. Demonstration of calcium uptake and release by sea urchin cortical endoplasmic reticulum. J. Cell Biol. 115: 1031–1037.

Vacquier, V. D. 1975. The isolation of intact cortical granules from sea urchin eggs: calcium ions trigger granule discharge. Dev. Biol. 43: 62–74.

Welter, K., M. Muller, and K. Mendgen. 1988. The hyphae of *Uromyces appendiculatus* within the leaf tissue after high pressure freezing and freeze substitution. Protoplasma 147: 91–99.

White, J. G., W. B. Amos, and M. Fordham. 1987. An evaluation of confocal versus conventional imaging of biological structures by fluorescence light microscopy. J. Cell Biol. 105: 41–48.

6 Spermatogenesis in the Ovotestes of the Solitary Ascidian *Boltenia villosa*

Michael J. Cavey

ABSTRACT Spermatocysts from the ovotestes of the solitary ascidian *Boltenia villosa* have been examined by light and electron microscopy. The spermatocyst wall is a discontinuous epithelium formed by the bodies and basolateral processes of accessory (somatic) cells. On the basis of ultrastructural features and morphometric analyses, the spermatogonia, primary and secondary spermatocytes, and spermatids can be identified. Spermatogonia are polygonal cells which occur singly or in small clusters near the periphery of a cyst. Primary spermatocytes, secondary spermatocytes, and spermatids are found in bundles (clones). Cloned cells, which are linked by cytoplasmic bridges, differentiate synchronously. It is estimated that a spermatogonium undergoes five mitotic divisions, creating a clone of 32 interconnected primary spermatocytes with the potential to produce 128 spermatozoa. Accessory cells project into the lumen of the spermatocyst and contact the cloned cells. Accessory cells do not form intercellular junctions with each other or with the spermatogenic cells. It is unlikely that accessory cells of the ascidian support or nourish the spermatogenic cells substantively or coordinate their differentiation. The ascidian accessory cells do perform a phagocytic function by engulfing malformed bundles of spermatids.

Introduction

There are appreciable gaps in our knowledge of ascidian spermatogenesis. Technical difficulties in preparing spermatozoa and spermatogenic cells for ultrastructural examination are partly responsible for this predicament, and the conflicting observations and interpretations in published reports make it difficult to draw valid comparisons with other animals. Most available studies on ascidian spermatogenesis have focused on spermiogenesis, as investigators have attempted to resolve controversial aspects of spermatozoon morphology.

The solitary (simple) ascidian *Boltenia villosa* (Class Ascidiacea, Family Pyuridae) was selected for a histological and ultrastructural examination of spermatogenesis. This pyurid is an oviparous species which relies on sexual reproduction. It is a simultaneous hermaphrodite that is self-sterile and dependent on external insemination (Cloney, 1987, 1989). The first ultrastructural description of its spermatozoon was provided by Cloney and Abbott (1980). They found that the gamete was divisible into a head and a tail and that it possessed a single mitochondrion, a perforated nuclear envelope, one centriole (representing the basal body of the flagellum), and a typical arrangement of flagellar microtubules. A putative acrosome, situated between the nuclear envelope and plasmalemma at the anterior end of the head was subsequently described by Fukumoto (1986). The spermatozoon of *B. villosa* is typical in most respects to the spermatozoa of

Michael J. Cavey, Division of Zoology, Department of Biological Sciences, The University of Calgary, Alberta, Canada T2N 1N4.

other solitary species (Franzén, 1983; Fukumoto, 1983, 1984, 1985, 1986; Rosati et al., 1985).

This chapter presents a brief description of the spermatocysts from the ovotestes of *B. villosa*; some additional observations on the fine structure of the spermatozoa; an ultrastructural survey and morphometric assessment of the spermatogenic cells; and a consideration of the associations and functions of the accessory (somatic) cells.

Materials and Methods

Adult specimens of *Boltenia villosa* Stimpson, 1864, were collected from floats at Jensen's Marina, Friday Harbor, Washington, and held in tanks supplied with flowing sea water at the Friday Harbor Laboratories of the University of Washington. Individuals were bisected in the sagittal plane with razor blades, and the two halves of the body were promptly flooded with fixative (see below). Segments of the ovotestes were removed with forceps and immersed in fresh fixative.

For light microscopy, the ovotestes were fixed with phosphate-buffered glutaraldehyde (Cloney and Florey, 1968) for 60–90 min at ambient temperature (see Fig. 6.1a). The specimens were dehydrated with graded ethanols and embedded in glycol methacrylate. Semithin sections (3 μm in thickness) were cut with Ralph glass knives, stained with hematoxylin and eosin, and examined and photographed with a Nikon Optiphot compound microscope equipped with plan-achromatic objective lenses and an HFX-II photomicrographic attachment.

For transmission electron microscopy, the ovotestes were fixed with phosphate-buffered glutaraldehyde (Cloney and Florey, 1968) for 45–60 min at ambient temperature (see Figs. 6.2a and 6.4d) or with cacodylate-buffered glutaraldehyde (Cavey and Cloney, 1972) for 90–120 min at 0–4°C (see Figs. 6.1b, 6.2b, d, e, 6.3a–d, 6.4a–c, and 6.6a–c). Aldehyde-fixed specimens were transferred, without rinsing, to bicarbonate-buffered osmium tetroxide (Wood and Luft, 1965) for 45–60 min at 0–4°C. Osmicated specimens were rinsed briefly with demineralized water, dehydrated with graded ethanols, transferred through propylene oxide, and embedded in epoxy resin. Ultrathin sections (70 nm in thickness) were cut with diamond knives, serially stained with aqueous solutions of uranyl acetate (saturated) and lead citrate (Reynolds, 1963), and examined and photographed with a JEOL JEM-100S transmission electron microscope at 60 kV.

For scanning electron microscopy, fragments of ovotestes were teased with forceps, fixed in phosphate-buffered glutaraldehyde (Cloney and Florey, 1968) for 30–45 min at ambient temperature (see Fig. 6.2c), dehydrated in graded ethanols, and dried by the critical-point method, using carbon dioxide as the transitional fluid. Specimens were sputter-coated with a gold-palladium film and examined and photographed with a JEOL JSM-35CF scanning electron microscope at 10 kV.

Transmission electron micrographs were used in morphometric analyses of the spermatogenic cells. The perimeters and areas of cells and nuclei were measured directly with Jandel Scientific's SigmaScan 3.9 software on a Zenith Z-316SX personal computer linked to a Hitachi HDG-1111C graphics tablet. Two sets of cellular and nuclear volumes were calculated in Borland International's Quattro Pro 4.0 software with the formula for the volume (V) of a sphere: $V = 4.189r^3$. Radius (r) was determined by substituting the measured perimeter for circumference (C) in the expression $C/2\pi$ or by inserting the measured area (A) into the expression $\sqrt{A/\pi}$. Cytoplasmic volume was derived by subtracting nuclear volume from cellular volume. Morphometric results were graphed with Software Publishing Corporation's Harvard Graphics 2.12 software.

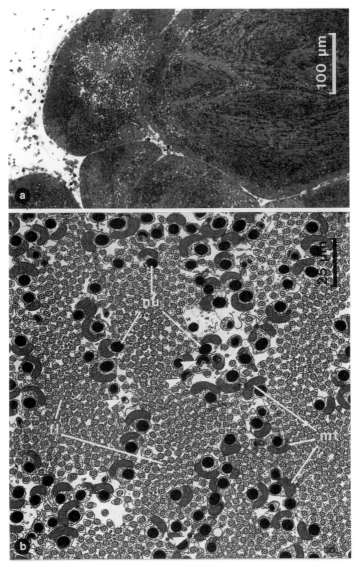

FIGURE 6.1. *a*. Spermatocysts from an ovotestis of *Boltenia villosa*. Pale areas in the larger cysts are the bundled tails of differentiating spermatozoa. *b*. Transverse section of clones of spermatozoa (fl, flagellum; mt, mitochondrion; nu, nucleus) in a spermatocyst. Helical tips of the nuclei are represented by arcuate and petite circular profiles (arrowheads).

Results

Spermatocysts

The ovotestes of *Boltenia villosa* are acinar organs situated on the left and right sides of the body. The thin-walled acini are dedicated to either oogenesis or spermatogenesis (Fig. 6.1a). Spermatogenic acini, or spermatocysts, contain small clusters of spermatogonia and bundles, or "clones" (Roosen-Runge, 1977), of differentiating gametes (Fig. 6.1b). Clones may consist of primary spermatocytes, secondary spermatocytes, or spermatids, and the differentiation of cells within a clone is synchronized. In this solitary ascidian, there are surprisingly few images of cells undergoing mitosis or meiosis, possibly indicating that the nuclear and cytoplasmic divisions occur rapidly or that elapsed time between divisions is relatively long.

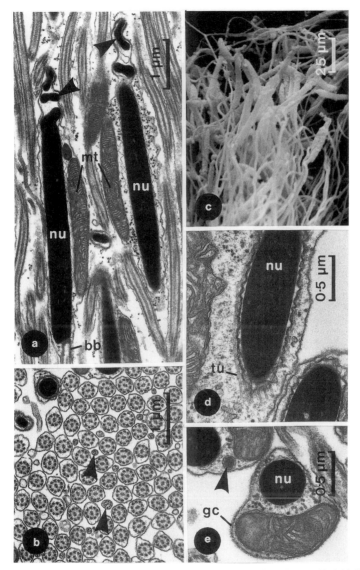

FIGURE 6.2. *a*. Longitudinal section of the heads of mature spermatozoa (mt, mitochondrion). Helical tips (arrowheads) of the electron-dense nuclei (nu) appear within loose-fitting extensions of the nuclear envelopes. A basal body (bb) occurs in the centriolar fossa of the nucleus. *b*. Transverse section of the tails of mature spermatozoa. Flagellar axonemes show the 9 × 2 + 2 pattern of microtubules. Subterminal and terminal segments (arrowheads) of the axonemes exhibit the 9 × 1 + 2 pattern of microtubules and variable numbers of microtubular singlets, respectively. *c*. Scanning electron micrograph of a clone of spermatozoa. *d*. Oblique section of the nucleus (nu) of a spermatozoon. Undulations of the nuclear envelope are apparent, and a sleeve of closely spaced microtubules (tu) flanks the outer membrane. *e*. Transverse section of the heads of spermatozoa. The nucleus (nu) overlies a shallow mitochondrial sulcus. Plasmalemmal segments in the head are invested by a glycocalyx (gc). In the upper spermatozoon, a small vacuole with granular contents (arrowhead) is evident.

Spermatozoa

The spermatozoon of *B. villosa*, as in all ascidians examined thus far, lacks a midpiece (Franzén, 1970; Cloney and Abbott, 1980). The plasmalemma of the gamete is invested by a glycocalyx which is well developed over the head but sparse or absent along the tail. The head encloses an electron-dense nucleus, a single mitochondrion, and a number of glycogen rosettes (Figs. 6.2a,e). The nucleus is surrounded by a perforated envelope, and the anterior end of the organelle has a helical ("corkscrew") configuration (Figs. 6.2a,c). The envelope is not appressed to the helical tip of the nucleus; instead, the nuclear gyres reside in a loose-fitting extension of the envelope. A sleeve of cytoplasmic microtubules, described previously by Franzén (1976) and Kubo et al. (1978), adjoins the outer membrane of the nuclear envelope (Fig. 6.2d). Undulations of the envelope are especially prevalent over the posterior two-thirds of the nucleus, and these segments are devoid of pores (Cloney and Abbott, 1980).

The nucleus is oriented parallel to a longitudinal sulcus in the mitochondrion (Fig. 6.2e). The mitochondrion stretches from a level immediately behind the helical tip of the nucleus to the level of the centriolar fossa. The basal body in the centriolar fossa of the nucleus organizes the flagellar microtubules of the tail. Membrane-bounded, multigranular bodies frequently adjoin the nucleus and mitochondrion, localizing behind the tip of the nucleus at a level where perforations exist in the nuclear envelope.

The basal body aligns parallel to the primary axis of the tail (Fig. 6.2a), and it displays the usual $9 \times 3 + 0$ arrangement of microtubules when viewed in transverse sections (Baccetti and Afzelius, 1976). There is no evidence of an anchoring apparatus between the microtubular triplets and the plasmalemma. The microtubular patterns along the tail of the ascidian spermatozoon are undistinguished (Fig. 6.2b). Flagellar microtubules span the entire length of the tail, and a typical $9 \times 2 + 2$ pattern of microtubules is found in the axoneme (Baccetti and Afzelius, 1976). In distal, subterminal sections through a tail, the axoneme may exhibit a $9 \times 1 + 2$ or a $9 \times 1 + 0$ pattern of microtubules. In terminal sections of a tail, fewer than nine microtubular singlets may be seen.

Spermatogenic Cells

Spermatogonia

Spermatogonia are large polygonal cells confined to positions near the peripheries of the cysts (Fig. 6.3a). Relatively few cells are encountered in any one location. The eccentric nucleus of a spermatogonium is bounded by a perforated envelope and contains a preponderance of dispersed chromatin. Some chromatin condensation is apparent in the peripheral nucleoplasm beneath the envelope. A prominent nucleolus is present, and intranuclear vacuoles of various sizes and shapes are common.

The cytoplasm of a spermatogonium is finely granular in appearance, owing to large numbers of free ribosomes and polysomes. The cytoplasm contains a few small mitochondria which are spherical and dispersed around the circumference of the nucleus. Mitochondrial cristae are poorly developed. A small Golgi body situates beside the nucleus (Fig. 6.3b), and electron-dense material within its cisternae is an indication of synthetic activity. A few cisternae of granular endoplasmic reticulum, arising from evaginations of the nuclear envelope, are present. Cytoplasmic bridges between the spermatogonia were not observed. Since cellular linkages typically arise during mitotic divisions that precede meiosis, it is likely that the bridges between spermatogonia were missed in those random sections examined with the electron microscope.

Primary Spermatocytes

The cytoplasmic organization of a primary spermatocyte is superficially similar to that of a spermatogonium. The cytoplasm has a granular appearance, but it now includes more granular and agranular cisternae of the endoplasmic reticulum (Fig. 6.3c and see Fig. 6.6a). Small spherical mitochondria encircle the nucleus, but there are signs of a mitochondrial shift toward the side of the nucleus opposite to the Golgi apparatus. Mitochondrial fusion is occurring in some cells, and paired centrioles are often observed in the cellular cortices. Slender cytoplasmic bridges link

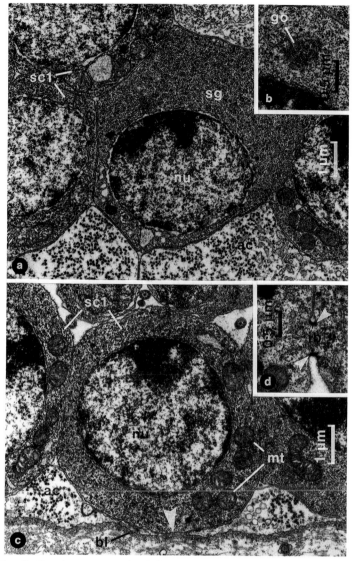

FIGURE 6.3. *a*. Section of a spermatogonium (sg), identified by its polygonal outline, eccentric nucleus (nu), and granular cytoplasm (ac, accessory cell; sc1, primary spermatocyte). *b*. Golgi body (go) in a spermatogonium. *c*. Section of primary spermatocytes (sc1), identified by their rounded profiles, central nuclei (nu), and granular cytoplasms. The plasmalemma of the central cell (arrowhead) contacts the basal lamina (bl) of the spermatocyst wall (ac, accessory cell; mt, mitochondrion). *d*. Intercellular bridge (ib) between primary spermatocytes. Plasmalemmal segments within the bridge (arrowheads) appear denser than outlying sectors.

the primary spermatocytes, and plasmalemmal segments within the bridges appear thickened (Fig. 6.3d), as previously noted in ascidians (Villa, 1981) and other animals (Burgos et al., 1970). The thickened appearance may be attributable to a grazing section of curved membranes and to a filamentous mat on the cytoplasmic aspect of the bridge.

Chromatin condensation is generally more pronounced in a primary spermatocyte than in a spermatogonium. Condensation is progressing toward central regions of the nucleoplasm, and a distinct nucleolus can still be identified. Intranuclear vacuoles are present, and synaptonemal complexes reside in the nuclei of zygotene and pachytene cells (cf. Villa and Tripepi, 1982).

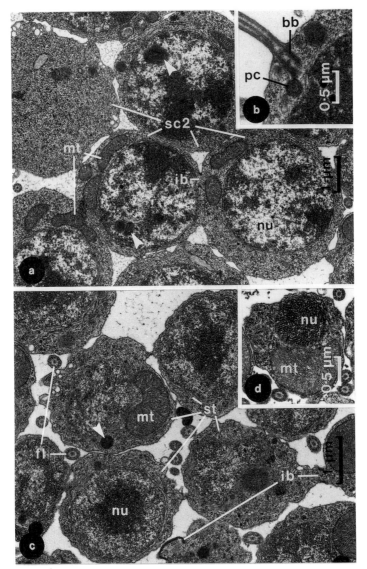

FIGURE 6.4. *a*. Section of secondary spermatocytes (sc2), identified by their rounded profiles, central nuclei (nu) with membranous inclusions (arrowheads), granular cytoplasms, and polarized mitochondria (mt). A dividing cell appears in the upper left corner of the field (ib, intercellular bridge). *b*. Basal body (bb), or distal centriole, and nascent flagellum of a secondary spermatocyte (pc, proximal centriole). *c*. Section of early spermatids (st), identified by their undulated margins, central nuclei (nu), indented mitochondria (mt), and multigranular vacuoles (arrowhead) (fl, flagellum; ib, intercellular bridge). *d*. Coarse chromatin strands in the nucleus (nu) of an intermediate spermatid (mt, mitochondrion).

Secondary Spermatocytes

The cytoplasmic bridges are easier to detect between the secondary spermatocytes than between the primary spermatocytes. The nuclear envelope and plasmalemma of a secondary spermatocyte can come into close proximity of one another (Fig. 6.4a). Chromatin condensation augments the mottling of the nucleoplasm. A single nucleolus is still evident in some nuclei, and heterochromatin probably masks the nucleoli in others. Mitochondria have now congregated at one pole of the nucleus. The larger size and more elongate shapes of the mitochondria can be attributed to continued fusion of the organelles. Paired centrioles localize in the cortical cytoplasm near the mitochondrial aggregation (Fig. 6.4b). One centriole of the pair is involved with the organization of microtubules for the flagellum. The microtubules in the nascent flagellum are virtually indistinguishable in terms of size and pattern from those in the flagellum of a mature spermatozoon. The microtubule-organizing centriole becomes the basal body of the flagellum, thus corresponding to the distal centriole in those spermatozoa of other animals which retain the pair. In the ascidian, the fate of the proximal centriole is unclear. It could simply disappear, or it could conceivably survive to organize the microtubules for the nonflagellated daughter cell of the secondary spermatocyte.

Spermatids

All spermatids are joined to other members of their clone by intercellular bridges, and these bridges persist until the end of spermiogenesis. Ascidian spermiogenesis is readily divisible into early, intermediate, and late phases (Kubo et al., 1978; Cotelli et al., 1980; Villa, 1981; Villa and Tripepi, 1982; Fukumoto, 1983, 1985). Since flagella begin to differentiate on the secondary spermatocytes, the events of spermiogenesis in ascidians are somewhat less dramatic than those in other organisms.

A large spherical or subspherical nucleus with filamentous strands of chromatin appears in an early spermatid (Fig. 6.4c). One large mitochondrion adjoins the nucleus, and both organelles alter their shapes in tandem during the ensuing phases of spermiogenesis.

Elongation of both the nucleus and the mitochondrion typifies the intermediate spermatid (Fig. 6.4d). The chromatin condenses into long, coarse strands which orient in the primary axis of the oblong nucleus. The change in nuclear shape coincides with the appearance of a sleeve of cytoplasmic microtubules outside the nuclear envelope. As the nucleus lengthens, it aligns with a shallow mitochondrial sulcus and associates posteriorly with the flagellar basal body, establishing the centriolar fossa. The Golgi apparatus of the spermatid remains active, and it is the likely source of membrane-bounded bodies of various sizes. Some Golgi vacuoles are electron-lucent, while others contain electron-dense granules.

The late spermatid shows further condensation of the chromatin, obscuring the coarse strands and other internal features. Coiling of the anterior end of the nucleus forms about one and a half gyres. The nuclear envelope does not remain firmly attached to the helical tip of the nucleus. Ascidian spermiogenesis involves no cytoplasmic sloughing by the spermatids.

The smallest bundles of late spermatids consist of 108–134 tails ($N = 10$). Approximately thirty-two primary spermatocytes would be required to yield the number of spermatids in this range. A spermatogonium would thus be expected to undergo five mitotic divisions before the two maturation divisions of the spermatocytes.

Morphometric Analyses

Volumetric calculations indicate that spermatogenic cells decrease in size during spermatogenesis, and there is a concomitant increase in the cellular fraction occupied by the nucleus (Table 6.1, Fig. 6.5). Since primary spermatocytes are approximately one-half the size of spermatogonia, some growth must logically take place after the mitotic divisions which establish the spermatogenic clones. Little, if any, growth would be expected after the first and second meiotic divisions. The secondary spermatocyte is indeed approximately one-half the size of a primary spermatocyte, but the calculated volume for a spermatid is considerably less than predicted if the contents of a secondary spermatocyte were distributed equally between daughter cells. The volumetric discrepancy might be explained if spermatids are more oblong than spherical.

Table 6.1. Morphometric analyses and volumetric calculations for the spermatogenic cells of *Boltenia villosa*

	Radius (mean ± SD, N = 50)		Volume (mean ± SD, N = 50)	
	Cell (μm)	Nucleus (μm)	Cell (μm³)	Nucleus (μm³)
Spermatogonium	3.1 ± 0.3[a]	2.1 ± 0.3[b]	131.7 ± 42.2	39.2 ± 15.7
	3.0 ± 0.3[c]	2.0 ± 0.3[d]	114.6 ± 32.6	36.0 ± 14.5
Primary spermatocyte	2.5 ± 0.2[a]	1.8 ± 0.2[b]	65.4 ± 15.5	25.3 ± 7.9
	2.4 ± 0.2[c]	1.7 ± 0.2[d]	56.3 ± 11.2	22.9 ± 7.2
Secondary spermatocyte	2.0 ± 0.2[a]	1.6 ± 0.2[b]	37.3 ± 12.4	18.7 ± 7.3
	2.0 ± 0.2[c]	1.6 ± 0.2[d]	33.6 ± 10.6	16.9 ± 6.6
Spermatid	1.3 ± 0.1[a]	0.9 ± 0.2[b]	9.1 ± 3.2	3.4 ± 1.9
	1.2 ± 0.1[c]	0.9 ± 0.2[d]	7.2 ± 2.4	2.9 ± 1.5

[a]Derived from direct measurements of cellular perimeter.
[b]Derived from direct measurements of nuclear perimeter.
[c]Derived from direct measurements of cellular area.
[d]Derived from direct measurements of nuclear area.

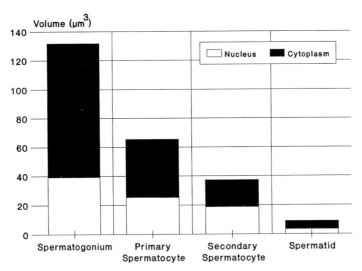

FIGURE 6.5. Mean volumes of the nucleus and cytoplasm in the spermatogonium, primary spermatocyte, secondary spermatocyte, and spermatid of *Boltenia villosa*. Direct measurements of cellular and nuclear perimeters provided the radii for volumetric calculations.

Accessory (Somatic) Cells

Tall, flagellated cells at the periphery of a spermatocyst are the only identifiable accessory cells. The pale cytoplasm of an accessory cell is endowed with a prominent Golgi apparatus and conspicuous glycogen rosettes (Fig. 6.6a). Short cisternae of the granular endoplasmic reticulum originate from evaginations of the nuclear envelope. The bodies of these cells and their basolateral processes form a discontinuous epithelium comprising the spermatocyst wall. A robust basal lamina underlies the epithelium, segregating the spermatocyst from the connective tissue.

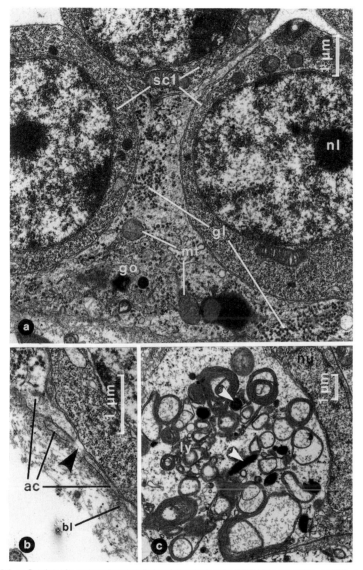

FIGURE 6.6. *a*. Section of a spermatocyst wall showing cytoplasmic detail of an accessory cell (gl, glycogen rosettes; go, Golgi apparatus; mt, mitochondrion). Three primary spermatocytes (sc1) closely approach the accessory cell, but there are no obvious intercellular junctions (nl, nucleolus). *b*. Discontinuity (arrowhead) of the spermatocyst wall where the processes of accessory cells (ac) are absent (bl, basal lamina). *c*. Section of a secondary lysosome (ly) within an accessory cell (nu, nucleus). Electron-dense profiles (arrowheads) are the nuclei of spermatozoa from an ingested clone.

The basolateral processes of the accessory cells are attenuated and overlapped (Fig. 6.6b). Intercellular junctions do not form between the processes. The bodies of accessory cells also project for short distances into the lumen of a spermatocyst, where they come into contact with the spermatogenic cells. Once again, there are no obvious intercellular junctions formed by the apposed plasmalemmata. Phagocytosis is the only role that can be confirmed for the accessory cells in the spermatocyst of *B. villosa*. Massive, membrane-bounded vacuoles containing the remnants of spermatozoa are frequently observed (Fig. 6.6c). The electron-dense nuclei of ingested spermatozoa are easily identified within some of these secondary lysosomes.

Discussion

It is sometimes preferable when referring to the male reproductive organs of invertebrates to employ a term other than *testes*, thus denoting a simpler organization than found in vertebrates and certain insects. Roosen-Runge (1977) recommends use of the term *spermatocyst*, referring to a spherical or ovoid body which measures 30–300 μm in diameter and contains a single clone of germ cells. A spermatogenic acinus of an ascidian ovotestis meets the prescribed shape of a spermatocyst, but it obviously does not contain just a single clone of germ cells.

The spermatocyst wall of *Boltenia villosa*, which is discontinuous and variable in thickness, is formed by the bodies and attenuated processes of the accessory cells. Ascidian accessory cells do not correspond to Sertoli cells which, by strictest definition, are somatic cells that associate simultaneously with more than one generation of spermatogenic cells (Fawcett, 1975; Roosen-Runge, 1977). The principal functions of Sertoli cells include support, coordination, and nutrition of spermatogenic cells and the phagocytosis of cytoplasm sloughed by spermatids during spermiogenesis. The only confirmed role of an ascidian accessory cell is that of phagocytosis. Entire clones of differentiating spermatozoa can apparently be engulfed for lysosomal digestion. Malformed clones, represented by several spermatidlike nuclei in a common mass of cytoplasm, are often encountered in the spermatocysts of *B. villosa*. Several workers, including Georges (1969) and Kubo et al. (1978), have also observed binucleate and multinucleate spermatids during ascidian spermiogenesis, and the latter investigators hypothesize a link between spermiogenic failure and the seasonality of spawning.

Spermiogenesis has garnered more attention than the earlier phases of spermatogenesis in ascidians (Franzén, 1976; Kubo et al., 1978; Cotelli et al., 1980; Villa, 1981; Fukumoto, 1983, 1985), because many investigators have attempted to resolve the controversy of whether or not an acrosome exists in the spermatozoon by studying the transformation of spermatids. Unfortunately, the controversy has not been settled. Single or multiple vesicles have been found near the tips of several spermatozoa. In *Ciona intestinalis*, these vesicles react during fertilization (De Santis et al., 1980; Fukumoto, 1984, 1988, 1990a,b; Rosati et al., 1985). In *Phallusia mammillata* (Honegger, 1986), on the other hand, there are no radical changes in the membranes, and intact vesicles can be observed in spermatozoa in the perivitelline space and in spermatozoa already fused to the oolemma. The putative acrosomes of ascidian spermatozoa are comparatively small, difficult to preserve, and midsagittal in placement (Fukumoto, 1983, 1984). They frequently lack the electron-dense contents that distinguish the acrosomes in other organisms. Lambert and Koch (1988) have compiled a list of prerequisites for conclusive identification of an acrosome and an acrosomal reaction, including ultrastructural identification of the organelle and confirmation that it is a discrete body; demonstration that the acrosome derives from the Golgi apparatus during spermatogenesis; detection of suitable proteases within the vesicle; and fusion of the plasmalemma and acrosomal vesicle when the spermatozoon contacts the vitelline envelope of the egg. Until such prerequisites are met, they recommend use of the term *apical vesicle(s)* instead of *acrosome, putative acrosome*, or *acrosomelike structure(s)*.

From the few ultrastructural studies that cover the gamut of spermatogenic events in ascidians (Georges, 1969; Cotelli et al., 1980; Villa and Tripepi, 1982), a pattern is beginning to emerge. Spermatogonia must undergo several rounds of mitosis to seed the spermatogenic clones, and primary spermatocytes are linked by cytoplasmic bridges that persist throughout the remainder of spermatogenesis. Flagella appear precociously on the secondary spermatocytes, and mitochondria of the spermatocytes merge into the single organelle serving the spermatid. A Golgi apparatus seems to be synthetically active in all spermatogenic cells. The condensed chromatin in the spermatid nucleus changes from a filamentous mass to a network of coarse strands. The nucleus and the mitochondrion assume an intimate relationship during spermiogenesis. Cytoplasmic sloughing is not a feature of ascidian spermiogenesis.

Spermiogenesis in *B. villosa* is generally consistent with published observations on other solitary species (Kubo et al., 1978; Cotelli et al., 1980; Villa and Tripepi, 1982; Fukumoto, 1983, 1985). In *B. villosa*, however, there is no convincing evidence of either an *apical blister* formed by the plasmalemma or a *dense plate* derived from the subjacent sector of nuclear envelope. The blister is the purported site of acrosomal formation, and the plate is the forerunner of a so-called acrosomal pedestal. The absence of an apical blister and a dense plate distinguishes spermiogenesis of *B. villosa* from that of other pyurids, including *Pyura haustor* (Fukumoto, 1983) and *Halocynthia roretzi* (Kubo et al., 1978).

Acknowledgments

Two undergraduate students, Ms. Vicki Whyte and Mr. Trevor Andrews, worked on aspects of this project, and their efforts are warmly acknowledged. This investigation was supported by Research Operating Grant No. OGP0000484 from the Natural Sciences and Engineering Research Council of Canada.

Literature Cited

Baccetti, B., and B. A. Afzelius. 1976. The Biology of the Sperm Cell. S. Karger, Basel.

Burgos, M. H., R. Vitale-Calpe, and A. Aoki. 1970. Fine structure of the testis and its functional significance. *In* A. D. Johnson, W. R. Gomes, and N. L. Vandemark, eds. The Testis, Vol. 1. Development, Anatomy, and Physiology. Academic Press, New York, pp. 551–649.

Cavey, M. J., and R. A. Cloney. 1972. Fine structure and differentiation of ascidian muscle. I. Differentiated caudal musculature of *Distaplia occidentalis* tadpoles. J. Morphol. 138: 349–374.

Cloney, R. A. 1987. Phylum Urochordata, Class Ascidiacea. *In* M. F. Strathmann, ed. Reproduction and Development of Marine Invertebrates of the Northern Pacific Coast, Data and Methods for the Study of Eggs, Embryos, and Larvae. University of Washington Press, Seattle, pp. 607–639.

———. 1989. Urochordata—Ascidiacea. *In* K. G. Adiyodi and R. G. Adiyodi, eds. Reproductive Biology of Invertebrates, Vol. 4, Pt. B. Fertilization, Development, and Parental Care. Oxford & IBH Publishing Company, New Delhi, pp. 391–451.

Cloney, R. A., and L. C. Abbott. 1980. The spermatozoa of ascidians: Acrosome and nuclear envelope. Cell Tissue Res. 206: 261–270.

Cloney, R. A., and E. Florey. 1968. Ultrastructure of cephalopod chromatophore organs. Z. Zellforsch. Mikrosk. Anat. 89: 250–280.

Cotelli, F., R. De Santis, F. Rosati, and A. Monroy. 1980. Acrosome differentiation in the spermatogenesis of *Ciona intestinalis*. Dev. Growth Differ. 22: 561–569.

De Santis, R., G. Jamunno, and F. Rosati. 1980. A study of the chorion and the follicle cells in relation to the sperm-egg interaction in the ascidian, *Ciona intestinalis*. Dev. Biol. 74: 490–499.

Fawcett, D. W. 1975. Ultrastructure and function of the Sertoli cell. *In* D. W. Hamilton and R. O. Greep, eds. Handbook of Physiology, Sect. 7 (Endocrinology), Vol. 5. Male Reproductive System. American Physiological Society, Washington, D.C., pp. 21–55.

Franzén, Å. 1970. Phylogenetic aspects of the morphology of spermatozoa and spermiogenesis. In B. Baccetti, ed. Comparative Spermatology. Academic Press, New York, pp. 29–46.

———. 1976. The fine structure of spermatid differentiation in a tunicate, *Corella parallelogramma* (Müller). Zoon 4: 115–120.

————. 1983. Urochordata. *In* K. G. Adiyodi and R. G. Adiyodi, eds. Reproductive Biology of Invertebrates, Vol. 2. Spermatogenesis and Sperm Function. John Wiley & Sons, Chichester, pp. 621–632.

Fukumoto, M. 1983. Fine structure and differentiation of the acrosome-like structure in the solitary ascidians, *Pyura haustor* aud [*sic*] *Styela plicata*. Dev. Growth Differ. 25: 503–515.

————. 1984. Fertilization in ascidians: Acrosome fragmentation in *Ciona intestinalis* spermatozoa. J. Ultrastruct. Res. 87: 252–262.

————. 1985. Acrosome differentiation in *Molgula manhattensis* (Ascidiacea, Tunicata). J. Ultrastruct. Res. 92: 158–166.

————. 1986. The acrosome in ascidians. I. Pleurogona. Int. J. Invert. Reprod. Dev. 10: 335–346.

————. 1988. Fertilization in ascidians: Apical processes and gamete fusion in *Ciona intestinalis* spermatozoa. J. Cell Sci. 89: 189–196.

————. 1990a. Morphological aspects of ascidian fertilization: Acrosome reaction, apical processes and gamete fusion in *Ciona intestinalis*. Invert. Reprod. Dev. 17: 147–154.

————. 1990b. The acrosome reaction in *Ciona intestinalis* (Ascidia, Tunicata). Dev. Growth Differ. 32: 51–55.

Georges, D. 1969. Spermatogenèse et spermiogenèse de *Ciona intestinalis* L. Observées au microscope électronique. J. Microsc. (Paris) 8: 391–400.

Honegger, T. G. 1986. Fertilization in ascidians: Studies on the egg envelope, sperm and gamete interactions in *Phallusia mammillata*. Dev. Biol. 118: 118–128.

Kubo, M., M. Ishikawa, and T. Numakunai. 1978. Differentiation of apical structures during spermiogenesis and fine structures of the spermatozoon in the ascidian *Halocynthia roretzi*. Acta Embryol. Exp. 3: 283–295.

Lambert, C. C., and R. A. Koch. 1988. Sperm binding and penetration during ascidian fertilization. Dev. Growth Differ. 30: 325–336.

Reynolds, E. S. 1963. The use of lead citrate at high pH as an electron-opaque stain in electron microscopy. J. Cell Biol. 17: 208–212.

Roosen-Runge, E. C. 1977. The Process of Spermatogenesis in Animals. Cambridge University Press, London.

Rosati, F., M. R. Pinto, and G. Casazza. 1985. The acrosomal region of the spermatozoon of Ciona intestinalis: Its relationship with the binding to the vitelline coat of the egg. Gamete Res. 11: 379–389.

Villa, L. 1981. An electron microscope study of spermiogenesis and spermatozoa of *Molgula impura* and *Styela plicata* (Ascidiacea, Tunicata). Acta Embryol. Morphol. Exp., N.S., 2: 69–85.

Villa, L., and S. Tripepi. 1982. An electron microscope study of spermatogenesis and spermatozoa of *Microcosmus sabatieri* (Ascidiacea, Tunicata). Acta Embryol. Morphol. Exp., N.S., 3: 201–215.

Wood, R. L., and J. H. Luft. 1965. The influence of buffer systems on fixation with osmium tetroxide. J. Ultrastruct. Res. 12: 22–45.

7 Test Cell Secretions and Their Functions in Ascidian Development

Richard A. Cloney

ABSTRACT Ascidian test cells differentiate in close association with oocytes in the ovary, but they have no manifest role in oogenesis or vitellogenesis. During embryogenesis they typically move about in the perivitelline space. The test cells of many species synthesize submicroscopic structures called *ornaments* and deposit them on the outer cuticular layer of the larval tunic during late development. In *Ciona*, *Ascidia*, and *Ascidiella* the test cells do not deposit ornaments, but they adhere firmly to the tunic. Test cells must also secrete amorphous substances in order to account for the binding of ornaments or the test cells themselves to the larval tunic. Test cells contain anionic glycosaminoglycans, and their secretory products are probably hydrophilic. What is the functional significance of test cell secretions? One clue comes from the fact that normal ascidian larvae have hydrophilic surfaces and do not float. But when the larvae of *Corella inflata* were experimentally deprived of test cell secretions during development they were hydrophobic and floated at air-water interfaces (Cloney 1990). It is proposed that the products of test cells render ascidian larvae wettable and prevent them from becoming trapped at the surface of the sea where they typically swim before settlement. *Molgula pacifica* is an unusual species that has no test cells and hatches as a tailless, benthic juvenile with strong adhesive properties. Since test cells are clearly not essential for oogenesis in this species, they may have been eliminated through evolution from ancestors that had swimming larvae.

Introduction

Ascidian test cells are nonepithelial, somatic cells that adjoin the surfaces of oocytes throughout oogenesis and reside in the perivitelline space of zygotes throughout embryogenesis. They were clearly described in association with the oocytes, embryos, and larval stages of *Phallusia mammillata* and *Ciona intestinalis* by Kowalevsky (1866). Kupffer (1870) identified them as motile cells which adhere to the larval tunic or "test" of *Ciona intestinalis*, and he named them Testazellen. Similar cells have been found in nearly all ascidians and in thaliaceans, but they are not found in larvaceans or other invertebrates.

The origin of the test cells is uncertain. Young oocytes become surrounded by an extension of the germinal epithelium called the primary follicle. The primary follicle cells divide and form an *outer follicular layer* and an *inner follicular layer*. The cells of the inner follicular layer proliferate by mitosis. Some of these cells are believed to dissociate from the inner follicular layer to form test cells. This is consistent with the findings of Bancroft (1899), Tucker (1942), Ermak (1976), Mukai and Watanabe (1976), and Sugino et al. (1987). But other investigators believe the test cells are derived from ameboid cells or lymphocytes which migrate to the surface of young oocytes and

Richard A. Cloney, Department of Zoology, University of Washington, Seattle, WA 98195.

differentiate into test cells (De Vincentiis, 1962; Kalk, 1963; Mancuso, 1965; Gianguzza and Dolcemascola, 1979).

Regardless or their origin, the test cells are incorporated into discrete pockets on the surface of young oocytes. A vitelline coat (formerly called the chorion; Cotelli et al., 1981) soon forms between the test cells and the inner follicular layer. The test cells codifferentiate with the oocytes with which they are associated, and near the time of ovulation they move out of the pockets and enter the perivitelline space. In late oogenesis and throughout embryogenesis the test cells have intense metachromatic-staining properties. With few exceptions, they contain acidic glycos-aminoglycans, usually sulfated, as determined by staining with toluidine blue and alcian blue methods (Cowden and Markert, 1961; Kessel and Kemp, 1962; De Vincentiis and Materazzi, 1964; Vitaioli and Materazzi, 1972; Mansueto and Villa, 1982; Monniot et al., 1992a,b).

Early ultrastructural studies of ovarian test cells of *Styela plicata* showed that they contain vacuoles with rodlike submicroscopic structures composed of spherical subunits (Kessel, 1962). Similar structures were found in *Styela* sp. (Kessel and Beams, 1965). These structures are electron-dense when fixed with osmium tetroxide, and they also stain with uranyl acetate. Cavey (1976) made a significant advance in understanding the role of test cells by showing that submicroscopic granules made in the test cells of *Distaplia occidentalis* are deposited on the larval tunic, a process he called "ornamentation." Later, the test cells of *Corella inflata* were also shown to synthesize ornaments and deposit them on the larval tunic (Cloney and Cavey, 1982). Elemental microanalyses of sections have recently been performed on ovaries of *Styela clava* by Monniot et al. (1992b). These investigators used a scanning electron microscope equipped with wavelength-dispersing spectrometers to demonstrate that the test cell granules contain *opal,* a form of silica. These investigators also demonstrated that the test cell granules (ornaments) have a reduced electron-density after pronase digestion and proposed that the siliceous material is embedded in a protein matrix.

In this chapter I compare test cells, their secretory products, and the larval tunics of many species. There has been a paucity of hypotheses related to the functions of test cell secretions. I discuss the bases for an hypothesis that the secretory products of test cells have a role in making the larval tunic hydrophilic and wettable. I also describe the ovarian oocytes and juvenile tunic of an unusual ascidian, *Molgula pacifica*, a species with no test cells and no swimming larval stage.

Materials and Methods
Collection of Adults and Cultures of Embryos

The larvae of ovoviviparous ascidians were collected directly from adult colonies. The larvae of oviparous ascidians were obtained by fertilizing eggs and culturing them through hatching. These procedures and all fixation methods for light and electron microscopy were carried out at the following laboratories: Hopkins Marine Station (*Clavelina huntsmani* Van Name, 1931; *Polyclinum planum* [Ritter and Forsyth] 1917); Harbor Branch Foundation, Florida (*Eudistoma olivaceum* [Van Name] 1902; *Ecteinascidia turbinata* Huntsman, 1880); Marine Biological Laboratory, Woods Hole, Massachusetts (*Styela partita* Stimpson, 1852; *Aplidium constellatum* [Verrill, 1871]); Friday Harbor Laboratories, Washington (*Chelyosoma productum* Stimpson, 1864; *Ascidia callosa* Stimpson, 1852; *Metandrocarpa taylori* Huntsman, 1912; *Boltenia villosa* Stimpson, 1864). *Molgula occidentalis* Traustedt, 1883, was obtained from Gulf Specimen Co., Panacea, Florida, and cultured in Seattle, Washington. *Molgula pacifica* Huntsman, 1912, a species with direct development, was collected at the Bamfield Marine Station, British Columbia, Canada, and cultured at the Friday Harbor Laboratories.

Light and Electron Microscopy

Fixative I. The gonads, embryos, and juveniles of *Molgula pacifica* and some larvae of other species were fixed in a primary fixative containing 2 percent glutaraldehyde, 0.2 *m* sodium cacodylate, 0.27 *M* sucrose, and 1 percent tannic acid. Specimens were postfixed in 2 percent osmium tetroxide buffered with 1.25 percent sodium bicarbonate (Torrence and Cloney, 1981).

Fixative II. Larvae were fixed in 2 percent osmium tetroxide buffered with 1.25 percent sodium bicarbonate at pH 7.4.

Fixative III. Larvae and gonads were fixed in phosphate-buffered glutaraldehyde followed by bicarbonate-buffered osmium tetroxide (Cloney, 1977).

All of the specimens were dehydrated, infiltrated and embedded in epoxy resins. Fixative I is superior to the others for cytological preservation. Unfortunately, at least in some cases, it removes the ornaments from the larval tunic although it preserves the ornaments in the test cells (Cloney and Cavey, 1982). The fixatives used for specific tissues are given in the figure legends.

Sections for light microscopy were stained with azure II and methylene blue. All but three electron micrographs illustrated in this chapter were made from sections that were stained with uranyl acetate and lead citrate. Other sections were examined to determine if the ornaments were electron-dense without staining (Figs. 7.1a, 7.2a and 7.2b). Figure 7.3b of the larval tunic of *Chelyosoma productum* was kindly provided by Dr. Steven A. Torrence.

Terminology

In this chapter the term *ornaments* refers to a variety of submicroscopic granules or particles and filaments that are deposited on or near the surface of the outer cuticular layer (C1) of ascidian larval tunics. All ornaments are believed to be secretory products of test cells. In some species, *putative ornaments* have been found on larval tunics. In these cases it remains to be determined if they are actually secreted by test cells. The smaller granules are called *punctate ornaments*. The larger granules typically form aggregates called *multigranular ornaments*. The individual granules in multigranular ornaments are called *subunits*. In some species filaments are associated with granules. In other species the only ornaments are filaments. Filamentous ornaments form aggregates in some species; in other species filamentous ornaments attach to the tunic without forming aggregates.

Some granular ornaments and subunits of multigranular ornaments have a property which I have called *constitutive electron density* (CED). Ornaments with CED had a high density when they were examined in unstained sections by electron microscopy. The electron density could be either *intrinsic* or caused by the osmium tetroxide used in fixatives, as indicated by the findings of Monniot et al. (1992b). They determined that the ornaments in the test cells of *Styela clava* were electron-dense, owing to the presence of silicon, after osmium was removed from sections with periodic acid. The tissue samples used in this chapter were prepared with protocols that included fixation with osmium tetroxide. The basis for the CED of the various ornaments described is unknown. Some of the granular ornaments and all of the filamentous ornaments that have been found lack CED.

The larval tunic consists of extracellular products that are secreted by epidermal cells, beginning at the tailbud stage (Mancuso, 1973, 1974). Two thin, transparent, cuticular layers and filamentous matrix are invariably formed. The outer cuticular layer (C1) forms the larval fins, and

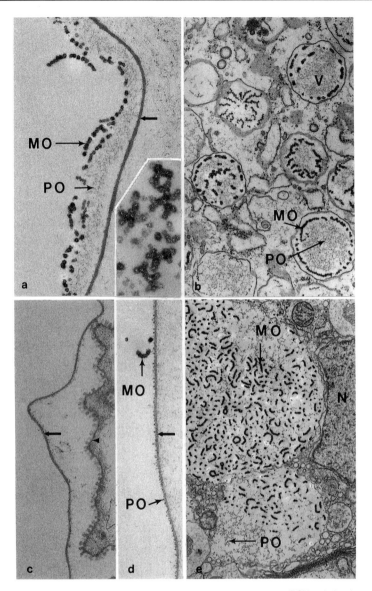

FIGURE 7.1. *a. Aplidium constellatum*, with a broad band of punctate ornaments (PO) bordering the outer cuticular layer of the larval tunic (arrow). Outside of this band are one or two layers of multigranular ornaments (MO). The subunits join to form linear or curvilinear strands. The subunits have electron-lucent cores (fixative III, 22,400×). Insert: *Polyclinum planum*. Unstained section illustrating the constitutive electron density of the subunits of multigranular ornaments. Notice that the subunits join to form a branching pattern rather than linear arrays (fixative III, 44,000×). *b. Aplidium constellatum*—cytoplasm of a test cell at the *tailbud stage*. Some vacuoles (V) contain central clusters of punctate ornaments (PO) and peripheral multigranular ornaments (MO) identical with those on the larval tunic (*a*). In some vacuoles the subunits of the multigranular ornaments are smaller than those on the larval tunic and may still be growing. The vacuoles are apparently fragile and many were ruptured in this preparation (fixative I, 18,300×). *c. Clavelina huntsmani*—larval tunic with ornaments lost from the outer cuticular layer (arrow) in processing the tissues. The inner cuticular layer (arrowhead) bears short protrusions (fixative III, 22,400×). *d. Clavelina huntsmani*—larval tunic, with the outer cuticular layer (arrow) bearing a few multigranular ornaments (MO) and numerous punctate ornaments (PO) (fixative II, 22,400×). *e. Clavelina huntsmani*—Test cell. The vacuoles in the cytoplasm contain both multigranular ornaments in the form of rods and rings (MO) and punctate ornaments (PO). These are identical to the ornaments found on the larval tunic (*d*) (fixative I, 10,000×).

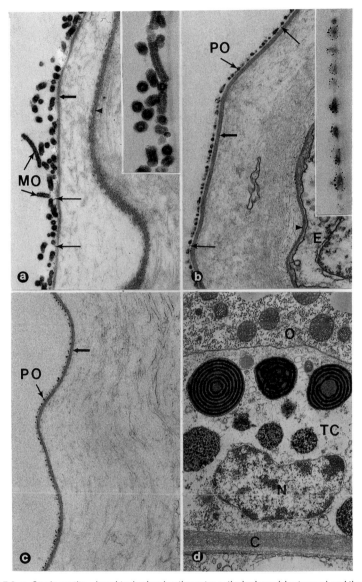

FIGURE 7.2. a. *Styela partita*—larval tunic showing the outer cuticular layer (short arrow) and the papillate, inner, cuticular layer (arrowhead). The multigranular ornaments (MO) are linear arrays of subunits near the surface of C1. They appear to be joined to thin, dense filaments (long arrows) (fixative II, 22,400×). Insert: Unstained section to show the constitutive electron density of the ornament subunits. Notice that the filaments are not visible (fixative II, 44,000×). b. *Metandrocarpa taylori.*— larval tunic and epidermis (E). Punctate ornaments (PO) form a single layer near the outer cuticular layer (short arrow); they are joined laterally by fine filaments (long arrows). The inner cuticular layer is indicated by the arrowhead. The structure in the compartment between the two cuticular layers is a process of a blood cell (fixative II, 22,400×). Insert: Unstained section of larval tunic to show the constitutive electron density of the punctate ornaments. The ornaments are composed of numerous, separate, electron-dense granules which surround larger structures of intermediate electron density. Notice that the filaments are not visible (fixative II, 44,000×). c. *Boltenia villosa*—larval tunic showing the outer cuticular layer (arrow) with attached ornaments. The punctate ornaments (PO) form a single thin layer (fixative II, 22,400). d. *Boltenia villosa*—ovarian oocyte (O), vitelline coat (C), and test cell (TC). Some vacuoles in the test cell contain concentric light and dark rings; others contain punctate, electron-dense granules. The punctate granules are somewhat larger at this stage of differentiation than those on the larval tunic (c) (fixative III, 8,800×).

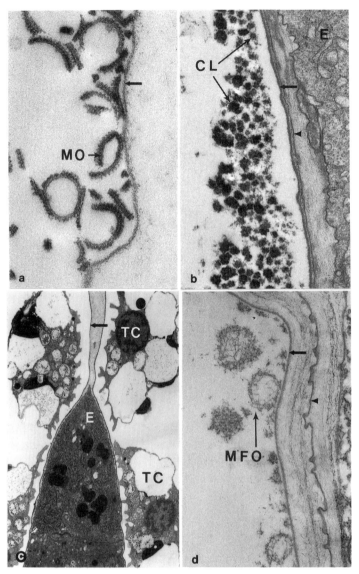

FIGURE 7.3. *a. Corella inflata*—larval tunic. Linear, curvilinear, and ring-shaped multigranular ornaments (MO) are found on and near the surface of the outer cuticular layer of tunic (arrow). Individual granules have electron-lucent cores (fixative II, 52,700×). *b. Chelyosoma productum*—larval tunic. Subunits apparently aggregate the form multigranular clusters (CL) of different sizes. Notice the stratification. There is a prominent gap between the outer cuticular layer (arrow) and the ornaments which probably contains an unresolved extracellular matrix. The epidermis (E) and the inner cuticle (arrowhead) are visible. This micrograph was provided by Dr. S. A. Torrence (fixative II, 22,400×). *c. Ascidia callosa*—larval tail showing epidermis (E) and dorsal fin. Four vacuolated test cells (TC) are visible. Most of the test cells processes do not make direct contact the outer cuticular layer (arrow). There are no ornaments in this species (fixative I, 5,000×). *d. Molgula occidentalis*—larval tunic. The ornaments consist of filaments. Some of these aggregate to form hollow multifilamentous spherical ornaments (MFO), and others form smaller clumps. The outer cuticular layer (arrow) is smooth, and the inner cuticular layer bears papillae or plications (arrowhead) (fixative II, 22,400×).

it is the only layer that can bear ornaments. The inner cuticular layer (C2) becomes the surface of the juvenile when C1 is cast off at metamorphosis.

Results

Ornaments, Test Cells, and Larval Tunics of Twelve Species

Aplidium constellatum *(Family Polyclinidae)*

In *Aplidium constellatum* the outer cuticular layer of the larval tunic is covered with two layers of ornaments. The inner layer consists of scattered 7–11-nm granules in a band with a width of 75–400 nm. These are punctate ornaments. They have very low CED (see Material and Methods) and are barely visible without staining. The outer layer consists mostly of multigranular ornaments in one or two layers. Most of these are linear strands of subunits, but occasionally they are branched. The subunits have a short axis of 39 nm and a long axis of 49 nm (Fig. 7.1a). The subunits have CED with electron-lucent cores. Both types of ornaments are evidently attached to the outer layer of tunic with an extracellular matrix that was not resolved.

The test cells were examined at the late tailbud stage. Many vacuoles contain ornaments that correspond closely with those found on the larval tunic. In most vacuoles the punctate ornaments are centrally located and are surrounded by a layer of multigranular ornaments (Fig. 7.1b) Some vacuoles contain multigranular ornaments with subunits that are smaller than those that were found on the larval tunic. No additional information is available on the stages of assembly of these ornaments.

Polyclinum planum *(Family Polyclinidae)*

In *Polyclinum planum* the larval tunic bears multigranular ornaments and punctate ornaments. The punctate ornaments have no CED and are 10–20 nm in diameter (not illustrated). They attach in a single layer close to the surface of C1. The subunits in the multigranular ornaments have a diameter of about 49 nm. Instead of joining in linear arrays or rings, most of the subunits join to form branched clusters (Fig. 7.1a). The subunits have CED with electron-lucent cores. The test cells of this species have not been examined.

Clavelina huntsmani *(Family Polycitoridae)*

I reported that the larva of *Clavelina huntsmani* had no ornaments (Cloney, 1990) based on specimens prepared with fixative III (Fig. 7.1c). Later, I found two types of ornaments on larvae that were prepared with fixative II. Punctate ornaments, 10–16 nm in diameter, form a single layer close to the surface of C1. A few linear, multigranular ornaments, composed of 49 nm subspherical subunits were found outside of the punctate layer (Fig. 7.1d) These have CED with electron-lucent cores.

The test cells were examined in both the ovary and in the late tailbud stage with fixative I. None of the vacuoles in the test cells in the ovary contained multigranular ornaments although they contained flocculent material. At the late tailbud stage the test cells were more fully differentiated. Their vacuoles contained multigranular ornaments, joined in linear arrays or rings, and punctate ornaments (Fig. 7.1e). The multigranular ornaments have CED with electron-lucent cores. The punctate ornaments lack CED and were barely visible in unstained sections.

Distaplia occidentalis *(Family Polycitoridae)*

There are two size classes of ornaments in *D. occidentalis*. Both of these have been previously described from specimens that were stained with uranyl acetate and lead citrate (Cavey and

Cloney, 1984). I have determined that the smaller punctate ornaments, 10–14 nm in diameter, have no CED and are not visible without staining. These form a single layer very close to the outer surface of C1. The larger granules are ovoid with a long axis of 38 nm and a short axis of 33 nm. They have CED with electron-lucent cores in unstained sections (not illustrated). The larger granules or subunits aggregate to form multigranular rings about 500 nm in diameter. Some of the rings are separated from C1 by a distance of several hundred nanometers.

The test cells in this species were examined during the middle and late period of embryonic development. The vacuoles in these cells contain multigranular ornaments, as previously reported (Cavey and Cloney, 1984). Upon reexamination of the test cell I found that they also contain abundant punctate ornaments, identical to those found on the surface of C1. The embryonic development of this colonial species requires about 28 days at 13°C. If the test cells are removed before the tailbud stage, ornaments do not appear on the larval tunic (M. J. Cavey, pers. comm., 1984).

Corella inflata *(Family Corellidae)*

In *Corella inflata* the outer cuticular layer of the larval tunic is covered with multigranular ornaments. All of the constituent subunits in these structures are similar. They have CED with electron-lucent cores in unstained sections. They are asymmetric, subovoid structures with a long axis of 35 nm and a short axis of 17.5 nm. They aggregate to form linear and curvilinear stands or rings (Fig. 7.3a). Some of the aggregates are separated from C1 by several hundred nanometers. I assume that they are held in place by an unresolved glycocalyx. No punctate ornaments were found in this species.

The ornaments found in vacuoles of the test cells in the embryos of *Corella inflata* match those that are found on C1. The embryonic development of this species is completed in about 30 h at 12°C. If the test cells are removed from an embryo before the tailbud stage, ornaments never appear on the larval tunic (Cloney and Cavey, 1982).

Ecteinascidia turbinata *(Family Perophoridae)*

In *Ecteinascidia turbinata* the outer cuticular layer of the larval tunic is covered with filaments with a diameter of 7 nm (not illustrated). These have only moderate electron density in stained sections. The majority of filaments are oriented perpendicular to the surface of the tunic. The larval tunic of *Perophora orientalis* bears similar structures (Terakado, 1970). Each filament appears to be attached at one end to C1. These structures must be regarded as putative ornaments until it can be demonstrated that they are present in the test cells.

Chelyosoma productum *(Family Corellidae)*

In *Chelyosoma productum* the outer cuticular layer of the larval tunic is covered with 3–12 stratified layers of multigranular putative ornaments in clusters of different sizes. These ornaments are quite different from those in any other species. The smallest subunits appear to be about 15 nm in diameter, with moderate electron density. It is not known whether these have CED. The largest clusters are about 150 nm in diameter. The smaller clusters are found nearest the surface of C1 (Fig. 7.3b). The entire layer of ornaments is separated from C1 by a prominent gap of 160–220 nm. I presume that this gap is occupied by some type of extracellular matrix that was either not preserved by the methods that were employed or lacks sufficient contrast to be detected. The test cells of this species have not been examined.

Ascidia callosa *(Family Ascidiidae)*

The test cells of *Ascidia callosa* adhere firmly to C1 beginning late in embryogenesis (Fig. 7.3c). They do not secrete ornaments. The test cells have many irregular fingerlike processes on their surfaces, but only a few make direct contact with the larval tunic. They apparently are suspended in an invisible extracellular matrix that adheres to the outer cuticular layer.

Molgula occidentalis *(Family Molgulidae)*

In *Molgula occidentalis* the outer cuticular layer is covered with 7-nm filamentous putative ornaments which form compact clumps and conspicuous, hollow spheres, 100–750 nm in diameter (Fig. 7.3d). Some spheres are separated from C1 by gaps as large as 1 μm, with no visible connections. These unusual structures were illustrated but not described in another study of this species by Torrence and Cloney (1981) before they were recognized as putative ornaments. The test cells of this species have not been examined.

Styela partita *(Family Styelidae)*

In *Styela partita* the outer cuticular layer of the larval tunic bears rodlike multigranular ornaments. Individual ornaments have a short axis of 49 nm and a long axis of 58 nm (Fig. 7.2a). They have CED with electron-lucent cores in unstained sections (Fig. 7.2a). Near the surface of the outer cuticular layer, a delicate reticular network of fine filaments with diameters of 3–6 nm joins the multigranular ornaments. These could not be detected in unstained specimens and thus have no CED (Fig. 7.2a). The multigranular ornaments closely resemble those described in the test cells of *Styela plicata*, *Styela* sp. (Kessel, 1962; Kessel and Beams, 1965), *Styela clava* (Monniot et al., 1992a), and *Polycarpa gracilis* (Botte et al., 1979).

Metandrocarpa taylori *(Family Styelidae)*

In *Metandrocarpa taylori* the outer cuticular layer is covered with a single layer of ovoid putative ornaments with diameters of 30–45 nm as seen in transverse sections. In some micrographs they appear elongated in the plane of C1 with lengths to 175 nm. They are joined laterally by a network of fine filaments without CED (Fig. 7.2b). The ovoid ornaments have an unusual CED in unstained sections. They are complex structures consisting of a central core of intermediate electron density surrounded by a single layer of electron-opaque granules, 1–20 nm in diameter (Fig 7.2b [insert]). The test cells of this species have not been examined.

Boltenia villosa *(Family Pyuridae)*

In *Boltenia villosa* the outer cuticular layer of the larval tunic is covered with a single layer of punctate ornaments with an average diameter of 15 nm (Fig. 7.2c). These are associated with fine filamentous material of low electron density.

Ovarian test cells at intermediate stages of differentiation contain vacuoles with alternating light and dense concentric lamellae, vacuoles with interconnecting light and dark layers, and vacuoles with electron-dense granules joined by fine filaments (Fig. 7.2d). At the late tailbud stage the vacuoles in the test cells contain dense granular and filamentous material, but no concentric lamella. The granules and filaments are identical to those attached to the tunic at this stage in stained sections. Only the granules have CED. In unstained sections they appear black, in contrast to the surrounding light gray tissues.

Oogenesis and Hatching in an Ascidian without Test Cells
Ovarian Oocytes of molgula pacifica

Young et al. (1988) reported that they could not find test cells in the embryos of *M. pacifica*, but they mentioned that I had found a few small structures which I assumed were test cells when I demembranated the eggs. To resolve this problem I examined the ovaries and embryos of this species by light and electron microscopy. I found no test cells in the ovaries or embryos. These results confirm the findings of Young et al. (1988) and those of Bates and Mallett (1991a).

Oocytes in late vitellogenesis have numerous light, spherical bodies at the periphery (Fig. 7.4a). When thin sections were examined by light microscopy, it appeared that they might be test cells or degenerate test cells. However, when they were examined by electron microscopy they were found to be simple invaginations of the oolemma which contained scattered, fine, filamentous material but no cellular components. I have called these pseudo–test cells (Fig. 7.4b).

Small oocytes, 6 μm in diameter, are surrounded by a layer of squamous cells, the outer follicular layer. Larger previtellogenic oocytes acquire a layer of inner follicle cells. The vitelline coat is secreted between these cells and the oocyte (Fig. 7.4a). In all other ascidians that have been examined, with rare exceptions (Monniot et al., 1992a), the test cells become embedded in the surface of the oocyte before the vitelline coat is secreted.

The smaller follicle cells of *M. pacifica* have a large, central, dense vacuole surrounded by a ring of smaller, dense vacuoles. The dense material consists of tightly packed arrays of filaments. The larger follicle cells have a single large, dense vacuole with numerous peripheral, smaller, light vacuoles and a few small, dense vacuoles near the vitelline coat (Fig. 7.4b). The follicle cells are known to secrete adhesives which anchor fertilized eggs to solid substrata (Young et al., 1988).

The vitelline coats around the larger oocytes in the ovary are mainly bipartite structures; each has an outer dense component, about 20 percent of the thickness, consisting of fine filaments, and an inner, loose component, about 80 percent of the thickness, consisting of coarser filaments. At this time the diameter of the larger ovarian oocytes, excluding the vitelline coat and follicle cells, is about 164 μm, and the overall thickness of the vitelline coat is about 1.2 μm (Fig. 7.4b). The inner coarse layer is secreted first, around previtellogenic oocytes. The outer layer is secreted later, after the follicle cells have grown considerably.

In the spaces between the bases of follicle cells there are a few loose filaments outside the dense component. This might constitute an incomplete third layer, as seen in other species (Honegger, 1986; Cloney, 1990). Previtellogenic oocytes and those in late vitellogenesis contain many vesicles filled with filamentous material that resemble those found in the vitelline coat (Fig. 7.4b). This is consistent with the findings of Cotelli et al. (1981) that oocytes secrete at least part of the vitelline coat. However, it remains possible that the outer, more dense, component, composed of fine filaments, is secreted by the follicle cells.

The Juvenile of Molgula pacifica

In *Molgula pacifica* development proceeds directly to the juvenile stage with no vestige of a tail. In preparation for hatching, the spherical embryo elongates and forms a nipplelike primary epidermal ampulla, which protrudes and eventually breaks through the vitelline coat (Cloney, 1987; Young et al., 1988) (Fig. 7.5b). This has been described in detail by Bates and Mallett (1991a,b). The juvenile has only one cuticular layer of tunic, about 11 nm in thickness. This layer is homologous to the inner cuticular layer (C2) of the larva of *Molgula occidentalis* (Torrence and

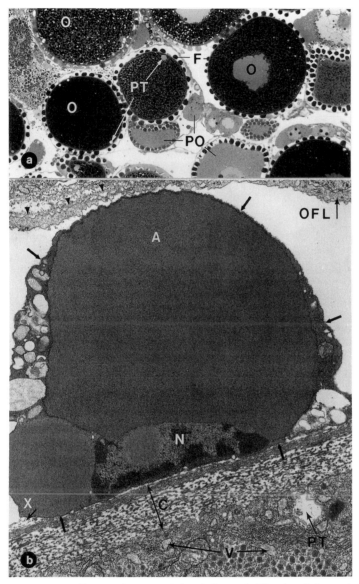

FIGURE 7.4. *a. Molgula pacifica*—ovarian oocytes in late vitellogenesis (O) and previtellogenesis (PO). The follicle cells (F) are attached to the vitelline coat. "Pseudo–test cells" (PT) are simple inpocketings of the oolemma, about the size of test cells. There are no true test cells around any of the oocytes (fixative I, 175×). *b. Molgula pacifica*—surface of an ovarian oocyte (O) in late vitellogenesis. The vitelline coat (C) consists of two major layers. Between follicle cells (X) small patches of filaments are found that may constitute a third layer. One follicle cell (ringed by arrows) and the outer follicular layer (OFL) with its basal lamina are visible (arrowheads). The follicle cell has a large, central, electron-dense vacuole containing filamentous adhesive material (A). It is surrounded by smaller light and dark vacuoles. The follicle cell nucleus (N) is near the vitelline coat. On the edge of the oocyte is a grazing section of a large invagination of the surface. This corresponds to the pseudo–test cells (PT) in *a*. The vesicles (V) contain filamentous material similar to material in the vitelline coat (fixative I, 12,000×).

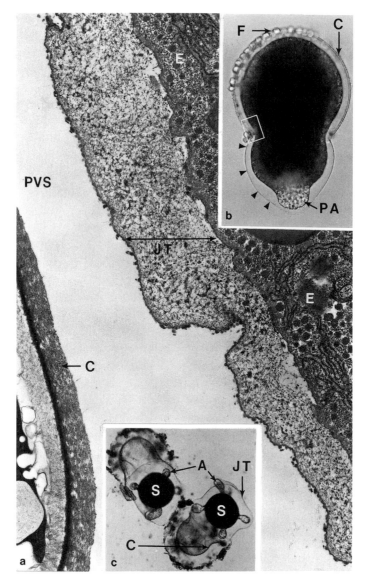

FIGURE 7.5. *a. Molgula pacifica*—Hatching juvenile showing the epidermis (E) tunic (JT) and perivitelline space (PVS). This section is from an area similar to the one in the white box in *b*. The inner compartment of the tunic beneath the cuticle is packed with filamentous material. To the left of the outer dense component of the vitelline coat is adhesive material from ruptured follicle cells (fixative I, 13,600×). *b. Molgula pacifica*—Fresh-mount of hatching juvenile. Notice the vitelline coat (C) with follicle cells (F) and the primary ampulla (PA) of the juvenile. The cuticular layer of the juvenile tunic is marked with arrowheads on the left side (150×). *c. Molgula pacifica*—Fresh-mount of two juveniles showing soma (S), ampullae (A), vitelline coat (C), and juvenile tunic (JT). The vitelline coat remains associated with juveniles for several days (60×).

Cloney, 1981). It bears short protrusions, giving it a beaded texture commonly found on C2 of other species (Figs. 7.1c, 7.2a, and 7.3d). The compartment between the epidermis and the cuticle is packed with a loose network of filaments and irregular clumps of filaments. The entire layer is 1–2 μm in width (Fig. 7.5a).

At hatching the vitelline coat is more compressed than it is in the ovary. Its thickness is reduced to 0.6 μm. Within several hours after hatching the juveniles form up to five epidermal ampullae (Cloney, 1987; Bates and Mallett, 1991a) In laboratory cultures the newly hatched juveniles adhere firmly to culture dishes and to the vitelline coats from which they emerged (Fig. 7.5c).

Discussion

Fate of the Secretory Products of the Test Cells

Do the test cells transport secretory products into developing oocytes? This function was proposed by Harvey (1927), Kessel and Kemp (1962), Kalk (1963), De Vincentiis and Materazzi (1964), Mancuso (1965), Materazzi and Vitaioli (1969), and Gianguzza and Dolcemascolo (1979). But, as pointed out by Kessel (1983), there is no direct evidence to support this view. Test cells are not the source of hatching enzymes as suggested by Knaben (1936); these are secreted by the embryos (Caggegi et al., 1974). To determine the function of the test cells it is necessary to follow the fate of their secretory products. Fortunately, in many species the secretory products are distinctive and easy to trace through development.

The test cells of *Distaplia occidentalis* and *Corella inflata* (Cavey, 1976; Cavey and Cloney, 1984; Cloney and Cavey, 1982), as well as those of *Aplidium constellatum*, *Clavelina huntsmani*, and *Boltenia villosa* (which are reported here for the first time), all contain secretory products with the same characteristics (size, shape, and electron density) as ornaments found on the larval tunics of these respective species. Although test cells of *Styela partita* were not examined, the ornaments on the larval tunic of this species are nearly identical to those found in the test cells of *S. plicata* and *S. clava* (Kessel, 1962; Monniot et al., 1992b). The observations on these species are especially significant because they firmly establish that test cells make distinct and recognizable products which adhere to the larval tunics of species in all three orders of ascidians. In those species in which the test cells have not yet been examined, the identification of structures on the larval tunics as ornaments remains to be confirmed.

Table 7.1 is a tentative classification of all of the known types of ornaments and putative ornaments that have been found on ascidian larvae. The distribution of various types of ornaments in ten families is summarized in Table 7.2. Six species in four families, representing three orders, have Type 1 multigranular ornaments. The subunits of these ornaments all have CED with central cores of low electron density. All of these are similar to the ornaments found in the ovarian test cells of *Styela clava*. In this species the ornaments have been shown to contain opal, a form of silica that is bound to proteins (Monniot et al., 1992b). The presence of opal is a major new finding which will probably generate renewed interest in test cell biochemistry.

Monniot et al. (1992a) have examined seventy-nine species of ascidians and found silica in the ovarian test cells of species in eight families. Five of the six genera listed in Table 7.1 with Type 1 ornaments contain silica according to their findings. It is interesting that Monniot et al. (1992a) found no silica in *Clavelina* ovarian test cells. I found that the test cells of *Clavelina huntsmani* had no Type 1 ornaments in the ovary, but they were present in the test cells of the late tailbud stage. This suggests that some of the species listed by Monniot et al. (1992a) as having no silica should be reexamined at a later state in development. Alternatively, Type 1 ornaments may represent a heterogeneous group in which ultrastructural similarity is not a reliable indication of the presence of silica.

The Type 1 ornaments of *Distaplia*, *Clavelina*, *Aplidium*, and *Polyclinum* coexist with Type 3 punctate ornaments (without CED). *Boltenia* and *Metandrocarpa* have a mixture of Type 2 punctate ornaments (with CED) and Type 5 cross-linking ornaments (without CED). In these species the test cells secrete at least two products with distinctive ultrastructural characteristics and chemical properties.

Table 7.1. Tentative ultrastructural classification of ornaments and putative ornaments found on ascidian larvae

Type	Structural characteristics of ornaments	Electron density when stained	Constitutive electron density	Generic examples
1	Subunits with electron-lucent cores join to form multi-granular rods, rings, or branched clusters	High	High	*Distaplia, Clavelina, Aplidium, Polyclinum, Corella, Styela*
2	Punctate granules	High	High	*Metandrocarpa, Boltenia*
3	Punctate granules	High	Very low	*Distaplia, Clavelina, Aplidium, Polyclinum, Diplosoma*
4	15-nm subunits form hetero-geneous clumps; stratified on tunic	Moderate	Unknown	*Chelyosoma*
5	Nonaggregating filaments; attach perpendicular to larval tunic	Moderate	Very low	*Ecteinascidia, Perophora*
6	Cross-linking filaments join to other types of ornaments	High	Very low	*Styela, Metandrocarpa*
7	7-nm filaments join to form clumps and hollow spheres	Moderate	Very low	*Molgula*

The multigranular, Type 4 ornaments of *Chelyosoma productum* are unique among the species that have been examined. The subunits form clusters rather than linear strands or rings and do not stain as intensely as Type 1 subunits. It is remarkable that two species within the same family (*Chelyosoma* and *Corella*) have such different ornaments. The ovarian test cells in both of these genera contain silica (Monniot et al. 1992a).

The filamentous, Type 5 ornaments, on the surface of swimming larvae of *Perophora orientalis*, were illustrated but not described by Terakado (1970). He also illustrated an embryonic stage in which the outer cuticular layer appears to have no filaments; this suggests that they are deposited late in development. In *Ecteinascidia turbinata* and *P. orientalis*, the filaments do not form aggregates, but attach separately, very close to and approximately perpendicular to the surface of C1. Neither of these genera contains silica in the ovarian test cells (Monniot et al. 1992a).

The multifilamentous clumps and spheres (Type 7) on larvae of *Molgula occidentalis* are often separated from C1 by relatively large gap. Neither of these classes of filaments stains intensely. Silica has not been found the ovarian test cells of molgulids (Monniot et al., 1992a).

Larvae in the families Cionidae and Ascidiidae have no ornaments, but all species that have been examined have test cells that firmly attach to the larval tunic (Table 7.2). The ovarian test cells of species in these families contain no silica (Monniot et. al. 1992a). It is not known why the attachment of test cells to C1 is correlated with the absence of ornaments. An essential part of the hypothesis presented in this chapter is that the test cells secrete a glycocalyx of some kind that holds the ornaments, or the test cells themselves, to the larval tunic. Without this postulated component it would be difficult to explain how the ornaments adhere to the tunic. A glycocalyx was not visible in electron micrographs with the techniques that were employed in this study. I

Table 7.2. Characteristic structures found on outer cuticular layer of ascidian larval tunics

Family and species	Occurrence and organization of ornaments and putative ornaments	Unit or subunit size (nm)	References
Order Phlebobranchia			
Cionidae			
Ciona intestinalis	ATC[a] only, no orn.[b]	—	Gianguzza and Dolcemascolo, 1984
Corellidae			
Corella inflata	Type 1 orn.; form rings and rods	17.5 × 35 15	Figure 7.3a
Chelyosoma productum	Type 4 granular orn. Orn. form clumps	15–150	Figure 7.3b Figure 7.3b
Ascidiidae			
Ascidia callosa	ATC only, no orn.	—	Figure 7.3c
Ascidia paratropa	ATC only, no orn.	—	Cloney and Cavey, 1982
Ascidiella aspersa	ATC only, no orn.	—	Lübbering and Goffinet, 1991
Perophoridae			
Ecteinascidia turbinata	Type 5 filamentous orn.	7–10 diam	
Perophora orientalis	Type 5 filamentous orn.	7–10 diam	Terakado, 1970
Order Aplousobranchia			
Polyclinidae			
Aplidium constellatum	Type 1 orn.; form rods	39 × 49	Figure 7.1a
	Type 3 punctate orn.	7–11	Figure 7.1a
Polyclinum planum	Type 1 orn.; form branched clusters	49	Figure 7.1a
	Type 3 punctate orn.	10–20	
Euherdmania claviformis	No ATC or orn.	—	
Polycitoridae			
Distaplia occidentalis	Type 1 orn.; form rings	33 × 38	Cavey and Cloney, 1984
	Type 3 punctate orn.	10–14	
Eudistoma olivaceum	No ATC or orn.	—	
Clavelina huntsmani	Type 1 orn.; form rods	49	Figure 7.1d
	Type 3 punctate orn.	10–16	Figure 7.1d
Didemnidae			
Diplosoma macdonaldi	Type 3 punctate orn.	15	

intend to look for a carbohydrate-rich coating using several techniques, including the alcian blue–ruthenium red staining method employed on other organisms by Dykstra and Aldrich (1978).

Eudistoma olivaceum and *Euherdmania claviformis* are exceptional species since neither ornaments nor test cells were found on the larval tunic (Table 7.2). The ornaments could have been lost in tissue processing of larvae. No conclusion can be made until the test cells are examined

Table 7.2. Continued

Family and species	Occurrence and organization of ornaments and putative ornaments	Unit or subunit size (nm)	References
Order Stolidobranchia			
Styelidae			
Styela partita	Type 1 orn.; form rods	49 × 58	Figure 7.2a
	Type 6 filamentous orn.	3–6 diam	Figure 7.2a
Metandrocarpa taylori	Type 2 punctate orn.	30–45	Figure 7.2b
	Type 6 filamentous orn.	3 diam.	Figure 7.2b
Pyuridae			
Boltenia villosa	Type 2 punctate orn.	15	Figure 7.2c
Halocynthia roretzi	Fibrous material	?	Lübbering et al., 1992
Molgulidae			
Molgula occidentalis	Type 7 filamentous orn.	7 diam	Figure 7.3d
	Orn. form clumps and hollow spheres	100–750	Figure 7.3d

[a]ATC = Attached test cells. Other characteristics of ornaments are listed in Table 7.1.
[b]orn. = Ornaments.

directly. I assumed that the test cells of *Clavelina huntsmani* made no ornaments when I could not find them on the larval tunic. Later, after I found ornaments in the test cells of embryos, I also found them on the larval tunic of specimens that were fixed with a different method. Even if ornaments are not secreted by the test cells of some species, it is still possible that they secrete an amorphous material with a similar function.

Comparative histological studies have established that the test cells synthesize a variety of products with distinctive submicroscopic characteristics. There is now substantial evidence that these are deposited on the larval tunic of many species. Within a species the ornaments on the larval tunic have a consistent range of sizes and shapes and a characteristic pattern of distribution. Differences between species are sometimes so striking that the ornaments might be useful, in conjunction with other morphological characters, for identifying larvae from the plankton when the parent is unknown.

Characteristics of the Larval Surface

The larvae of solitary ascidians, when normally reared in laboratory cultures, are never sticky. Particulate matter does not adhere to them, they do not clump together, even in crowded cultures, and they do not adhere to culture dishes. However, when embryos are demembranated (vitelline coat, follicle cells, and test cells removed) at the *neurula stage* or earlier, and the embryos are allowed to continue development in relatively large volumes of sea water (10 ml), they form both cuticular layers of the tunic, but fins are not formed. These larvae cannot swim effectively, they are sticky, the tunic is fragile, and larvae tend to clump together, although they can metamorphose (Cloney and Cavey, 1982; Cloney, 1990).

In other experiments demembranated neurulae were cultured in groups ranging in size from 2 to 80 in 1-ml volumes of sea water. In these experiments, the mean lengths of the caudal fins increased in proportion to group size. Many embryos in the larger groups develop fins that are as long as controls! Larvae with well-developed fins in these experiments were not sticky as in the previous experiments with large volumes of sea water (Cloney 1990). The absence of test cells alone apparently does not cause stickiness as might be suggested to explain the unusual stickiness of the juveniles of *Molgula pacifica* (see above).

It is probable that the ornaments and amorphous secretions of the test cells have an important role at the larval stage when swimming begins. But until recently there seemed to be no basis for suggesting their functional significance. When I deprived the larvae of *Corella inflata* of test cells and their secretion, by demembranation at the neurula stage, I found that they could develop normal fins if cultured in large groups, even though they were deprived of ornaments (Cloney, 1990). These larvae had different surface characteristics than controls, which developed with all extraembryonic structures. When they were transferred from a culture dish to a drop of sea water on a slide, they floated, and it was difficult to force them under the surface. This was not a problem with control larvae. They never became trapped at the air-water interface. This suggested that the test cell secretions make the larval tunic hydrophilic (Cloney, 1990). Most ascidian larvae are negatively geotactic and positively phototactic at some period after hatching. This behavior often brings them to the surface of the sea. If larvae were hydrophobic it might be hazardous for them to swim close to the surface, where they could be trapped. Although this hypothesis has been tested only on *Corella inflata* and must be evaluated on other species, it appears to be compatible with what is known about the histochemical characteristics of test cell. The test cells of many species are known to contain acidic glycosaminoglycans. These negatively charged molecules would attract water molecules if they formed a coating on the larval tunic. The intense staining properties of ornaments, with positively charged uranyl and lead ions, also indicate that they are anionic.

Oogenesis and Development without Test Cells

Nearly all ascidians have test cells. Several molgulids with direct development may lack test cells (Berrill, 1931), and three species in the family Agneziidae lack them (Monniot et. al., 1992a), but nothing is known about the development of these ascidians. *Molgula pacifica* is a rare exception in which the development has been studied. The absence of test cells in this species is consistent with the view that these cells have no essential role in oogenesis. The juveniles of this species settle on rocky substrates and adhere firmly, probably with epidermal secretions as in other molgulids (Cloney, 1978; Torrence and Cloney, 1981). They have no outer cuticular layer of tunic, and they have no swimming larva. I suggest that when the test cells no longer had an important role in coating the larval tunic they were eliminated. The pockets that were found in the periphery of oocytes are very puzzling. Is it possible that this species has retained vestigial pockets (pseudo–test cells) which once contained test cells? This is a question that Christopher G. Reed would have enjoyed discussing.

Acknowledgments

This chapter is dedicated to my close friend, former graduate student, and collaborator, Christopher G. Reed. His interests in biology were boundless, and he was an honored teacher in the field. Many of his publications on bryozoan developmental biology will endure as models of fine scholarship. He would have been elated with this book and with his many friends who celebrated his wit, gamesmanship, collegiality, and academic excellence at the Reed Symposium in 1992.

Literature Cited

Bancroft, F. W. 1899. Ovogenesis in *Distaplia occidentalis* Ritter (MS.) with remarks on other species. Bull. Mus. Comp. Zool. Harvard. 4: 1–111.

Bates, W. R., and J. Mallett. 1991a. Anural development of the ascidian *Molgula pacifica* (Huntsman). Can. J. Zool. 69: 618–627.

———. 1991b. Ultrastructure and histochemical study of anural development in the ascidian *Molgula pacifica* (Huntsman). Roux's Arch. Dev. Biol. 200: 193–201.

Berrill, N. J. 1931. Studies in tunicate development. II. Abbreviation of development in the Molgulidae. Philos. Trans. Roy. Soc. London B 218: 37–78.

Botte, L., S. Scippa, and M. de Vincentiis. 1979. Content and ultrastructural localization of transitional metals in ascidian ovary. Dev. Growth Differ. 21: 483–491.

Caggigi, S., A. Flugy, E. Puccia, and G. Reverberi. 1974. On the hatching of ascidian larvae. Rend. Acc. Naz. Lincei. 5: 803–807.

Cavey, M. J. 1976. Ornamentation of the larval ascidian tunic by test cells. J. Ultrastruct. Res. 55: 297–298.

Cavey, M. J., and R. A. Cloney. 1984. Development of the larval tunic in a compound ascidian: Morphogenetic events in the embryos of *Distaplia occidentalis*. Can. J. Zool. 62: 2392–2400.

Cloney, R. A. 1990. Larval tunic and the function of the test cells in ascidians. Acta Zool. 71: 151–159

———. 1977. Larval adhesive organs and metamorphosis in ascidians. 1. Fine structure of the everting papillae of *Distaplia occidentalis*. Cell. Tissue. Res. 183: 423–444.

———. 1987. Phylum Urochordata, class Ascidiacea. *In* M. F. Strathmann, ed. Reproduction and Development of Marine Invertebrates of the Northern Pacific Coast. University of Washington Press, Seattle, pp. 607–639.

Cloney, R. A., and M. J. Cavey. 1982. Ascidian larval tunic: Extraembryonic structures influence morphogenesis. Cell Tissue Res. 222: 547–562.

Cotelli, F., F. Andronico, R. DeSantis, A. Monroy, and F. Rosati. 1981. Differentiation of the vitelline coat in the ascidian *Ciona intestinalis*: An ultrastructural study. Roux's Arch. Dev. Biol. 190: 252–258.

Cowden, R., and C. Markert. 1961. A cytochemical study of the development of *Ascidia nigra*. Acta Embryol. Morphol. Exp. 4: 142–160.

De Vincentiis, M. 1962. Ulteriori indagini istospettrograpfiche e citochimiche su alcuni aspetti dell' ovogenesi di *Ciona intestinalis*. Atti. Soc. Peloritana Sc. Fis. Mat. Nat. 8: 189–198.

De Vincentiis, M., and G. Materazzi. 1964. Indagine citochimica sui mucopolisaccaridi delle cellule testacee di ovociti di *Ciona intestinalis* L. nel corso dell'ovogenesi. Riv. Biol. 57: 301–325.

Dykstra, M. J., and H. C. Aldrich. 1978. Successful demonstration of an elusive cell coat in amebae. J. Protozool. 25: 38–41.

Ermak, T. H. 1976. Renewal of the gonads in *Styela clava* (Urochordata: Ascidiacea) as revealed by autoradiography with tritiated thymidine. Tissue & Cell. 8: 471–478.

Gianguzza, M., and G. Dolcemascolo. 1979. On the ultrastructure of the test cells of *Ascidia malaca* during oogenesis. Acta Embryol. Morphol. Exp. 1979: 173–189.

———. 1984. Formation of the test in the swimming larva of *Ciona intestinalis*: An ultrastructural study. J. Submicrosc. Cytol. 16: 136–162.

Harvey, L. A. 1927. The history of cytoplasmic inclusions of the egg of *Ciona intestinalis* (L.) during oogenesis and fertilization. Proc. Roy. Soc. London B 150: 136–162.

Honegger, T. G. 1986. Fertilization in ascidians: Studies on the egg envelope, sperm and gamete interactions in *Phallusia mammillata*. Dev. Biol. 118: 118–128.

Kalk, M. 1963. Cytoplasmic transmission of a vanadium compound in a tunicate oocyte, visible with electron microscopy. Acta Embryol. Morphol. Exp. 6: 289–303.

Kessel, R. G. 1962. Fine structure of pigment inclusions in the test cells of the ovary of *Styela*. J. Cell Biol. 12: 637–640.

————. 1983. Urochordata. *In* K. G. Adiyodi and R. G. Adiyodi, eds. Reproductive Biology of Inverte-brates, Vol. 1, Oogenesis, Oviposition and Oosorption. John Wiley & Sons, Chichester, England, pp. 655–734.

Kessel, R. G., and H. W. Beams. 1965. An unusual configuration of the Golgi complex in pigment-producing "test" cells of the ovary of the tunicate *Styela*. J. Cell Biol. 25: 55–67.

Kessel, R. G., and N. E. Kemp, 1962. An electron microscopic study of the oocyte, test cells and follicular envelope of the tunicate *Molgula manhattensis*. J. Ultrastruct. Res. 6: 57–76.

Knaben, N. 1936. Über Entwicklung und Funktion der Testazellen bei *Corella parallelogramma* Mull. Bergens Mus. Arb. Nat. Rek. 1: 5–33.

Kowalevsky, A. 1866. Enwickelungsgeschichte der einfachen Ascidien. Mem. Acad. St. Petersbourg. 7: 1–19.

Kupffer, C. 1870. Die Stammverwandtschaft zwischen Ascidien und Wirbelthieren. Arch. Micr. Anat. 6: 115–172.

Lübbering, B., and Goffinet G. 1991. Ultrastructural survey of tunic morphogenesis in the larval and young adult ascidian *Ascidiella aspersa* (Tunicata, Ascidiacea). Belg. J. Zool. 121: 39–53.

Lübbering, B, T. Nishikata, and G. Goffinet. 1992. Initial stages of tunic morphogenesis in the ascidian *Halocynthia*: A fine structural study. Tissue & Cell 24: 121–130.

Mancuso, V. 1965. An electron microscopic study of the test cells and follicle cells of *Ciona intestinalis* during oogenesis. Acta Embryol. Morphol. Exp. 8: 239–266.

————. 1973. Changes in the fine structure associated with the test formation in the ectoderm cells of *Ciona intestinalis* embryo. Acta Embryol. Exp. 3: 247–257.

————. 1974. Formation of the ultrastructural components of *Ciona intestinalis* tadpole test by the animal embryo. Experientia 30: 1078.

Mansueto, C., and L. Villa. 1982. The test cells of *Ciona intestinalis* and *Ascidia malaca* during embryonic development. Acta Embryol. Morphol. Exp. N.S. 2: 107–116.

Materazzi, G., and Vitaioli, L. 1969. Further cytochemical studies of the mucopolysaccharides of the test cells of *Ciona intestinalis* L. Histochemie. 19: 58–63.

Monniot, F., R. Martoja, and C. Monniot. 1992a. Silica distribution in ascidian ovaries, a tool for systemat-ics. Biochem. Systematics Ecol. 20: 541–552.

Monniot, F., R. Martoja, M. Truchet, and F. Fröhlich. 1992b. Opal in ascidian: A curious bioaccumulation in the ovary. Mar. Biol. 112: 283–292.

Mukai, H., and H. Watanabe, H. 1976. Studies on the formation of germ cells in a compound ascidian *Botryllus primigenis* Oka. J. Morph. 148: 337–361.

Sugino, Y. M., A. Tominaga, and Y. Takashima. 1987. Differentiation of the accessory cells and structural regionalization of the oocyte in the ascidian *Ciona savignyi* during early oogenesis. J. Exp. Zool. 242: 205–214.

Terakado, K. 1970. Tunic formation in the larva of an ascidian *Perophora orientalis*. Sci. Reports Saitama Univ. B. 5: 183–191.

Torrence, S. A., and R. A. Cloney. 1981. Rhythmic contractions of the ampullar epidermis during meta-morphosis of the ascidian *Molgula occidentalis*. Cell Tissue Res. 216: 293–312.

Tucker, G. H. 1942. The histology of the gonads and development of the egg envelopes of an ascidian (*Styela plicata* Lesueur). J. Morph. 70: 81–113.

Vitaioli, L., and G. Materazzi. 1972. Comparative cytochemistry studied on the test cells of several Ascidiacea. Riv. Biol. 65: 145–151.

Young C. M., R. F. Gowan, J. Dalby, C. A. Pennachetti, and D. Gagliardi. 1988. Distributional conse-quences of adhesive eggs and anural development in the ascidian *Molgula pacifica* (Huntsman, 1912). Biol. Bull. 174: 39–46.

8   Ultrastructural Evidence That Somatic "Accessory Cells" Participate in Chaetognath Fertilization

George L. Shinn

ABSTRACT Transmission electron microscopy was used to evaluate controversial classical reports that chaetognath fertilization occurs before ovulation by passage of sperm through specialized somatic accessory cells to which differentiating oocytes are attached. The accessory cells occur in pairs and form part of the wall of the oviducal complex in which sperm received during mating are stored. The more laterally located cell of each pair is traversed by a long, coiled, extracellular "fertilization canal." Under most circumstances, the canal is completely occluded by a complementary cytoplasmic process arising from the more medially located cell. A single oocyte is attached to the soma of each medial cell. Shortly before fertilization, the cytoplasmic process of the medial cell disappears from the canal by an as yet unknown process. This allows sperm to pass into the fertilization canal and thereby reach the associated oocyte prior to ovulation.

Introduction

The Chaetognatha is a distinctive phylum of small, predaceous, typically planktonic marine worms. As far as is known, all chaetognaths are hermaphroditic and have internal fertilization. These two features are considered to be adaptations allowing reproductive success at low population densities (Ghiselin, 1961).

The mechanism of fertilization in chaetognaths has been a subject of controversy ever since the phenomenon was first described by Stevens (1903, 1910). She reported that fertilization in *Sagitta bipunctata*, *S. minima*, and *S. elegans* is mediated by pairs of specialized somatic cells, which she called accessory fertilization cells (AFCs). The latter are serially arranged along the medial side of the elongate, blindly ending oviducal complex, which is situated laterally in each ovary. A single differentiating oocyte is attached to the outer, medial side of each pair of AFCs. Sperm received at mating pass into the oviducal complex, where they are stored until fertilization. Stevens (1903, 1910) determined that fertilization occurs while oocytes are still within the oogenic compartment of the ovary by passage of sperm through a minute canal traversing the adjoining AFCs. The AFCs degenerate after fertilization, thereby creating small openings in the medial wall of the oviducal complex through which ovulating zygotes move into the oviducal complex.

Although several subsequent investigators have provided similar descriptions of fertilization for other species of chaetognaths (reviewed and documented with light micrographs by Ghirardelli, 1968), serious doubts about the involvement of AFCs in chaetognath fertilization were

George L. Shinn, Division of Science, Northeast Missouri State University, Kirksville, MO 63501.

expressed by Buchner (1910), Pierrot-Bults (1975), and Alvarino (1983a,b, 1990). The latter two authors suggested that fertilization occurs after ovulation.

In this study, transmission electron microscopy has been used to document the morphology of AFCs in *Ferosagitta hispida* and to determine whether these somatic cells participate in sperm transfer during fertilization. *Ferosagitta hispida* occurs along the eastern coast of North America and is relatively easy to keep alive in laboratory cultures (Reeve and Walter, 1972a). Consequently, several aspects of its reproductive biology have been fairly extensively studied, including mating (Reeve and Walter, 1972b), ovulation and egg laying (Conant, 1896; Reeve and Lester, 1974), and the ultrastructure of the male and female reproductive organs (Bergey, 1991; Shinn, 1992, respectively). Reeve and Lester (1974) reported that differentiating oocytes are attached to the oviducal complex by pairs of somatic accessory cells (i.e., AFCs). They presumed the latter cells to be involved in fertilization but did not investigate their role in any detail. Shinn (1992) described all somatic ovarian tissues except the AFCs.

Materials and Methods

Ferosagitta hispida were collected near the Fort Pierce Inlet into the Indian River Lagoon, Florida (27°14′ N; 80°90′ W; 0–5 m depth). This was accomplished by lowering a 300-μm-mesh plankton net into tidal currents flowing beneath a public fishing pier. In the laboratory, specimens were kept in small dishes of sea water (room temperature) and fed microcrustaceans that had been collected in the same tows as the chaetognaths.

The primary fixative for electron microscopy was a paraformaldehyde-glutaraldehyde mixture (Karnovsky, 1965) in 0.2 M phosphate buffer (20°C, pH 7.4, 1–2 h). One or two minutes after immersion in the fixative, specimens were bisected with a razor blade. Worms were postfixed in 2 percent OsO_4 in 0.2 M phosphate buffer (20°C, pH 7.4, 1–2 h). They were then serially dehydrated in ethanol to 100 percent, transferred to propylene oxide, and embedded in Epon 812 (Tousimis). Thin sections were stained in uranyl acetate and lead citrate (Reynolds, 1963) and examined with a Zeiss EM9-S2 transmission electron microscope.

The following description of AFC morphology is based upon those cells associated with the largest oocytes in worms preserved at about 3:00 P.M. Had the worms remained alive, the AFCs presumably would have participated in fertilization approximately 15 h later. For observations concerning fertilization, specimens were preserved just prior to egg-laying, which occurred predictably at about 6:00 A.M. Beginning about 5:00 A.M., mature specimens were transferred to petri dishes containing sea water and then observed with a dissecting microscope. The status of oocytes in the ovaries was clearly visible due to the transparency of the chaetognath body. Specimens were flooded with primary fixative immediately upon germinal vesicle breakdown, which occurs 15–30 min prior to ovulation and coincides with fertilization (Stevens, 1903).

Results
General Morphology of the Ovary

The cylindrical ovaries of *F. hispida* contain up to five size classes of primary oocytes (Shinn, 1992; Figs. 8.1 and 8.2). In live specimens, only those AFCs associated with the largest size class of oocytes are conspicuously visible (Figs. 8.1 and 8.2). The oocytes protrude medially from the oviducal complex into a fluid-containing ovarian space. The latter is bounded and separated from the trunk coelom by a thin ovary wall.

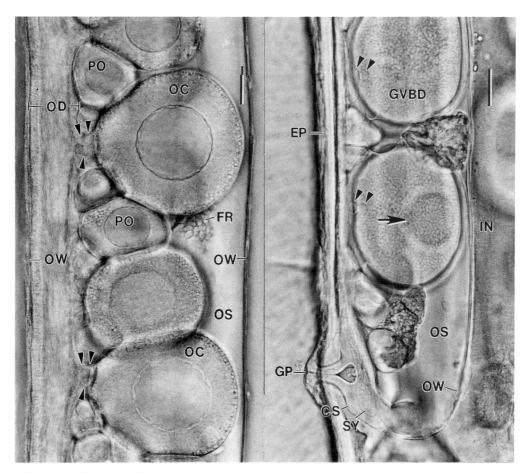

FIGURE 8.1 *(left)*. Mid-level of left ovary dissected out of a glutaraldehyde-fixed worm. Dorsal view, light micrograph, specimen preserved at 3:00 P.M. Arrowheads indicate the borders of accessory fertilization cells that attach the largest oocytes (OC) to the oviducal complex (OD). The central location of oocyte nuclei is typical except just prior to ovulation (FR, follicular reticulum; PO, previtellogenic oocyte; OS, ovarian space; OW, ovary wall; scale = 25 μm). FIGURE 8.2 *(right)*. Posterior end of left ovary in a live specimen about to commence ovulation. Dorsal view, light micrograph. The upper oocyte (GVBD) appears to lack a nucleus as a result of nuclear envelope disintegration. This occurs a few seconds after migration of the nucleus to the medial side of the egg (arrow in lower oocyte). Arrowheads indicate the location of accessory fertilization cells (CS, cellular sheath of oviducal complex; GP, female gonopore; IN, intestine; OS, ovarian space; OW, ovary wall; SY, syncytium of oviducal complex; scale = 50 μm).

Morphology of Accessory Fertilization Cells

Each differentiating oocyte is associated with two highly specialized oviducal sheath cells—the accessory fertilization cells (Fig. 8.3). These are in turn bordered by an undetermined number of "supporting cells" which differ from other surrounding sheath cells in their more columnar shape (Fig. 8.3).

Within each pair of AFCs, the more laterally located cell, referred to by Stevens (1910) as AFC1, is approximately spheroidal in overall shape. A coiled, convoluted, extracellular canal traverses this cell but, rather than being "empty" or containing extracellular fluids as implied by Stevens (1910), the canal encloses an elongate, membrane-bounded cytoplasmic process extending from the more medially located AFC2 (Figs. 8.3–8.5). The canal in AFC1 appears to consist of a deep, complex invagination of the plasmalemma. AFC1 contains a single nucleus, scattered mitochondria, inactive Golgi bodies, ribosomes, and small profiles of endoplasmic reticulum.

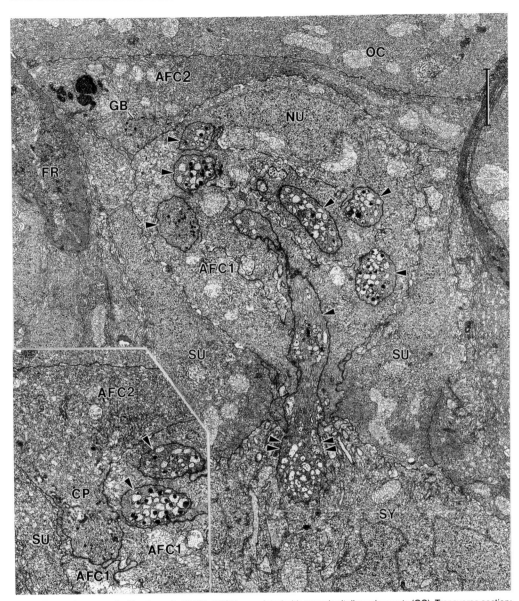

FIGURE 8.3. Accessory fertilization cells (AFC1, AFC2) associated with an early vitellogenic oocyte (OC). Transverse section; medial at top, lateral at bottom; transmission electron micrograph (TEM). Arrows indicate profiles of the elongate, membrane-bounded cytoplasmic process that extends from the soma of accessory fertilization cell 2 (AFC2) through the fertilization canal traversing accessory fertilization cell 1 (AFC1). The end of the cytoplasmic process (paired arrowheads) protrudes into the syncytium (SY) of the oviducal complex (FR, follicular reticulum in ovarian space; GB, Golgi body; SU, supporting cells of cellular sheath of oviducal complex; scale = 2.5 μm). Inset: Origin of cytoplasmic process (CP) from the soma of AFC2.

The soma of AFC2 covers the medial side of AFC1, completely separating the latter from the associated oocyte (Fig. 8.3). In addition to a single flattened nucleus, the soma contains mitochondria, numerous ribosomes, and scattered profiles of smooth and rough endoplasmic reticulum. Although the soma of AFC2 appears compressed at this stage of differentiation (Fig. 8.3), it eventually becomes spheroidal and occupies a micropylelike gap in the forming egg coat (Shinn, unpublished electron micrographs). The cytoplasmic process of AFC2 is uniform in width, approximately 2 μm in diameter. It is dominated by spheroidal, membrane-bound vesicles and by

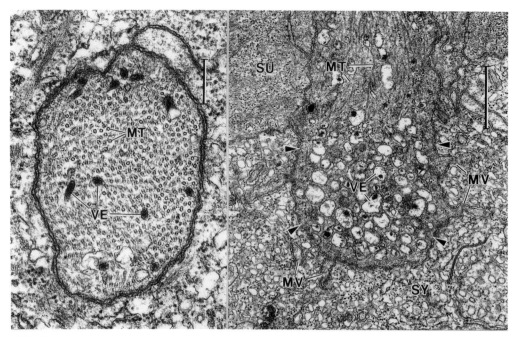

FIGURE 8.4 *(left)*. Cross-section through cytoplasmic process of AFC2, showing microtubules (MT) and vesicles (VE) (scale = 0.3 μm). FIGURE 8.5 *(right)*. Apical tip (arrowheads) of cytoplasmic process of AFC2 protruding into syncytium (SY) of oviducal complex (MT, microtubules; MV, microvilli; SU, support cells of cellular oviducal sheath; VE, vesicles; scale = 1.0 μm).

microtubules whose long axes conform to the convoluted long axis of the process (Figs. 8.3–8.5). The vesicles range from 0.1 μm to 0.4 μm in diameter and contain one or more small electron-dense granules in an electron-lucent matrix. The vesicles are evidently produced by a small Golgi body located in the soma near the origin of the process. The apical tip of the cytoplasmic process protrudes a short distance beyond AFC1 into a complementary invagination of the oviducal syncytium (Fig. 8.5). Like more typical cells of the oviducal complex, AFC2 has microvilli that extend into the syncytium. The soma of AFC2 has no extracellular channel like that found in AFC1. Thus, at this stage of AFC differentiation and until fertilization (see below), there is no open canal in the AFCs through which sperm might pass from the oviducal complex to the oocytes.

Fertilization

Fertilization coincides with germinal vesicle breakdown (GVBD) (Stevens, 1910; Figs. 8.6–8.9). Specimens preserved a few minutes after GVBD show the following transformations involving the accessory fertilization cells: (1) Supporting oviducal sheath cells have retracted from the sides of AFC1, giving AFC1 a much broader apical surface and allowing this cell to bulge into the syncytium (Fig. 8.6). (2) Numerous spermatozoa have moved from the oviducal syncytium into the extracellular canal of AFC1 (Fig. 8.6). The fate of the cytoplasmic process of AFC2, which no longer occludes the canal in AFC1, remains unresolved (see Discussion). (3) The soma of AFC2 has been displaced from the micropylelike gap in the egg envelope by a lobe of oocyte cytoplasm that protrudes through the gap (Fig. 8.7). (4) One or more sperm cells have entered the oocyte cytoplasm, near the micropylelike opening in the egg envelope (Figs. 8.7–8.9). The oolemma abuts one of the AFCs, presumably AFC2, whose cytoplasm is filled with large irregularly shaped

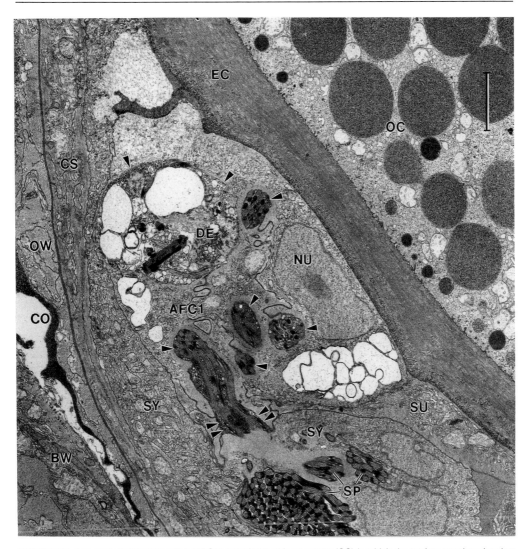

FIGURE 8.6. Accessory fertilization cell 1 (AFC1) associated with an oocyte (OC) in which the nuclear envelope has just disintegrated (see Figure 8.2). Transverse section near but not at micropylar opening in extracellular egg coat (EC); TEM. Arrowheads indicate sectional profiles of the fertilization canal traversing AFC1. Note that smaller-diameter parts of the canal are occupied by numerous closely packed spermatozoa; larger-diameter parts contain cellular debris (DE). Paired arrowheads indicate the lateral opening of the fertilization canal to the syncytium (SY) of the oviducal complex (BW, body wall; CS, lateral wall of cellular sheath of oviducal complex; NU, nucleus of AFC1; OW, ovary wall; SU, supporting cell of cellular oviducal sheath; SP, spermatozoa in lumina of oviducal syncytium; scale = 3 μm).

vacuoles (Fig. 8.7). Due to ambiguities in my sections, I have not yet been able to provide a detailed description of the changes that AFC2 undergoes during fertilization.

Discussion

Morphology of Accessory Fertilization Cells

To my knowledge, no previous authors have described the intricate association that I have found between the cytoplasmic process of AFC2 and the "fertilization canal" in AFC1. In *S. elegans*, one of the species studied by Stevens (1910), the AFCs have a similar arrangement (Shinn, unpublished electron micrographs). Whether this is typical of the phylum remains to be deter-

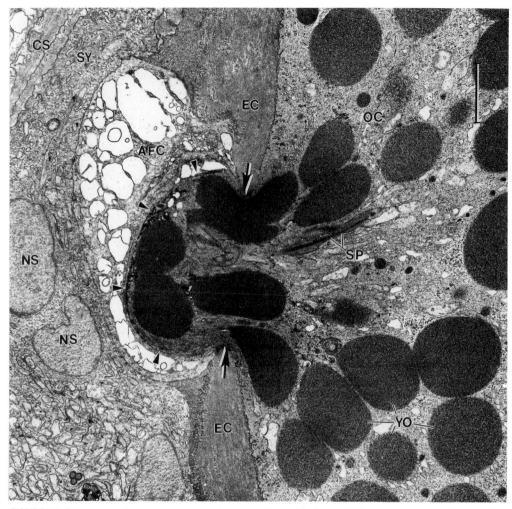

FIGURE 8.7. Transverse section through same oocyte as that in Figure 8.6, showing fertilizing spermatozoan (SP) embedded within the oocyte cytoplasm (OC) and lobe of oocyte (arrowheads) protruding through micropylelike opening (between arrows) in extracellular egg coat (EC) (AFC, vacuolated cytoplasm of an accessory fertilization cell, presumably AFC2; CS, cellular sheath of oviducal complex; NS, nucleus of oviducal syncytium (SY); YO, yolk; scale = 2.5 μm).

mined by additional electron microscopical studies. Features described by several other chaeto-gnath researchers probably correspond to sectional profiles through the cytoplasmic process of AFC2, however. These include: (1) in Buchner (1910) the "Faden" (attachment cord) extending laterally from the nucleus of AFC2 to the "Netz" (net) that underlies the oviductal epithelium of an unspecified *Sagitta*; (2) in Stevens (1910), the granular, nonstaining, spindle-shaped body in the fertilization canal of *S. elegans*; (3) in Ghirardelli (1968: 299) the "vacuoli" or "particular states of the cytoplasm" of the accessory cells in *Sagitta bipunctata*, *S. inflata*, and *S. minima*; and (4) in Pierrot-Bults (1975), either the "nonstaining areas" thought to correspond to the vacuoles of Ghirardelli (1968) or the "homogeneously red staining substance" in AFC1 of *S. planctonis*. Because the canal is not permanently open, my observations justify Buchner's (1910) and Ghirar-delli's (1968) skepticism that a fertilization canal exists. Stevens (1903, 1910) was incorrect in describing the canal as a persistent space.

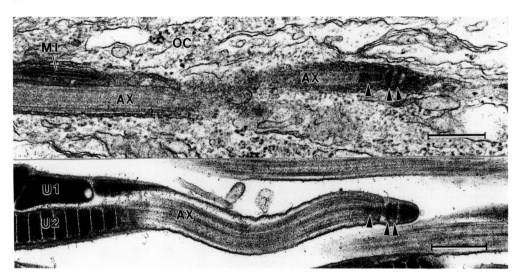

FIGURE 8.8 *(top)*. High-magnification view of fertilizing spermatozoan within oocyte cytoplasm (OC). The anterior tip of the spermatozoan contains a characteristic basal-body-like structure (arrowheads) from which the axoneme (AX) arises. The two offset images of the sperm axoneme presumably belong to the same sperm cell which curves out of and then back into the plane of section. Alternatively, the egg may have been fertilized by more than one spermatozoan (MI, mitochondrial derivative within sperm; scale = 0.5 μm). FIGURE 8.9 *(bottom)*. Anterior end of a spermatozoan in seminal vesicle of *F. hispida*, demonstrating the characteristic basal-body-like structure (arrowheads) and supporting the conclusion that the image shown in Figure 8.6 is a spermatozoan. The axoneme (AX) extends the full length of the sperm, but in this micrograph passes out of the plane of section. U1 and U2 indicate sperm organelles of unknown function (scale = 0.5 μm).

I cannot yet account for Stevens' (1910) observation that AFC2, like AFC1, encloses a fertilization canal. According to Stevens (1903), this part of the canal appears relatively late during differentiation of the AFC complex (exact timing not specified). In specimens of *F. hispida* preserved 1 h prior to the expected time of fertilization there is still no canal through AFC2 (Shinn, unpublished electron micrographs). Thus, formation of this part of the canal may coincide with fertilization.

The structure that I have found to be the protruding apical tip of AFC2 was interpreted by Ghirardelli (1968) and Pierrot-Bults (1975) to be part of AFC1 "pouring" into the syncytium. Noting that sperm in the syncytium accumulate at these sites, Ghirardelli (1968) hypothesized that these contents serve to attract sperm for fertilization. My observations provide no new information for evaluating whether sperm are attracted to the tip of the cytoplasmic process. Pierrot-Bults (1975) hypothesized that the protruding materials are indicative of metabolite exchange between AFCs, the oviducal syncytium, and oocytes. Although the metabolic activities of the AFCs remain unexplored, I have found no conspicuous morphological evidence of bulk transport of materials between the syncytium and oocytes. Specifically, I have found no evidence that the vesicles in the cytoplasmic process are formed by endocytosis or that they exocytose their contents toward either the syncytium or associated oocyte. Rather, the cytoplasmic process of AFC2 appears to accumulate vesicles that are produced within that cell. Perhaps the vesicles participate in the dramatic morphological changes that AFC2 presumably undergoes during fertilization.

Role of AFCs in Fertilization

This study confirms the observations of Stevens (1903, 1910) and Ghirardelli (1968) with regard to two major points: (1) fertilization precedes ovulation and (2) the so-called accessory fertilization cells mediate access of sperm to oocytes. Stevens (1905, 1910) was correct in stating that spermatozoa enter a continuous coiled canal in AFC1 during fertilization. Important questions about the role of AFC2 in fertilization remain unresolved: (1) What changes in AFC2 are responsible for opening the fertilization canal in AFC1? and (2) How do sperm traverse the region that, prior to fertilization, is occupied by the soma of AFC2? The opening of the fertilization canal presumably involves either the disintegration of the cytoplasmic process of AFC2 or its active withdrawal from the canal. The former scenario would be consistent with Buchner's (1910) report that AFC2 begins disintegrating prior to fertilization. The disappearance of the cytoplasmic process of AFC2 apparently occurs fairly rapidly at the onset of fertilization because, in gravid specimens preserved at the expected time of GVBD but in which GVBD has not yet begun, the cytoplasmic process of AFC2 is still in place and intact (Shinn, unpublished electron micrographs). Thus, it seems likely that changes in the morphology of AFC2 directly regulate the occurrence of fertilization. Resolution of this problem requires further electron microscopical analysis.

In recent reviews of chaetognath reproduction, Alvarino (1983a,b, 1990) stated that reports of the participation of AFCs in fertilization (e.g., Ghirardelli, 1968) are incorrect. Noting that cleavage does not begin until after ovulation, she concluded that fertilization could not occur prior to ovulation. Without providing morphological documentation, Alvarino (1990) asserted that fertilization occurs just prior to release of eggs from the body as ovulated oocytes move through the "vagina" and past the posterior end of the oviducal complex in which sperm are stored. Prior to the present study, the reasoning behind Alvarino's model for fertilization applied as well to *F. hispida* as to any other chaetognath species. Because her model is incorrect for *F. hispida*, there appears to be no basis for expecting it to apply to other species of chaetognaths. Furthermore, Alvarino's premise that cleavage necessarily commences immediately after fertilization is not valid. Temporary delay of the onset of cleavage occurs in diverse organisms having internal fertilization (e.g., polyclad flatworms: Patterson and Wieman, 1912; Kato, 1940; phoronids: Zimmer, 1972; Emig, 1977; bryozoans: Temkin, 1988). In these organisms, as in chaetognaths, cleavage commences when the zygotes contact sea water. Although there may be several different modes of fertilization in the Chaetognatha, their discovery awaits careful electron microscopical studies of well-preserved specimens.

Acknowledgments

This research was initiated at the Harbor Branch Oceanographic Institution. I am grateful for advice and assistance provided by Kevin Eckelbarger and Pamela Blades-Eckelbarger. The project was completed at Northeast Missouri State University, which provided financial assistance through Faculty Research Grants, and at the Friday Harbor Laboratories. I am especially indebted to my wife, Anne Bergey, who helped collect and prepare specimens and who, along with two anonymous reviewers, provided suggestions for improving the manuscript. This chapter is dedicated to the memory of Christopher G. Reed, a close friend with whom I enjoyed sharing many hours indulging our mutual interests in the natural world.

Literature Cited

Alvarino, A. 1983a. Chaetognatha. *In* K. G. Adiyodi and R. G. Adiyodi, eds. Reproductive Biology of Invertebrates, Vol. 1. Oogenesis, Oviposition, and Oosorption. John Wiley, New York, pp. 585–610.

———1983b. Chaetognatha. *In* K. G. Adiyodi and R. G. Adiyodi, eds. Reproductive Biology of Invertebrates, Vol. 2. Spermatogenesis and Sperm Function. John Wiley, New York, pp. 531–544.

————1990. Chaetognatha. *In* K. G. Adiyodi and R. G. Adiyodi, eds. Reproductive Biology of Inverte-brates, Vol. 4. Pt. B. Fertilization, Development, and Parental Care. John Wiley, New York, pp. 255–282.

Bergey, M. A. 1991. Ultrastructure of the male reproductive system of the chaetognath *Ferosagitta hispida*. Master's thesis, Northeast Missouri State University, Kirksville.

Buchner, P. 1910. Keimbahn und Ovogenese von *Sagitta*. Anat. Anz. 35: 433–443.

Conant, F. S. 1896. Notes on the chaetognaths. Ann. Mag. Nat. Hist. 6: 201–204.

Emig, C. C. 1977. Embryology of Phoronida. Am. Zool. 17: 21–37

Ghirardelli, E. 1968. Some aspects of the biology of chaetognaths. Adv. Mar. Biol. 6: 271–375.

Ghiselin, M. T. 1961. The evolution of hermaphroditism among animals. Q. Rev. Biol. 44: 189–208.

Karnovsky, M. H. 1965. A formaldehyde-glutaraldehyde fixative of high osmolarity for use in electron microscopy. J. Cell Biol. 27: 137A.

Kato, K. 1940. On the development of some Japanese polyclads. Japan. J. Zool. 8: 537–574.

Patterson, J. T., and H. L. Wieman. 1912. The uterine spindle of the polyclad *Planocera inquilina*. Biol. Bull. 23: 271–293.

Pierrot-Bults, A. C. 1975. Morphology and histology of the reproductive system of *Sagitta planctonis* Steinhaus, 1896 (Chaetognatha). Bijdr. Dierkd. 45: 225–236.

Reeve, M. R., and B. Lester. 1974. The process of egg-laying in the chaetognath *Sagitta hispida*. Biol. Bull. 147: 247–256.

Reeve, M. R., and M. A. Walter. 1972a. Conditions of culture, food-size selection and the effects of temperature and salinity on growth rate and generation time in *Sagitta hispida* Conant. J. Exp. Mar. Biol. Ecol. 9: 191–200.

————1972b. Observations and experiments on methods of fertilization in the chaetognath *Sagitta hispida*. Biol. Bull. 143: 207–214.

Reynolds, E. S. 1963. The use of lead citrate at high pH as an electron opaque stain in electron microscopy. J. Cell Biol. 17: 208–214.

Shinn, G. L. 1992. Ultrastructure of somatic tissues in the ovaries of a chaetognath (*Ferosagitta hispida*). J. Morphol. 211: 221–241.

Stevens, N. M. 1903. On the ovogenesis and spermatogenesis of *Sagitta bipunctata*. Zool. Jahrb. Anat. 18: 227–240.

————1905. Further studies on the ovogenesis of *Sagitta*. Zool. Jahrb. Anat. 21: 243–252.

————1910. Further studies on reproduction in *Sagitta*. J. Morphol. 21: 279–319.

Temkin, M. H. 1988. Location of fertilization in *Membranipora membranacea*. Am. Zool. 29: 169A.

Zimmer, R. L. 1972. Structure and transfer of spermatozoa in *Phoronopsis viridis*. *In* C.J. Arceneaux, ed. Thirtieth Ann. Proc. Electron Micros. Soc. Am. at Los Angeles, pp. 108–109.

9 Mechanisms for Enhancing Fertilization Success in Two Species of Free-Spawning Invertebrates with Internal Fertilization

Richard L. Miller

ABSTRACT Some marine organisms spawn sperm externally but have internal fertiization. These organisms should possess adaptations that increase the chances of sperm locating the sequestered eggs to counteract the high dilution capacity of the marine environment. Laminar flow past the large polyp of the hydroid *Tubularia* produces a downstream cylindrical eddy with a core of reversed flow that carries small particles upstream toward the oral tentacles. This probably allows the polyp to capture food particles that pass through the whorl of long capture tentacles. Sperm may also be carried toward the oral area where a low flow zone exists close to the gonophores, which release sperm attractant. In this manner, the flow regime around the polyp may enhance fertilization success. In salps, sperm swimming behavior after spawning appears intermittent, with substantial time spent drifting. This presumably saves energy and increases sperm longevity. The sperm, entering the anterior siphon of a female stage salp with the normal water flow, pass into the atrial cavity, where a fertilization duct contacts the inner wall. Some of these sperm activate and aggregate close to the wall where the duct is located and penetrate the duct to fertilize the egg in the ovary. Sperm remain actively motile close to the duct entrance for hours. These observations imply that both of these suspension feeders utilize water flow both for feeding and transport of sperm to particular sites, increasing the chances of fertilization.

Introduction

In situ studies of fertilization success and models based on estimates of sperm and egg interaction rates in laminar and turbulent flows suggest that fertilization success in free-spawning organisms is low, even when the two sexes are in close proximity (Denny, 1988; Denny and Shibata, 1989; Pennington, 1985). Presumably the rate would be even lower if the sperm were spawned externally but the eggs sequestered and limited to internal fertilization. Many marine organisms are pre-adapted both anatomically and behaviorally for the capture and manipulation of diluted food particles in suspension. They probably can capture sperm as well. If the sperm can be selected and moved close to the eggs, enough successful fertilization might occur to compensate for the highly adverse conditions brought about by the combination of high sperm dilution and internal fertilization.

Benthic and planktonic suspension feeders extend a modified portion of the body into the current or deflect a current into a chamber within their bodies for passive or active trapping of small particles, which may then be manipulated and sorted. Since sperm are small particles, these

Richard L. Miller, Department of Biology, Temple University, Philadelphia, PA 19122, and Friday Harbor Laboratories, The University of Washington, Friday Harbor, WA 98250.

organisms may be able to capture and manipulate sperm. Although the sperm may be treated like any other particle and be either discarded or digested, they could also be used to fertilize eggs, particularly when fertilization is internal. The important role of female or reproductive apparatus morphology in inducing the flow field to encourage the capture of drifting gametes has already been demonstrated in some conifers and models of primitive seed plants, in which the male gametes (pollen grains) are windborne and selectively collected as a result of flow eddies created by the structure of the cones (Niklas, 1981; Niklas and U, 1982). The normal operation of the filter-feeding apparatus in marine invertebrates, coupled with the appropriate positioning of the eggs, might operate to bring the sperm to a location where fertilization is possible or even encouraged.

Sperm chemotaxis is relatively common in marine invertebrates (Miller, 1985). The attractants are low-molecular-weight substances (Coll and Miller, 1992) that activate the motility of spawned but nonmotile sperm and direct the sperm paths toward the source of a gradient of the attractant. In some cases, sperm are attracted to egg-bearing structures which they must penetrate to fertilize the eggs (Miller, 1966, 1974). Although sperm chemotaxis to the egg usually operates over a few hundred microns at most, it increases the chances of nearby sperm striking the egg (Miller and King, 1983). Sperm chemotaxis could have evolved in part to counteract the sperm dilution normally experienced by free-spawning organisms which shed both eggs and sperm.

Preliminary study of sperm behavior in two suspension-feeding organisms has suggested ways in which anatomical adaptations to the flow environment over relatively small scales may also act to ensure that fertilization takes place under conditions of low sperm numbers in low to moderate flow. The benthic, passive suspension-feeding hydroid, *Tubularia,* relies on collision of drifting prey with its long tentacles. The active, planktonic suspension-feeding urochordate, *Thalia democratica*, uses a highly efficient mucus net in conjunction with jet-propelled swimming movements to capture small particles indiscriminately. The observations described here suggest that mechanisms exist for the capture of relatively rare sperm by both of these organisms under laminar flow-field conditions.

Materials and Methods

Tubularia larynx was collected in late winter and early spring at Shark River, New Jersey, on the sides of floating docks. *T. marina* was found in April to June on the floating docks at Friday Harbor, Washington. East Coast animals were maintained in a cold room (5°C) at Temple University in a tank of rapidly flowing sea water and fed local plankton or newly hatched *Artemia* larvae. Observations were also made on West Coast animals at Friday Harbor Labs in the summers of 1984 and 1989 using a flow tank attached to the lab sea water supply at temperatures from 8° to 10°C. The animals remained healthy enough to be of use for up to a week, but did not thrive, possibly for lack of food. The anatomy of the two species was very similar, and the results obtained were identical.

Single polyps with a long stem were suspended vertically from a rod, the lower end of which was submerged in a flow tank (Vogel, 1983). Several observations were made of the same animal each day in currents ranging from 1 to 10 cm/s. The flow field around the animal was revealed using thin streams of fluorescein dye or the motion of dextran particles added to the water. Flash photographs were taken to capture the shape of the dye streams or the positions of the particles.

Thalia democratica was collected daily in the Wistari Channel close to the west-southwest reef edge of Heron Island, Australia, in mid-December of 1989 and 1990. The animals were usually found in swarms at the surface before 9:00 A.M. and were captured using hand-held scoops. When animals were not at the surface, they could be obtained by towing a plankton net with a 2-g Plexiglas cod-end at 2–8 m depth. Every effort was made to minimize pouring the animals from one container into another. For example, the cod-ends were dismounted from the net and used as aquaria in the lab to avoid transferring the animals.

Newly released aggregates were placed in sea water under a tilted cover slip open at both ends and observed with low-power objectives of a compound microscope using Nomarski optics. The surfaces in contact with sea water were coated with 1 percent polyethylene glycol (>10 kD) to prevent sperm or salps sticking to the glass. Usually the animals wedged themselves loosely toward the narrow end of the cover slip, where they continued to pump water through their siphons. The transparency of the animals permitted detailed observation of the position and form of the female reproductive apparatus and the behavior of nearby sperm.

Diluted sperm were added through the wide end of the tilted cover slip, and excess water was removed from the narrow end. The same was done to refresh the sea water and add food. Adding water through the narrow end washed the animals out through the wide end, permitting them to be recovered and placed in larger culture containers for later observation. The salps were fed a mixture of microalgae to supplement the natural sea water in the culture dishes. Observations of sperm within the salp body were captured on videotape using a GBC low-light video camera and a Toshiba VCR. A motor-driven camera with a flash attachment fitted to the microscope base was used to make 35-mm still photographs. Over twenty examples of sperm aggregation at the oviduct were observed during the research periods.

Observations and Results

Tubularia—A Passive Suspension Feeder

Members of the *Tubularia* genus of large, subtidal, athecate, colonial or solitary, hydroid polyps are distributed widely throughout temperate and boreal oceans. The colonial species often occur in huge numbers where current flow is substantial and can monopolize available space under such circumstances (Ricketts and Calvin, 1968). Individuals of the solitary species rarely monopolize space but may be widely scattered over both hard and soft substrata or under boulders in the intertidal zone (Ricketts and Calvin, 1968) where surge or tidal currents can exceed speeds of several centimeters per second.

The *Tubularia* body plan is the same regardless of the location or species (Fig 9.1a). The polyp sits on the end of a stiff stalk that rises vertically from the substrate. It is attached to the stalk by a flexible neck which permits it to bend in almost any direction relative to the stalk. The polyp bears two whorls of tentacles and has a cone-shaped body that bears the mouth at its narrow, distal end (Mackie, 1966). A whorl of long tentacles, which form an array that resembles the spokes of an open umbrella, is found at the base of the body above its junction with the stalk. These tentacles rapidly fold over the exposed oral surface of the polyp when the animal is disturbed. A ring of smaller tentacles exists just below the apex of the cone, closely surrounding the mouth (Fig. 9.1a). Small planktonic Crustacea are captured when they strike the surface of one of the large tentacles, which bends inward to transfer the prey to the whorl of short tentacles, which then pass it into the mouth.

The reproductive system of *Tubularia* consists of a ring of grapelike clusters of medusoids (Fig. 9.1b) surrounding the base of the body just above the whorl of large tentacles. The number of medusoids is often so great that the clusters fill most of the space between the large tentacles and the polyp body. The sexes are separate, and both fertilization and embryonic development take place within the female medusoid. Fertilizable eggs are present at all times in mature females.

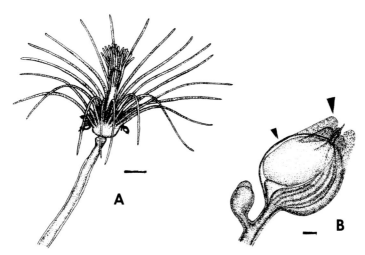

FIGURE 9.1. *a*. Drawing of a polyp of *Tubularia crocea* (scale bar = 1.0 mm). *b*. Drawing of a female reproductive medusoid showing the bladelike rudimentary tentacles (large arrow) surrounding the opening through which sperm pass to fertilize the egg (small arrow) (scale bar = 10 μm).

Males appear to release sperm in small bursts ("trickle spawning") (Miller, 1976). To fertilize the egg, sperm have to locate and penetrate a small pore at the apex of a medusoid bearing one or more ripe eggs. A sperm attractant is released through this pore. As a result, sperm actively turn to approach and aggregate at the pore and will penetrate it to fertilize the egg within the medusoid (Miller, 1976, and unpub. obs.).

In a strong current, a *Tubularia* polyp bends at the neck area so that the mouth points downstream (Fig. 9.2). The long tentacles extend outward radially, permitting water to pass between them, particularly at their outer ends. These tentacles remain surprisingly stiff even in current velocities over 10 cm/s (Miller, unpub. obs.). The polyp at this time resembles an everted umbrella in a strong wind, and the tentacle array may still act as a passive capture device.

Observation of dye streams and particle motions revealed that the polyp body sets up a vortex, which is radially symmetrical and extends 2–3 polyp diameters downstream (Fig. 9.2). The vortex induces a backflow such that water close to the geometric center of the polyp flows in a direction opposite to that of the main current (Figs. 9.2a, and b). The backflow stops close to the reproductive structures, and a substantial volume of still water exists at this location (Fig. 9.2b). Particles that passed between or around the capture tentacles close enough to enter the reverse flow zone were invariably carried back toward the polyp mouth, into the zone of still water. Small, independently motile food items might then swim into contact with the short, oral tentacles (Fig. 9.2b). Motile or nonmotile sperm would follow a similar path into the still water zone, where the sperm attractant released by the medusoids could form a gradient. The attractant gradient would direct them closer to the site of fertilization.

Thalia democratica—An Active Suspension Feeder

Salps are highly efficient suspension feeders that use forward swimming to pass water through an internal mucus net that can capture particles down to 1 μm in size (Alldredge and Madin, 1982). The ability of salps to reproduce asexually is prodigious, and large swarms or blooms may occur under favorable conditions of mixing and food supply (Wiebe et al., 1979). Salps are protogynous hermaphrodites which fertilize the egg within the ovary of newly released "aggregate" or "fe-

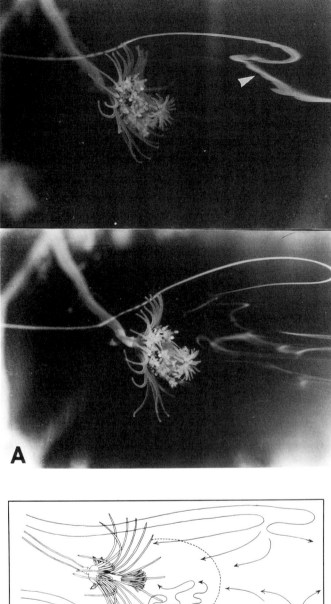

FIGURE 9.2 *a*. Two photos, about 5 s apart, showing the movement of a stream of dye flowing past a *Tubularia* polyp in a current of 5 cm/s. Note that the parcel of dye indicated by the arrow in the upper frame moves rapidly toward the polyp during the 5-s interval. *b*. Diagram of the approximate flow pattern in the vicinity of a *Tubularia* polyp based on the behavior of dye streams in a laminar flow regime of 5 cm/s. The diagram represents a vertical section through the approximately cylindrical flow vortex downstream of the polyp. The dotted line represents the volume within which particle velocity approaches zero.

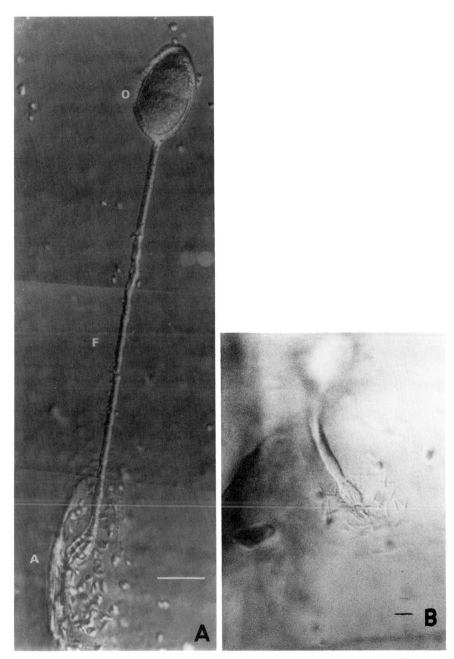

FIGURE 9.3 *a*. The anatomy of the salp female reproductive apparatus in a representative salp, *Cyclosalpa affinis*. (o = ovary; f = fertilization duct; a = area where duct attaches to the atrial wall; scale bar = 75 μm). *b*. An aggregation of motile sperm at the tip of the fertilization duct in *T. democratica*. The duct in *T. democratica* has a blunt end at the attachment zone rather than the elaborate one shown in *a*. The ovary is the light oval that is out of focus at the top of the picture (scale bar = 10 μm).

male" phase individuals. The formation of sperm during the "male" phase occurs later in reproductive life, at about the time the brooded embryo (called a "solitary") of the previously "female" individual has been released. An individual bearing an egg is therefore incapable of self-fertilization (Brooks, 1876a). Sperm are spawned freely into the water by the older "male" phase aggregates and are probably picked up by newly released "female" aggregates during normal swimming and feeding behavior.

The female system in salps consists of one ovary containing a single egg, connected to the inner wall of the atrium (posterior or excurrent cavity) by a long fertilization duct (Fig. 9.3a). The whole system is suspended within a blood sinus on the right posterio-ventral surface of the atrium. Because fertilization occurs in the ovary and the embryo is subsequently brooded, the anatomy presents a series of problems for the fertilizing sperm. Spawned externally during the male phase of an older aggregate from a previous generation, the sperm must gain entry to the atrium, where the fertilization duct is located. After avoiding the mucus feeding net that divides the anterior respiratory cavity from the atrium, the sperm must locate the tip of the fertilization duct . Once in the atrium, the sperm must also avoid being swept past the fertilization duct by the strong atrial currents. The sperm must then penetrate up the fertilization duct (Leloup, 1929) to reach the ovary. A single published figure in the older literature suggests that unusual sperm behavior occurs in the aggregate stage before fertilization (Brooks, 1876b), but it is diagrammatic and lacks verbal commentary.

Individual male-phase aggregates of *T. democratica* have been observed to shed sperm in the early morning of the day after collection. Spawning of any particular aggregate began suddenly and was often over in less than 5 min. Solitaries carrying chains released them about an hour later. Newly released chains consist of up to fifty aggregates (each about 2 mm long), all of which are in fertilizable condition (Brooks, 1876b). Spawned *Thalia* sperm have poor motility. Each cell alternated for at least an hour after spawning between 1 and 2-min periods of active forward motility and inactivity (Miller, 1991, unpub. obs.).

Sperm entering the anterior siphon of a female-stage aggregate can reach the fertilization duct by passing through the mucus net (Miller, unpub. ob.). How this is accomplished is not yet clear. Once in the atrial cavity, motile sperm were observed to hover around the tip of the fertilization duct in aggregations (Fig. 9.3b) that lasted for up to several hours. Sperm in the aggregations remained continuously motile in contrast to those free in the atrial volume. No comparable sperm aggregations were seen anywhere else inside the salp. Similar sperm aggregations were seen at the tips of the fertilization ducts in newly released aggregate chains collected early in the morning in the field (Miller, unpub. obs.).

Discussion

Several recent studies have documented fertilization success of benthic, free-spawning organisms such as sea urchins in the field (Pennington, 1985; Levitan, 1988, 1991; Yund, 1990; Levitan et al., 1991, 1992). All were done in the subtidal under relatively benign current regimes. The results indicate surprisingly low levels of fertilization, even a short distance from the source of the sperm. Mathematical models suggest that less than 0.01 percent fertilization of eggs occurs beyond a few meters downstream from spawning individuals even if they are close together in turbulent conditions equivalent to vigorous surf (Denny, 1988; Denny and Shibata, 1989). Organized spawning behavior, coordinated behavioral interactions between the sexes (Miller, 1989; Minchin, 1987; Tyler et al., 1992; Young et al., 1992), or other reproductive adaptations as well as appropriate flow conditions are required if substantial numbers of eggs are to be fertilized (Levitan, 1991; Petersen et al., 1992; Sewell and Levitan, 1992).

Further impediments to fertilization success arise when the sperm are spawned but the eggs are held in place or are negatively or positively buoyant, or when fertilization is internal. All these situations potentially restrict sperm access to the eggs, although egg bouyancy, coupled with coordinated spawning behaviors and favorable environmental conditions, has been shown to enhance fertilization success in a sea cucumber (Sewell and Levitan, 1992). Examples are scattered across the marine phyla, including the cnidarians (Yund, 1990; Babcock, 1984), annelids (Daly and Golding, 1977), bivalve molluscs (O'Foighil, 1985), crinoids (Holland, pers. comm.), and some compound urochordates (Bishop and Ryland, 1991). A substantial fraction of these organisms are hermaphroditic, but despite restriction of sperm access to eggs, selfing is apparently rare (Cohen, 1990; Daly and Golding, 1977; Ryland and Bishop, 1990).

Large numbers of sperm can be moved long distances while remaining undiluted in a spermatophore, a method of sperm storage and movement often associated with internal fertilization. Spermatophores are used by some members of the following phyla: Annelida (Hsieh and Simon, 1990), Mollusca (O'Foighil, 1985, 1989), Bryozoa (Silen, 1966), and Phoronida (Zimmer, 1972) (see Mann [1984] for a review of spermatophore use in animals). Lacking these sorts of adaptations, some marine animals with internal fertilization still depend on freely spawned individual sperm, in defiance of the supposed adverse effects of high dilution and wide adult dispersion (Bishop and Ryland, 1991; Eckelbarger et al., 1989).

Enhancing sperm longevity after spawning may increase the chances of fertilization in the deep sea and possibly elsewhere. A very long, motile, and fertilizable life is likely for the sperm of echinothurid echinoids in the deep sea (Eckelbarger et al., 1989; Tyler et al., 1992). In warmer surface waters, sperm longevity could be enhanced in a number of ways. Because sperm life span is directly related to the amount of energy used during swimming (Rothschild and Cleland, 1952; Anderson and Personne, 1975), prolonged inactivation after spawning will increase the dispersal of sperm while retaining their ability to swim and fertilize eggs (Havenhand, 1991). Some sort of signal would be required to activate the sperm when in the vicinity of an egg. Suitable signals, such as the release of sperm chemo-attractants by eggs (Miller, 1978), exist in a number of marine invertebrate phyla (Miller, 1985).

Consider the possible path of a *nonmotile* sperm that enters the eddy created by a female *Tubularia* polyp. It will be carried against the ambient current flow and toward the polyp body, reaching the zone of almost still water close to the clusters of reproductive medusoids. The zone of calm water would enable attractant gradients to form, so a sperm entering this area would activate rapid motility upon perception of the attractant and follow the gradient to the pore area of a ripe medusoid. Given these circumstances, a *Tubularia* sperm may have to swim only a few millimeters during its entire postspawning existence to fertilize an egg, even though it may have been carried as an inert object for many meters by currents after release from the male. This description of sperm capture by *Tubularia* is largely hypothetical because the study of fertilization success versus sperm concentration in *Tubularia* is still to be carried out. It does suggest that appropriate exposure of body form to the local water mass is likely to be important for the proper function of sexual reproductive processes in this marine invertebrate.

Suggestions that the shedding of downstream vortices by filtering arrays or by whole animals enhances particle capture rates have been made by a number of workers (Shimeta and Jumars, 1991). This mechanism may also occur in gorgonians (Sponaugle, 1991; Leversee, 1976). It has been called "downstream vortex feeding" by Leonard (1992), who states that it does not occur in crinoids. LeClair (pers. comm.) has reported generation of vortices downstream of the tips of artificial vertical arrays representing the erect feeding arms of groups of suspension-feeding ophiuroids. The vortex created by an upstream arm may enhance particle capture by the down-stream arms, demonstrating the enhancement of feeding of individuals within a group relative to the individuals feeding in isolation (Merz, 1984; Okamura, 1984, 1985). A parallel array of vertical, stiff arms is a useful but less than realistic model of the organisms involved. In the case of *Tubularia*, the reverse curvature of the radial array of tentacles directed at right angles to the flow, the fusiform body of the animal, and even the thin stem holding the animal up in the current, may contribute to a radial downstream vortex that retains particles close to feeding and reproductive portions of the polyp for a substantial length of time.

The literature contains a number of examples of situations similar to that described in *Tubularia*. The flow patterns around the polyps of the gorgonian *Pseudopterogorgia acerosa* at intermediate velocities (5–10 cm/s) (Sponaugle, 1991) are similar to those described here. The frequency of successful food capture is highest at intermediate-flow velocities (Sponaugle and LaBarbera, 1991), the same velocities where the reverse flow vortex is best maintained by *Tubularia*. Similar flow patterns probably occur in the vicinity of actinians and scleractinians in which fertilization is known to be internal (Spaulding, 1974; Strathmann, 1987). Sperm capture and storage occurs in the operculum of the suspension-feeding annelid *Spirorbis* (Daly and Golding, 1977), and it has been described in the bryozoan *Electra posidonae* (Silen, 1966).

Planktonic species such as salps, which actively process large volumes of water for food capture, may also have their anatomy and behavior specialized for the capture of sperm. Purcell and Madin (1991) reported that various species of oceanic salps vertically migrated at different times during the 24-h day. The male phase aggregates that were captured in surface waters had empty guts and motile sperm in their testes. By not feeding at the surface they presumably avoid depleting the sperm after they have spawned. The absence of the feeding net may not be required for sperm to reach the atrial cavity, since *Thalia* sperm have been observed to penetrate the mucus feeding web and enter the atrial cavity (Miller, unpub. obs.).

There are no observations on whether the spawning male distributes the sperm along its spiral path through the water, though this almost surely occurs. However, the periodic active-inactive motility of spawned sperm will increase the volume of space occupied by the sperm population without excessive use of the energy stores available for flagellar motility. The sperm's fertilizable and motile life span will be extended by this behavior, and the chances of being picked up by a passing female aggregate stage increased.

Thalia democratica sperm are "captured" within the atrial cavity near the tip of the fertilizing duct (Brooks, 1876; Miller, 1991, unpub. obs.). Sperm remain in this location even though they are more actively motile than in the external sea water or the water within the atrial cavity. Because little is known about the anatomical details of the area where the sperm aggregate, it is not clear what is responsible for initiation and maintenance of the aggregation. Sperm may be swept there by some local current deflection, or a mucus mass may exist near the end of the duct that traps sperm that randomly enter it but does not prevent flagellar motility, or a sperm attractant may be released from the duct or cells near the duct. No ciliated bands can be seen in this area. Besides the *Spirorbis* situation mentioned earlier, the ascidian *Diplosoma listerianum* also stores sperm prior to fertilization (Bishop and Ryland, 1991). The storage mechanism is unknown in both species, but sperm behavior similar to that observed in salps may be taking place.

In conclusion, a suite of adaptations probably occurs together to enhance the fertilization success of free-spawning, brooding invertebrates, each increasing by some unknown increment the overall fertilization success of the organisms in question. They will not be the same for all marine invertebrates, but are likely to be similar for those that manipulate currents for particle capture. These adaptations may include:

1. Increasing the longevity of the sperm during dilution in the sea, probably by suppression of motility after spawning;
2. Increasing the sperm concentration in the vicinity of the adult by capture of diluted sperm using the suspension feeding apparatus or organized currents;
3. Reactivation of sperm motility or initiation of sperm chemotaxis when sperm are close to the reproductive adult or the gonads or eggs;
4. Storing the captured sperm for later use in fertilization.

Acknowledgments

I thank Samson Xiong, who took and prepared the photographs of the dye streams in the vicinity of the *Tubularia* polyp, and Ms. Debi Milham, who helped with the salp studies in Australia. The staffs of Heron Isand Research Station (Dr. Ian Lawn, Director) and Friday Harbor Labs (A. O. D. Willows, Director) provided support during portions of the research effort. The salp work was supported in part by National Science Foundation grants INT-871403 and OCE-9121355, Temple University Research Incentive Fund, and personal funds.

Literature Cited

Alldredge, A., and L. P. Madin. 1982. Pelagic tunicates: Unique herbivores in the marine plankton. Bioscience. 32: 655–663.

Anderson, W. A., and P. Personne. 1975. The form and function of spermatozoa: A comparative view. In B. Afzelius, ed.The Functional Anatomy of the Spermatozoon. Pergamon Press, Oxford, pp. 3–14.

Babcock, R. C. 1984. Reproduction and distribution of two species of *Goniastrea* (Scleractinia) from the Great Barrier Reef Province. Coral Reefs 2: 187–195.

Bishop, J. D. D., and J. S. Ryland. 1991. Storage of exogenous sperm by the compound ascidian *Diplosoma listerianum*. Mar. Biol. 108: 111–118.

Brooks, W. K. 1876a. The development of *Salpa*. Bull. Mus. Comp. Zool. Harvard 3: 291–384.

————1876b. A remarkable life history and its meaning. Am. Nat. 10: 1–16.

Cohen, S. 1990. Outcrossing in field populations of two species of self-fertile ascidians. J. Exp. Mar. Biol., Ecol. 140: 147–158.

Coll, J. C., and R. L. Miller. 1992. The nature of sperm attractants in corals and starfish. *In* B. Baccetti, ed. Comparative Spermatology 20 Years After. Serono Symposia Publications, Raven Press, New York, pp. 129–134.

Daly, J. M., and D. W. Golding. 1977. A description of the spermatheca of *Spirorbis spirorbis* (L.) (Polychaeta: Serpulidae) and evidence for a novel mode of sperm transmission. J. Mar. Biol. Assoc. 57: 219–227.

Denny, M. 1988. Biology and the Mechanics of the Wave-Swept Environment. Princeton University Press, Princeton, N. J.

Denny, M. W., and M. F. Shibata. 1989. Consequences of surf-zone turbulence for settlement and external fertilization. Am. Nat. 134: 859–889.

Eckelbarger, K. J., C. M. Young, and J. L. Cameron. 1989. Modified sperm ultrastructure in four species of soft-bodied echinoids (Echinodermata: Echinothuriidae)) from the bathyal zone of the deep sea. Biol. Bull. 177: 230–236.

Havenhand, J. 1991. Fertilization and the potential for dispersal of gametes and larvae in the solitary ascidian *Ascidia mentula* Muller. Ophelia 33: 1–16.

Hsieh, H.-L., and J. L. Simon. 1990. The sperm transfer system in *Kinbergonuphis simoni* (Polychaeta: Onuphidae). Biol. Bull. 178: 85–93.

Leloup, E. 1929. La maturation et la fecondation de l'oeuf de *Salpa fusiformis* Cuv. Bull. Cl. Sci. Acad. Roy. Belg. 15: 461–478.

Leonard, A. B. P. 1992. The biomechanics, autecology and behavior of suspension-feeding in crinoid echinoderms. Ph.D. dissertation, University of California, San Diego.

Leversee, G. J. 1976. Flow and feeding in fan-shaped colonies of the gorgonian coral, *Leptogorgia*. Biol. Bull. 151: 344–356.

Levitan, D. R. 1988. Factors influencing fertilization success and fecundity in the sea urchin *Diadema antillarum* Phillipi. Am. Zool. 28: 139A.

———1991. Influence of body size and population density on fertilization success and reproductive output in a free-spawning invertebrate. Biol. Bull. 181: 261–268.

Levitan, D. R., M. A. Sewell, and F.-S. Chia. 1991. Kinetics of fertilization in the sea urchin *Strongylocentrotus franciscanus*: Interaction of gamete dilution, age and contact time. Biol. Bull. 181: 371–378.

———1992. How distribution and abundance influence fertilization success in the sea urchin *Strongylocentrotus franciscanus*. Ecology 73: 248–254.

Mackie, G. 1966. Growth of the hydroid *Tubularia* in culture. Symp. Zool. Soc. London 16: 397–412.

Mann, T. 1984. Spermatophores. Development, Structure, Biochemical Attributes and Role in the Transfer of Spermatozoa. Springer-Verlag, New York.

Merz, R. A. 1984. Self-generated versus environmentally produced feeding currents: A comparison for the sabellid polychaete *Eudistylia vancouveri*. Biol. Bull. 167: 200–209.

Miller, R. L. 1966. Chemotaxis during fertilization in the hydroid *Campanularia*. J. Exp. Zool. 162: 22–45.

———1974. The role of the gonomedusa and gonangium in the sexual reproduction (fertilization) of the Hydrozoa. Publ. Seto Mar. Biol. Sta. 20: 367–400.

———1976. Some observations on sexual reproduction in *Tubularia*. *In* G. O. Mackie, ed. Coelenterate Ecology and Behaviour. Plenum Press, New York, pp. 299–308.

———1978. Site-specific sperm agglutination and the timed release of a sperm chemo-attractant by the egg of the leptomedusan, *Orthopyxis caliculata*. J. Exp. Zool. 205: 385–402.

———1985. Sperm chemo-orientation in the Metazoa. *In* C. B. Metz, Jr., and A. Monroy, eds. Biology of Fertilization. Academic Press, New York, pp. 275–337.

———1989. Evidence for the presence of sexual pheromones in free-spawning starfish. J. Exp. Mar. Biol. Ecol. 130: 205–221.

———1991. Spawning and sperm approach to the egg in the salp *Thalia democratica* (Forskal, 1775). Am. Zool. 31: 138A.

Miller, R. L., and K. King. 1983. Sperm chemotaxis in *Oikopleura dioica* (Urochordata: Larvacea). Biol. Bull. 165: 419–428.

Minchin, D. 1987. Sea water temperature and spawning behaviour in the seastar *Marthasterias glacialis*. Mar. Biol. 95: 139–143.

Niklas, K. J. 1981. Simulated wind pollination and airflow patterns around ovules of some early seed plants. Science. 211: 275–277.

Niklas, K. J., and K. T. Paw U. 1982. Pollination and airflow patterns around conifer ovulate cones. Science. 217: 442–444.

O'Foighil, D. 1985. Sperm transfer and storage in the brooding bivalve *Mysella tumida*. Biol. Bull. 169: 602–614.

———1989. Role of spermatozeugmata in the spawning ecology of the brooding oyster *Ostrea edulis*. Gam. Res. 24: 219–228.

Okamura, B. 1984. The effects of ambient flow velocity, colony size and upstream colonies on the feeding success of Bryozoa. I. *Bugula stolonifera* Ryland, an arborescent species. J. Exp. Mar. Biol. Ecol. 83: 179–193.

———1985. The effects of ambient flow velocity, colony size and upstream colonies on the feeding success of Bryozoa. II. *Conopeum reticulum (Linnaeus)* an encrusting species. J. Exp. Mar. Biol. Ecol. 89: 69–80.

Pennington, T. 1985. The ecology of fertilization of echinoid eggs: The consequences of sperm dilution, adult aggregation and synchronous spawning. Biol. Bull. 169: 417–430.

Petersen, C. W., R. R. Warner, S. Cohen, H. C. Hess, and A. T. Sewell. 1992. Variable pelagic fertilization success: Implications for mate choice and spatial patterns of mating. Ecology. 73: 391–401.

Purcell, J. E., and L. P. Madin. 1991. Diel patterns of migration, feeding and spawning by salps in the subarctic Pacific. Mar. Ecol. Progr. Ser. 73: 211–217.

Ricketts, E. F., and H. Calvin. 1968. Between Pacific Tides, Stanford University Press, Stanford, Calif.

Rothschild, L., and K. W. Cleland. 1952. The physiology of sea urchin spermatozoa: The nature and location of endogenous substrate. J. Exp. Biol. 29: 66–71.

Ryland, J. S., and J. D. D. Bishop. 1990. Prevalence of cross-fertilization in the hermaphroditic compound ascidian *Diplosoma listerianum*. Mar. Ecol. Prog. Ser. 61: 125–132.

Sewell, M. A., and D. R. Levitan. 1992. Fertilization success during a natural spawning of the dendrochirote sea cucumber *Cucumaria miniata*. Bull. Mar. Sci. 51: 161–166.

Shimeta, J., and P. A. Jumars. 1991. Physical mechanisms and rates of particle capture by suspension feeders. Oceanogr. Mar. Biol. Ann. Rev. 29: 191–257.

Silen, L. 1966. On the fertilization problem in the gymnolaematous Bryozoa. Ophelia 3: 113–140.

Spaulding, J. G. 1974. Embryonic and larval development in sea anemones (Anthozoa: Actiniaria). Am. Zool. 14: 511–520.

Sponaugle, S. 1991. Flow patterns and velocities around a suspension-feeding gorgonian polyp: evidence from physical models. J. Exp. Mar. Biol. Ecol. 148: 135–145.

Sponaugle, S., and M. LaBarbera. 1991. Drag-induced deformation: A functional feeding strategy in two species of gorgonians. J. Exp. Mar. Biol. Ecol. 148: 121–134.

Strathmann, M. F. 1987. Anthozoa. *In* M. F. Strathmann, ed. Reproduction of Marine Invertebrates of the Northern Pacific Coast. University of Washington Press, Seattle, pp. 83–104.

Tyler, P. A., C. M. Young, D. S. M. Billett, and L. A. Giles. 1992. Pairing behaviour, reproduction and diet in the deep sea holothurian genus *Paroriza* (Holothurioidea: Synallactidae). J. Mar. Biol. Assoc. 72: 447–462.

Vogel, S. 1983. Life in Moving Fluids. The Physical Biology of Flow. Princeton University Press, Princeton, N. J.

Wiebe, P. H., L. P. Madin, L. R. Haury, G. R. Harbison, and L. M. Philbin. 1979. Diel vertical migration by *Salpa aspera* and its potential for large-scale particulate organic matter transport to the deep sea. Mar. Biol. 53: 249–255.

Young, C. M., P. A. Typer, J. L. Cameron, and S. G. Rumrill. 1992. Seasonal breeding aggregations in the low density populations of the bathyal echinoid *Stylocidaris lineata*. Mar. Biol. 113: 603–612.

Yund, P. O. 1990. An *in situ* measurement of sperm dispersal in a colonial marine hydroid. J. Exp. Zool. 253: 102–106.

Zimmer, R. L. 1972. Structure and transfer of spermatozoa in *Phoronopsis viridis*. *In* C. J. Arceneaux, ed. Thirtieth Ann. Proc. Electron Microscopy Soc. Am. at Los Angeles, pp. 108–109.

10 Cellular Dynamics during the Early Development of an Articulate Brachiopod, *Terebratalia transversa*

Corey Nislow

ABSTRACT Scanning electron microscopy was used to re-examine the early developmental stages of embryogenesis of the articulate brachiopod, *Terebratalia transversa*. These ultrastructural images provide novel views of cellular rearrangements from both inside and outside the embryo. A description of the changes in cell shapes that occur during cleavage, blastulation, and gastrulation is presented. In addition, separation of embryos at the two-cell stage produces a normal, half-sized larva, suggesting that development in this species proceeds by a regulative pathway.

Introduction

There are approximately 280 species of living brachiopods, which represent a small fraction of the 30,000 fossil species in this phyla that once flourished in the Paleozoic and Mesozoic eras (Barnes, 1980). Extant brachiopods are divided into two classes, the Inarticulata, whose valves are joined only by muscles, and the Articulata, whose valves are connected by a hinge apparatus as well as muscles (Hyman, 1959). Although there are several excellent descriptions of development of both inarticulate and articulate brachiopods (e.g., Yatsu, 1902; Conklin, 1902; Long, 1964), these studies are limited by the fact that they relied on light microscopic observations of living, intact specimens, which are typically yolky and opaque. To circumvent this problem, several studies have analyzed sectioned material (Long, 1964). This approach, while allowing a view of internal changes in structure, does not permit a three-dimensional perspective of embryonic development.

In this investigation, scanning electron microscopy (SEM) is used to re-examine the embryological development of one of the most thoroughly studied articulate brachiopods, *Terebratalia transversa*. These observations represent the first SEM views of prelarval stages, and are intended to complement previous light microscopic analyses of brachiopod development.

Materials and Methods

Adult specimens of *Terebratalia transversa* were collected by SCUBA divers from several subtidal locations near San Juan Island, Washington. Care was taken to avoid separating the animals from their substrates. Specimens were maintained at 12°C in artificial sea water (ASW, Marine Biological formula; 410 mM NaCl, 8 mM KCl, 10 mM CaCl$_2$, 23 mM MgCl$_2$, 25 mM MgSO$_4$, and 2 mM NaHCO$_3$) which was changed every three days. Gametes were collected according to Reed (1987). Adult shells were opened by severing their adductor muscles with a fine scalpel

Corey Nislow, Department of Molecular, Cellular and Developmental Biology, University of Colorado, Boulder, CO 80309–0347.

blade to expose the gonads. Ovaries were macerated and then passed several times through a 250-μm Nitex mesh into filtered (0.22-μm) ASW. Washed eggs were left as a monolayer in fresh ASW for 10–16 hr to allow them to shed their follicle cells. Immediately prior to fertilization, eggs were washed once with ASW. Sperm were prepared by macerating testes in a small volume of ASW. Large clumps of testes were removed, and an aliquot of this sperm mixture was adjusted to pH 10 by the addition of 10 N NaOH. As soon as the sperm appeared motile, several milliliters were added to 200 ml of ASW containing a monolayer of ovulated eggs. Fifteen minutes after the addition of sperm, the fertilized eggs were washed with several changes of fresh ASW and maintained at 12°C. To avoid contamination, the ASW was changed daily.

For scanning electron microscopy (SEM), embryos and larvae were fixed in 2 percent OsO_4 in ASW for 1 h at room temperature. Fixed specimens were washed twice in ASW, dehydrated through a graded series of ethanol, transferred to acetone, and critical point dried with CO_2 as the transitional fluid. Dried specimens were prepared for SEM observation by mounting them onto an SEM stub by means of double-sided tape. To fracture samples and reveal their internal structures, a second SEM stub containing double-sided tape but no embryos was gently brought down on top of the mounted embryos (Morrill 1986). Mounted embryos were coated with gold/palladium and viewed with a JEOL JSM-35 scanning electron microscope. Specimens to be sectioned were fixed as above, embedded in Epon, cut into 0.5-μm sections, and stained with 1.0 percent toluidine blue.

To separate two-cell stage blastomeres, a section of dental floss was teased apart into individual filaments which were then used to ligate embryos. Ligated embryos were finally separated with a mounted eyelash and then cultured separately in ASW.

Results

At 12°C, *Terebratalia transversa* takes three days to develop from fertilized egg to mature larva. By the fourth day, larvae are competent to undergo metamorphosis. The developmental stages discussed here are depicted in Figure 10.1. For further details, the reader is referred to Long (1964) and Long and Stricker (1991).

Early Cleavages

Each *T. transversa* oocyte is shed with its germinal vesicle intact (Fig. 10.2a) and measures about 150 μm in diameter. They are covered with a layer of small (ca. 15-μm) follicle cells which interdigitate to envelop the entire oocyte. The follicle cells posses long, thin cellular processes that penetrate through the dense lawn of microvilli and anchor the follicle cells to the surface of the oocyte. Over the course of several hours the follicle cells gradually detach from the oocyte surface.

Approximately 3 h postinsemination (PI), the first cleavage division results in the formation of a two-cell embryo. Although this first division usually forms two equal-sized blastomeres, occasional differences in the size of the two blastomeres are seen. This was also noted to occur in *Terebratulina septentrionalis* (Conklin, 1902). It is not clear how the plane of first cleavage relates to the future animal-vegetal axis of the embryo, but it is presumed that the first cleavage plane extends through the animal and vegetal poles (Long, 1964). At this stage, a thick vitelline envelope covers the entire embryo. The vitelline envelope is extremely resistant to both mechanical and chemical removal, making microsurgical manipulations difficult. Polar bodies can occasionally be seen protruding beneath the vitelline membrane (Fig. 10.2b).

The second cleavage occurs about 4 h PI and bisects the embryo into four approximately

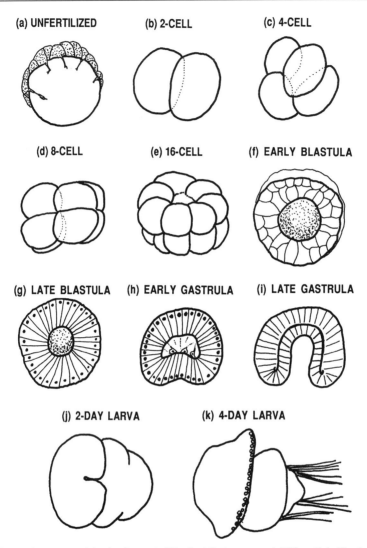

(a) UNFERTILIZED　　(b) 2-CELL　　(c) 4-CELL

(d) 8-CELL　　(e) 16-CELL　　(f) EARLY BLASTULA

(g) LATE BLASTULA　(h) EARLY GASTRULA　(i) LATE GASTRULA

(j) 2-DAY LARVA　　(k) 4-DAY LARVA

FIGURE 10.1. A summary of the development of *Terebratalia transversa* at 12°C. *a*. Unfertilized egg before insemination; *b*. Two-cell stage (3 h postinsemination [PI]); *c*. Four-cell stage (4 h PI). *d*. Eight-cell stage (5 h PI); *e*. Sixteen-cell stage (7 h PI); *f*. Early blastula (10 h PI); *g*. Late blastula (16 h PI); *h*. Early gastrula (20 h PI); *i*. Late gastrula (26 h PI); *j*. Two-day larva; *k*. Four-day larva.

equal-sized blastomeres. Typically, the second cleavage plane follows parallel to, but at right angles with, the first cleavage plane, to yield a ring of four cells. In most cultures, however, a small percentage of the embryos exhibit a second cleavage plane that appears up to 45° off the axis of the first (Fig. 10.2c). Such embryos nevertheless develop into normal gastrulae and larvae.

The third cleavage occurs about 5 h PI and usually divides the embryo into two tiers of eight equal-sized blastomeres (Fig. 10.2d). Occasionally, however, a ring of eight cells is formed by a third meridional cleavage. Such a "discrepancy" is compensated for at the next cleavage, and these embryos proceed to develop normally.

Fourth cleavage occurs 7 h PI to form a sixteen-cell embryo composed of two tiers of eight cells each. The embryo is wider along its equatorial axis, giving it a donutlike appearance with a depression at the animal and vegetal poles (Fig 10.3a). This depression has been interpreted as a

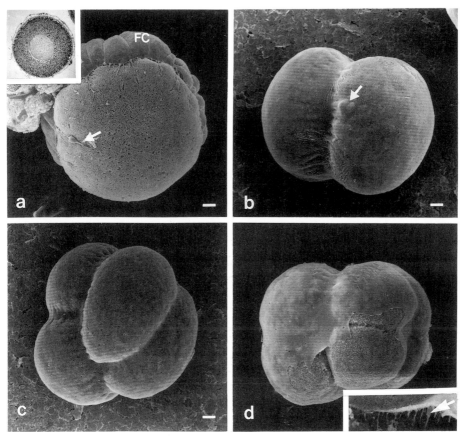

FIGURE 10.2. Early cleavages of *Terebratalia transversa* reared at 12°C (scale bars = 10 μm). *a.* Unfertilized egg with several adherent follicle cells (FC). Long cellular processes (arrow) anchor these cells to the egg surface. Inset: Section through an unfertilized oocyte showing the prominent germinal vesicle (GV). *b.* Two-cell embryo 3 h PI. Polar bodies can be seen (arrow) beneath the vitelline envelope. *c.* Four-cell embryo 4 h PI with aberrantly oriented blastomeres. Such embryos constitute a minor fraction of those in any given culture. *d.* Eight-cell embryo 5 h PI. In this embryo the vitelline envelope has been reflected off several blastomeres. Inset: Higher magnification of the vitelline envelope. Thin strands attach it to the surface of the blastomere (arrow).

pore by Hirai and Fukushi (1960), but it is not clear if there is actually a space between blastomeres. By the sixteen-cell stage, a loss of mitotic synchrony is already evident, as some blastomeres have cleaved completely whereas others have not.

Approximately 8 h PI, the blastocoelic cavity first appears (Fig. 10.3b). Early on the cavity is quite small (ca. 70 μm in diameter), but several rounds of cleavage reduce the size of blastomeres and increase the surface of the epithelium until the blastocoel reaches a maximum diameter of 140 μm (Fig. 10.3c). At this point the blastula wall is composed of a unilayered epithelium composed of nearly cuboidal blastomeres. The external surfaces of these cells are still covered by the vitelline envelope while their blastocoelic faces are quite smooth and apparently devoid of any basal lamina.

A few hours later (16 h PI), the cells of the blastula wall undergo a dramatic change in shape, elongating along their apicobasal axes to become extremely columnar. This change in blastomere morphology reduces the size of the blastocoel considerably (Fig. 10.3d). Each blastomere possesses a single cilium, and the embryo begins to rotate slowly due to ciliary beating. A fibrillar extracellular matrix (ECM) and a thin basal lamina are first observed in the late blastula.

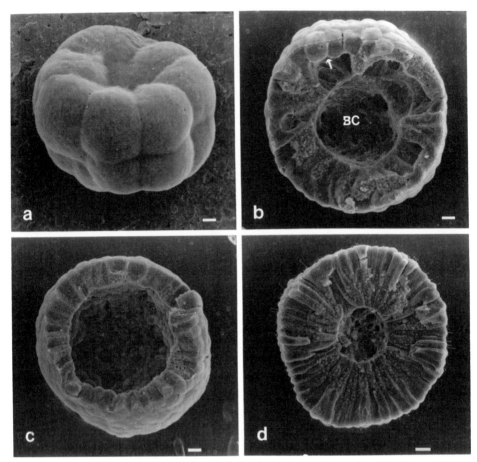

FIGURE 10.3. *a*. Sixteen-cell embryo 7 h PI. *b*. Early blastula 9 h PI showing the formation of the blastocoel (BC). *c*. Early- to mid-stage blastula 10 h PI. By this stage the blastocoel has reached its maximum diameter. *d*. Late blastula 16 h PI. The blastomeres have become columnar, and each contains a single apical cilium. (Scale bars = 10 μm.)

Gastrulation

Gastrulation begins approximately 18 h PI as 10–15 cells within the vegetal plate become bottle shaped. The bulbous ends of these cells extend into the blastocoel, and the thin apical ends often lose contact with other cells in the vegetal plate (Fig. 10.4a,b). Immediately following this shape change in the vegetal cells, the entire vegetal plate begins to invaginate (Fig. 10.4c,d). As invagination proceeds, a fibrous ECM extends from the basal surfaces of the cells at the tip of the archenteron to the animal pole.

The cells of the late gastrula exhibit several changes in shape. Once the archenteron reaches halfway across the blastocoel, cells at the archenteron tip become elongate and extend broad lamellipodia. These processes often contact the animal pole (Fig. 10.5a,b). Whether or not these contacts exert a pulling force that aids invagination (Gustafson and Wolpert, 1967) is not clear. As gastrulation proceeds the blastocoel is gradually obliterated (Fig. 10.5c). Cells detach from the archenteron tip and migrate into the shrinking blastocoel. Eventually, the putative mesenchyme cells between the presumptive ectoderm and presumptive endoderm anastomose to form a thin layer that fills the space formerly occupied by the now obliterated blastocoel (Fig. 10.5d–f).

Following gastrulation at about 38 h PI, the embryo elongates along its equatorial axis. The

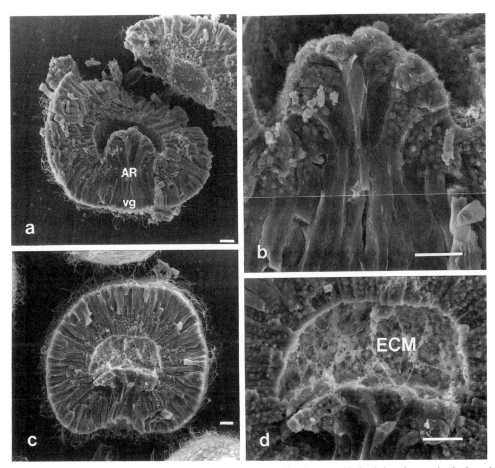

FIGURE 10.4. *a*. Early gastrula at 18 h PI. Cells of the archenteron (AR) at the vegetal (vg) pole have become bottle shaped and extend into the blastocoel. *b*. Higher magnification of the archenteron in *a*. *c*. Early gastrula 20 h PI. The cells at the vegetal pole have just begun to invaginate. A fibrillar extracellular matrix (ECM) is apparent within the blastocoel (arrow). *d*. Higher magnification of the blastocoel in *c*. Note how the ECM extends from the archenteron to the animal pole. (Scale bars = 10 μm.)

blastopore also becomes stretched in this direction and thus appears as an oval (Fig. 10.6a). The archenteron likewise becomes distended, with its anterior end terminating in a blind pouch. Soon thereafter (ca. 40–48 h PI) the larval mesoderm begins to form, presumably arising from the cells that emigrated from the archenteron tip during gastrulation. This single layer of presumptive mesoderm is now positioned at the anterior end of the larva (Fig 10.6c,d). The lateral and posterior growth of this sheet of cells results in formation of a coelomic space that flanks the archenteron laterally and dorsally. As the coelom is forming, the embryo undergoes several external changes. The oval blastopore eventually closes to form a slit running anterior to posterior within the incipient apical lobe at the anterior end of the larva. At this stage a second constriction that will eventually separate the mantle and pedicle lobes is first observed as a furrow encircling the larva (Fig. 10.6b).

The Mature Larva

The mature larva develops within three to four days. The larva, which measures about 200 μm in length, possesses three distinct lobes (Fig. 10.7a). The antero-most lobe is the apical lobe, which

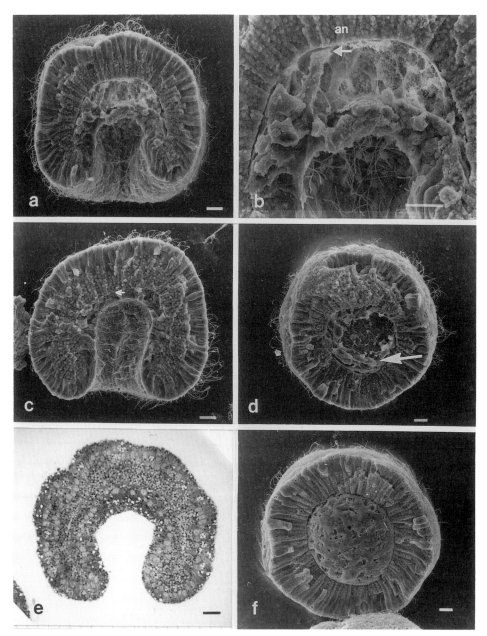

FIGURE 10.5. *a*. View of a mid-stage gastrula 24 h PI in longitudinal view (scale bars = 10 μm). *b*. Higher magnification of the embryo in *a*. Note that the cells at the tip of the archenteron extend lamellipodia (arrow) that contact the animal pole (an). *c*. A late gastrula 26 h PI. The archenteron has now completely obliterated the blastocoel (arrow). *d*. Animal pole view of a mid-late gastrula. Several mesenchyme cells can be seen migrating at the archenteron tip (arrow). *e*. Section along the animal-vegetal axis of a late gastrula. *f*. An animal pole view of a late gastrula, by this stage the mesenchyme cells along the archenteron anastomose to form a sheet.

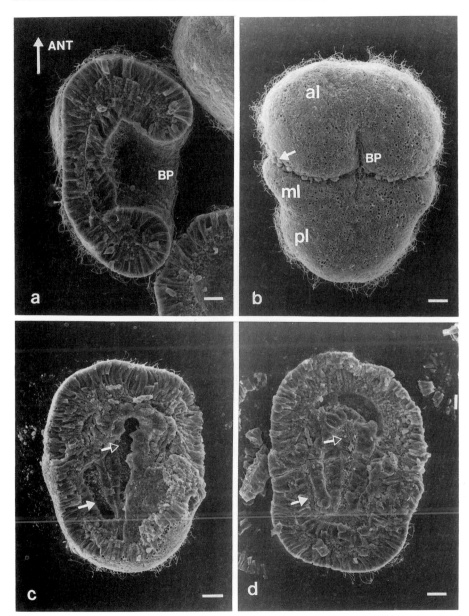

FIGURE 10.6. *a*. Longitudinal view of an embryo 38 h Pl. The archenteron and blastopore (BP) have elongated along the future anterior-posterior axis of the larva. *b*. External view of an early larva 48 h Pl. The blastopore (BP) has closed to form a slit in the apical lobe (al). A second ectodermal constriction is beginning to separate the mantle lobe (ml) from the pedicle lobe (pl). *d*. Fractures of larvae 48 h Pl. A sheet of mesoderm has grown both laterally and posteriorly to separate the coelomic cavity (white arrows) from the elongated archenteron (black arrows). (Scale bars = 10 μm.)

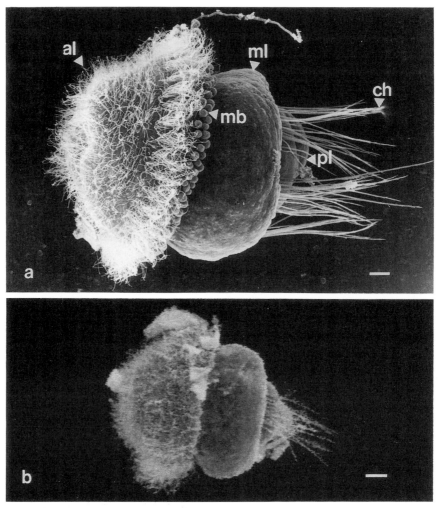

FIGURE 10.7. *a*. Mature larva 72 h PI. The apical lobe (al), mantle lobe (ml), and pedicle lobe (pl) are now clearly differentiated. Spherical multivesiculated bodies (mb) are seen at the posterior margin of the apical lobe. Four sets of chaetae (ch) extend posteriorly from the mantle lobe. *b*. Miniature larva reared from one blastomere of a two-cell stage embryo at 72 h PI. Except for size, this miniature larva is morphologically similar to the normal one shown in *a*. (Scale bars = 10 μm.)

is covered by cilia over its entire surface. At the posterior margin of the apical lobe there is a band of longer cilia as well as a collection of non-nucleated cytoplasmic protrusions that are termed *multivesiculated bodies*. The exact functions of both these features are not known. The middle lobe of the larva is the mantle lobe, which is devoid of cilia except for a narrow band on the ventral surface. Four groups of long (80–100-μm) chaetae extend posteriorly from the mantle lobe. The postero-most lobe is the pedicle lobe, which serves an important role during metamorphosis. For details concerning the later stages of larval development and metamorphosis the reader is referred to Stricker and Reed (1985a).

Development of isolated blastomeres

To determine if *T. transversa* development follows a mosaic or a regulative path, blastomeres were isolated from two-cell stage embryos and cultured for several days. As seen in Figure 10.7b, such manipulations yield a morphologically normal miniature larva. Furthermore, both the timing of morphogenesis and the swimming behavior of these larvae appeared to be normal.

Discussion

In this study, the embryology of the articulate brachiopod *Terebratalia transversa* has been analyzed by means of SEM. For the most part, the observations presented here are in agreement with earlier light microscopic studies. New findings revealed by SEM include the fact that the cells of the embryo undergo dramatic shape changes during blastulation and gastrulation. In addition, the transformation of the circular blastula into a gastrula bears striking similarity to echinoderm gastrulation. Much attention has been paid toward elucidating the mechanisms of archenteron elongation during echinoderm gastrulation, including the roles of cytoskeletal elements, cell shape and adhesivity changes, and changes in cell positioning (Nislow and Morrill 1988). Given the apparent morphological similarities between echinoderm and brachiopod gastrulation, it would be valuable to ask if similar morphogenetic "motors" are at work during development in these two phyla. Since large quantities of synchronously developing *Terebratalia transversa* embryos can be easily obtained, such comparative studies should be possible.

It should be noted that this study has focused on a single species of articulate brachiopod. To be able to draw general conclusions regarding the fine structural details of brachiopod development, comparable studies on other articulate species, as well as inarticulate species, are needed.

The observation that blastomeres isolated from two-cell embryos can form normal miniature larvae, combined with observations of Long (1964) on *Hemithyris psittacea* and the observation by Conklin (1902) of naturally occurring one-half size larvae of *Terebratulina septentronialis*, suggests that regulative, not mosaic, development is a common feature of embryogenesis in articulate brachiopods.

Acknowledgments

This work was conducted when I was a first-year graduate student in Chris Reed's laboratory at Dartmouth College. I would like to dedicate this report to his memory. During my stay at Dartmouth I learned a great many things from Chris, both about developmental biology and about the development of oneself. I am very lucky, indeed, to have spent the short time that I did with such a remarkable individual.

Literature Cited

Barnes, R. D. 1980. Invertebrate Zoology. Holt, Rinehart and Winston, Philadelphia.

Conklin, E.G. 1902. The embryology of a brachiopod, *Terebratulina septentrionalis* Couthouy. Proc. Am. Philos. Soc. 41: 41–76.

Gustafson, T., and L. Wolpert. 1967. Sea urchin morphogenesis. Biol. Rev. 42: 442–498.

Hirai, E., and T. Fukushi. 1960. The development of two species of lamp shells, *Terebratalia coreanica* and *Coptothyris grayi*. Bull. Mar. Biol. Sta. Asamushi 10: 77–80.

Hyman, L. H. 1959. The Invertebrates: Smaller Coelomate Groups, Vol. 5. McGraw-Hill, New York.

Long, J. A. 1964. The embryology of three species representing three superfamilies of articulate Brachiopods. Ph.D. dissertation, University of Washington, Seattle.

Long, J. A. and S. A. Stricker. 1991. Brachiopoda. *In* A. C. Giese, J. S. Pearse, and V. B. Pearse, eds. Reproduction of Marine Invertebrates, Vol. 6. Echinoderms and lophophorates. The Boxwood Press, Pacific Grove, Calif., pp. 47–85.

Morrill, J. B. 1986. Scanning electron microscopy of embryos. *In* T. Schroeder, ed. Methods in Cell Biology, Vol. 27. Academic Press, Orlando, pp. 263–294.

Nislow, C., and J. B. Morrill. 1988. Regionalized cell division during sea urchin gastrulation contributes to archenteron formation and is correlated with the establishment of larval symmetry. Dev. Growth Differ. 30: 483–499.

Reed, C. G. 1987. Phylum Brachiopoda. *In* M. F. Strathmann, ed. Reproduction and Development of Marine Invertebrates of the Northern Pacific Coast. University of Washington Press, Seattle pp. 486–493.

Stricker, S. A. and C. G. Reed. 1985a. The ontogeny of shell secretion in *Terebratalia transversa* (Brachiopoda: Articulata). I. Development of the mantle. J. Morph. 183: 223–250.

———1985b. The ontogeny of shell secretion in *Terebratalia transversa* (Brachiopoda: Articulata). II. Formation of the protegulum and juvenile shell. J. Morph. 183: 251–271.

———1985c. Development of the pedicle in the articulate brachiopod *Terebratalia transversa* (Brachiopoda: Inarticulata). Zoomorph. 105: 253–264.

Yatsu, N. 1902. On the development of *Lingula anatina*. J. Coll. Sci. Imp. Univ. Tokyo 17: 1–112.

11 Developmental Regulation of Ciliogenesis and Ciliary Length in Sea Urchin and Surf Clam Embryos

R. E. Stephens

ABSTRACT Sea urchin embryos generate cilia by the limited production of key structural elements of the 9 + 2 axoneme. These coassemble with the major building blocks, tubulin and dynein, to a length determined by their limited availability. Tektin-A, an integral component of the outer doublet microtubules, and certain other associated proteins are synthesized in "quantal" or limiting amounts; most of the other architectural proteins involved in maintaining ninefold symmetry are coordinately regulated, and tubulin, present in the most abundant pool, is autoregulated. Hyperciliation can be induced by theophylline treatment of zinc-arrested sea urchin embryos, accompanied by the replay of protein synthetic events in proportion to the length of cilia that are to be deployed. This frugal, limited-synthesis mechanism should be advantageous to actively swimming larvae, which must be able to replace cilia efficiently while carrying on an uninterrupted program of development. A comparative study of ciliary regeneration in the far more rapidly developing surf clam embryo reveals that all proteins of the axoneme, including tubulin, appear to be coordinately regulated, and all occur in limited, stoichiometric pools. Ciliary length in this case is probably determined in an equilibrium fashion, heretofore described only in protists.

Introduction

Aside from the mitotic apparatus, which is assembled and disassembled with each cell division, cilia are the first relatively permanent structures that a developing sea urchin embryo produces. Ciliogenesis also appears to represent the first truly zygotically programmed morphogenetic event, although the production of hatching enzyme mRNA is also zygotic and probably begins before the formation of cilia (Lepage and Gache, 1990). The number of cell divisions that must precede ciliogenesis, although precise, varies considerably with both the species and the cell lineage (Masuda, 1979). Consequently, after ciliogenesis, some ciliated blastomeres continue to divide. In doing so, each cell rapidly resorbs its single cilium and undergoes mitosis, and then the two daughter cells each reproduce one cilium. The basal body and daughter centriole from the initially resorbed cilium serve as a diplosome for the next cell division and, after division, the mother centriole in each cell is transformed into a basal body, reinitiating ciliogenesis (Masuda and Sato, 1984).

Embryos may be deciliated by high salt treatment and will rapidly regrow cilia (Auclair and Siegel, 1966), requiring only limited protein synthesis to do so (Burns, 1979). Subjected to successive deciliations, embryos will repeatedly reciliate, with identical growth kinetics, while

R. E. Stephens, Marine Biological Laboratory, Woods Hole, MA 02543; Present address: Department of Physiology, Boston University School of Medicine, Boston, MA 02118.

normal development proceeds undelayed (Stephens, 1977). Interestingly, ciliogenesis and regeneration require multicellularity, and presumably gap junctions (cf. Spiegel and Howard, 1983), since blastomeres dissociated immediately before ciliogenesis will not generate cilia, nor will ciliated blastomeres regenerate cilia if deciliated and then dissociated. In both cases, cilia formation resumes when the cells reaggregate (Amemiya, 1971).

At a molecular level, cilia formation has been studied mainly in terms of tubulin gene expression. A number of β-tubulin genes are expressed differentially throughout early development (Alexandraki and Ruderman, 1985; Harlow and Nemer, 1987a), and a subset of these genes is coordinately and selectively expressed in conjunction with both ciliogenesis and regeneration (Harlow and Nemer, 1987b). The expression of tubulin genes is autogenously regulated at the level of mRNA stability by the level of free tubulin, and the loss of cilia results in the increased transcription of tubulin genes (Gong and Brandhorst, 1987, 1988). Deciliation of animalized embryos bearing long cilia does not further increase an already high β-tubulin transcription rate but does increase message accumulation, indicating that the induction of β-tubulin mRNA by deciliation is regulated transcriptionally in normal embryos but posttranscriptionally in animalized embryos (Harlow and Nemer, 1987b).

In the sea urchin embryo, the major ciliary building blocks, tubulin and dynein, preexist in fairly large pools, evidently even in the unfertilized egg (Bibring and Baxandall, 1981; Foltz and Asai, 1990). Ciliogenesis itself is marked by the limited *de novo* production of a multitude of coassembling 9 + 2 architectural proteins (Stephens, 1972). Experimental deciliation triggers the resynthesis of these architectural components, with a subset of them being made in a "quantal" or strictly limiting amount (Stephens, 1977). The most prominent of these quantal proteins is tektin-A, a member of a unique class of highly insoluble protein associated with the junctional protofilaments of the nine outer doublet microtubules (Linck and Langevin, 1982; Linck and Stephens, 1987). An important implication of this limited synthesis is that the length of the doublet microtubule must be strictly limited by the available amount of coassembling junctional proteins. Deciliation also stimulates supplemental synthesis of tubulin and dynein, refilling the relatively large pools of these major axonemal building blocks.

These observations led to a limiting key component model for ciliogenesis and regeneration wherein it was proposed that an embryo need only synthesize *limited* or *limiting* amounts of certain minor components for the rapid assembly of the organelle from a warehouse already full of the major structural and enzymatic components (Stephens, 1977). Additional synthesis, stimulated by pool depletion, replenishes the store of tubulin and dynein. Since ciliary regeneration can be repeated many times without disrupting the normal process of development, regeneration (and its initial counterpart, ciliogenesis) may be viewed as a subroutine in the program of development. As such, the regenerating embryo represents a convenient system with which to investigate inducible coordinate gene expression and certain aspects of autoregulation. One logical function for such a subroutine would be to provide an embryo with an efficient mechanism for making new cilia after each cell division, or for regenerating lost cilia, without interrupting the main program of development to do so.

This chapter reviews the characteristic patterns of protein synthesis during ciliary regeneration in sea urchin embryos, investigates these same aspects in surf clam embryos, and then presents information on the coordinate control of protein synthesis in proportion to ciliary length in an "inducible" sea urchin system.

Regeneration of Cilia in Sea Urchin Embryos

To illustrate the protein synthetic patterns of ciliary regeneration, fractionation of isolated cilia into defined components is particularly useful. Cilia can be released from sea urchin embryos, and most other marine embryos as well, by a 1–2 min treatment at the growth temperature with sea water containing twice the normal amount of NaCl. The cilia are then isolated by differential centrifugation, and the embryos are returned to normal sea water for regeneration (Stephens, 1986). The ciliary membrane may be solubilized with nonionic detergents to produce a membrane plus periaxonemal matrix fraction, while the 9 + 2 axoneme may be further fractionated by gentle heating at low ionic strength into tubulin plus tubule-associated proteins and an insoluble remnant fraction consisting of the ninefold symmetric array of architectural proteins with which the tubulin and associated proteins originally coassembled (Stephens et al., 1989). This basic fractionation scheme is illustrated in Figure 11.1.

After high salt deciliation, 66-h mid-gastrula-stage embryos of the sea urchin *Strongylocentrotus droebachiensis* were pulse labeled at 7.5°C for 4 h with 3 μCi/ml ^{3}H-leucine, followed by an equivalent chase period with unlabeled 0.25 percent leucine during the initial regeneration of cilia. The embryos were then deciliated again and allowed to regenerate two additional crops of cilia, utilizing protein pools labeled during the first regeneration but supplemented with proteins synthesized from unlabeled leucine after the chase. The cilia were fractionated as outlined in Figure 11.1 and subjected to sodium dodecylsulfate polyacrylamide gel electrophoresis (SDS-PAGE) and fluorography. The results of three such regenerations are illustrated in Figure 11.2.

Several facts are evident from an inspection of the fluorogram, and these facts can be verified further by video densitometry (Table 11.1). First, the protein tektin-A (arrowheads, Fig. 11.2), so heavily labeled during the initial regeneration, is depleted by more than 80 percent upon a second regeneration, showing that its synthesis is quantal and thus potentially limiting in amount. Previous work, wherein embryos were animalized to produce cilia of twice normal length, demonstrated that the synthesis of this protein is directly proportional to length and also that its synthesis is coincident with elongation (Stephens, 1989a). There are also a number of other less-abundant proteins, particularly obvious in the lower-molecular-weight region of the R_1 fraction, that are label depleted to the same extent as tektin-A. This is also true of one minor membrane/matrix protein (asterisks, Fig. 11.2). Second, most of the remaining architectural proteins, both tubule- and remnant-associated, are derived from pools of fairly limited capacity since the labeled fraction drops rapidly in successive regenerations, averaging better than a 60 percent reduction. Furthermore, this decrease is uniform among most of the architectural proteins, suggesting that they are all present in roughly stoichiometric amounts, that is, all have similar molar pool sizes, in contrast to tektin-A and certain other quantal proteins whose labeled representatives are substantially depleted after only one regeneration. Third, the relative specific activity of tubulin decreases only slightly upon successive regenerations, consistent with the well-established point that it is present in a relatively large pool. The decrease in specific activity of about 22 percent through replenishment with unlabeled, newly synthesized protein suggests a pool of tubulin nearly five times larger than needed to produce one crop of cilia. The pool of the other major protein, dynein, is label depleted more than 35 percent per regeneration, also indicative of a considerable surplus, although less than tubulin. These observations are thus consistent with the limiting key component model for cilia formation, namely, that ciliogenesis involves the quantal production of strictly limiting amounts of certain key structural elements, limited amounts of most other architectural components, and the proportionate refilling of the major building block pools.

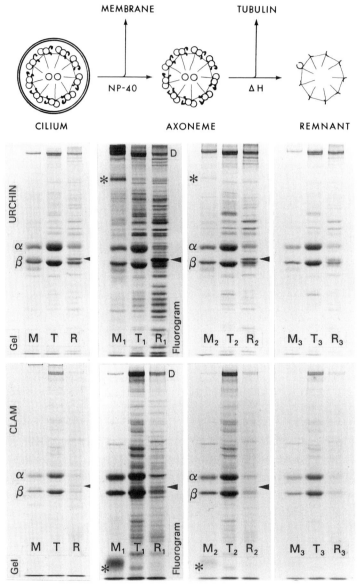

FIGURE 11.1 *(top)*. Ciliary fractionation scheme. The membrane is removed with NP-40 to yield a soluble membrane plus periaxonemal matrix fraction and an insoluble 9 + 2 axoneme fraction. Heating the axonemes at low ionic strength depolymerizes most of the tubulin, yielding a soluble tubulin plus tubule-associated protein fraction and an insoluble ninefold symmetric ciliary remnant fraction. The latter contains all of the tektins and many of the other architectural proteins. A depolymerizing A-tubule is depicted in the remnant to indicate its original location. (Based on Stephens et al. 1989.) FIGURE 11.2 *(middle)*. SDS-PAGE and fluorographic analysis of sea urchin *(Strongylocentrotus droebachiensis)* embryonic ciliary proteins synthesized during regeneration and then utilized in two successive regenerations. The membrane plus periaxonemal matrix fraction (M), the soluble tubulin and tubule-associated protein fraction (T), and the ciliary remnant (R) fraction are stoichiometrically loaded; subscripts in the second, third, and fourth panels (fluorograms) indicate the number of regenerations. Each regeneration contains equivalent amounts of protein, a representative Coomassie blue-stained gel of which is shown in the first panel (α, β = α and β tubulins; D = dynein heavy chains; arrowheads = tektin-A; asterisks = a quantal minor membrane protein). FIGURE 11.3 *(bottom)*. SDS-PAGE and fluorographic analysis of surf clam *(Spisula solidissima)* embryonic ciliary proteins synthesized while pulse-chase labeling during regeneration and then utilized in two successive regenerations. Designations as in Figure 11.2.

Table 11.1. Decrease in relative specific activities of axonemal components between successive regenerations and estimated approximate pool sizes (reciprocal of percentage decrease), determined by video densitometry with Jandel Scientific's JAVA system.

(These analyses used the integrated optical density, minus background, of individual proteins [dynein and tubulin] or blocks of components [architecturals, i.e., the remaining 9 + 2 proteins] from fluorograms, normalized by protein load from each gel sample [cf. Stephens 1991a, 1992].)

	Regeneration (%)			
	2d	3d	Average	Pool
Strongylocentrotus droebachiensis				
Dynein	−27	−42	−34.5	3 ×
Tubulin	−21	−23	−22.0	5 ×
Architecturals	−55	−68	−61.5	1.6 ×
Spisula solidissima				
Dynein	−32	−47	−39.5	2.5 ×
Tubulin	−33	−43	−38.0	2.6 ×
Architecturals	−45	−43	−44.0	2.3 ×

The tubulin found in the membrane plus periaxonemal matrix compartment has a specific activity equal to that of the axoneme in second- and third-regeneration cilia, consistent with its being a precursor to axonemal tubulin. Its labeling is somewhat higher in first-generation cilia, an observation consistent with the necessary early assembly of part of the axoneme with tubulin from the yet-unlabeled pool (cf. Stephens, 1991a). That the periaxonemal matrix must contain proteins destined for assembly into the axoneme is a widely accepted view, and the behavior of membrane/matrix tubulin is certainly consistent with this viewpoint. However, *most other heavily labeled axonemal proteins, particularly tektin-A, are not detected in the periaxonemal matrix compartment*, an observation that is true even in cilia that are isolated during partial regrowth from embryos heavily labeled prior to deciliation and regeneration (Stephens, 1992). This leaves us with the challenging question of how ciliary proteins traverse the distance from the cytoplasm, where they are synthesized, to the tip of the cilium, where they are known to assemble, without being detected in the membrane/matrix compartment.

Regeneration of Cilia in Surf Clam Embryos

Evolutionary arguments concerning comparative development in protostomes versus deuterostomes prompted the undertaking of parallel studies on ciliary regeneration in molluscan embryos. For these studies, the surf clam *Spisula solidissima* was chosen since this species has been extremely useful in analyses of cell cycle control and centriole formation (e.g., Palazzo et al., 1992). Unlike early sea urchin embryos, which have fairly uniform ciliation, veliger larvae of the far more rapidly developing clam are covered with cilia of fairly uniform length, but they have a very prominent apical tuft of bundled long cilia which accounts for an appreciable fraction of the mass of isolated cilia. Following deciliation and subsequent regeneration, the individual cilia reach their final length well before the apical tuft fully regenerates. Consequently, due to time constraints in the experimental design required to match the sea urchin example in Figure 11.2, the illustration which follows represents full regeneration of most of the cilia but only partial regeneration of the apical tuft. Figure 11.3 shows a 1-h pulse/1-h chase labeling of deciliated 18-h *Spisula* veligers, followed by two successive 2-h regenerations, all carried out at 24°C.

Compared to sea urchins, some very interesting qualitative similarities but quantitative differences are seen in *Spisula* embryonic ciliary regeneration. The parallel decrease in specific activities of architectural proteins, characteristic of limited stoichiometric pools, is clearly seen here also. The depletion of label, averaging nearly 45 percent, is more significant when one considers that these samples represent only partial regrowth; that is, nearly half of the label is replaced at each regeneration even though regeneration is not strictly complete over the entire surface of the embryo. The major building blocks, tubulin and dynein, are present in similarly limited pools (Table 11.1). For tubulin, this is a marked difference from what is seen in sea urchin embryos. The specific activities for both tubulin and dynein decrease by about one-third in the second regeneration and by nearly one-half again in the third regeneration.

In spite of the fact that the molluscan equivalent of sea urchin ciliary tektin-A has been demonstrated in *Spisula* sperm flagella and gill cilia (Stephens and Prior, 1991) and can be unequivocally identified in embryonic cilia by immunoblotting (data not shown), there is no evidence for the *de novo* quantal synthesis of tektin-A in these embryos. Rather, tektin-A appears to be coordinately regulated along with all of the other architectural and building block proteins, as evidenced by the uniform decrease in virtually all of the labeled axonemal components. Only a single, unidentified membrane component is significantly depleted after one generation (asterisks, Fig. 11.3). Thus the limiting key component model does not apply to these molluscan embryos since there are neither limiting axonemal protein components nor a warehouse disproportionately full of the major building blocks. Rather, ciliary regeneration in surf clam embryos is somewhat like flagellar regrowth in *Chlamydomonas* where relatively small pools of most structural components may limit growth in an equilibriumlike fashion, with deflagellation triggering protein synthesis to refill the depleted pools uniformly (Rosenbaum et al., 1969; Remillard and Witman, 1982).

As with sea urchin embryos, the detergent-solubilized fraction of surf clam cilia is dominated by tubulin. This tubulin of the membrane plus periaxonemal matrix compartment has essentially the same specific activity as tubulin of the axoneme in second- and third-regeneration cilia, which is consistent with a role as an axonemal precursor. Made even more striking by the fact that these preparations must contain cilia that are only partially regrown, no heavily labeled axonemal proteins can be detected in the membrane/matrix compartment, again raising the question of how newly synthesized proteins transit from the cytoplasm to the growing ciliary tip. Secondarily, these and the above data argue against any appreciable amount of axonemal protein appearing in the membrane/matrix fraction as a consequence of axonemal breakdown during isolation and detergent treatment.

In a final point of comparison, the relative degree of axonemal protein labeling in sea urchin embryos versus surf clam embryos is quite similar. For the experiments illustrated in Figures 11.2 and 11.3, the initial volume of embryos was essentially the same, as were the labeled and unlabeled leucine levels; only the time and growth temperature differed. The ^{3}H-leucine uptake, measured by depletion of counts from the culture medium, was 79 percent for the sea urchin culture but only 42 percent for the surf clam culture. The yield of cilia was about twice as great for sea urchin as for surf clam embryos. Reflecting this approximate fourfold difference in expected counts, the sea urchin fluorograms were exposed for 2 days while those for surf clams were exposed for 10 days, giving very similar intensities.

Induced Elongation of Cilia in Sea Urchin Embryos

Animalizing agents enhance ectodermal structures at the expense of entomesodermal structures (Lallier, 1975), typically resulting in the production of very long cilia, characteristic of the apical tuft, over most of the surface of the sea urchin embryo. Perhaps the most reliable of the animalizing agents are zinc ions which, when present from fertilization onward, arrest the embryo at the hatched blastula stage (Lallier, 1959). The Indo-Pacific sea urchin *Tripneustes gratilla* has been reported occasionally to yield unusually hyperciliated embryos but not in response to zinc ions (Burns, 1973). Recently, Riederer-Henderson (1988), using embryos of the sand dollar *Dendraster excentricus*, demonstrated that theophylline could be added to the growing repertoire of animalizing agents that can act after hatching. Because of its easily controlled dosage and application, theophylline was tested on *Tripneustes* embryos. Theophylline rapidly arrested development of hatched blastulae but it, like zinc ions alone, induced only a minimal long-cilia phenotype. Interestingly, when zinc-arrested blastulae (Fig. 11.4) were subsequently treated with theophylline at any time up until control embryos gastrulated, the uniformly short cilia elongated to approximately three times their normal length (Fig. 11.5) with kinetics exceeding ciliary regeneration and lacking the lag period characteristic of regeneration. After induction, hyperciliated embryos regenerated the long-cilia phenotype with kinetics similar to regeneration in noninduced embryos (Fig. 11.6). Thus *Tripneustes* embryos can provide a convenient inducible system with which to study ciliary length control (Stephens, 1989b).

This induction raised the question of whether theophylline brought about a replay of the subroutine of development described above. To test this, steady-state, zinc-arrested embryos with normal length cilia were pulse-chase labeled in parallel with zinc-arrested embryos treated with theophylline to induce the long-cilia phenotype. Since ciliary length increases approximately threefold, one might expect to see a threefold quantal production of tektin-A and the up-regulation of other architectural proteins over that seen in steady-state, nonregenerating control cilia. However, when examined at equal protein loadings, the pattern of incorporation of newly synthesized proteins into virgin control and induced hyperlong cilia was essentially the same (Fig. 11.7).

The high degree of incorporation of structural proteins into the cilia of zinc-arrested control embryos, averaging about two-thirds that of actively regenerating cilia (data not shown), is due to cell division, ciliary replacement, and an apparent dynamic exchange of ciliary proteins with the labeled cytoplasmic pool, as recently reported for a different sea urchin species (Stephens, 1991a,b). The proportionate degree of labeling of all components, including tektin-A, seen in cilia elongating during theophylline induction strictly parallels the steady-state (or regeneration) labeling pattern. The fact that these two populations of cilia are labeled to approximately the same degree is somewhat coincidental since two-thirds of the unlabeled proteins are replaced by labeled axonemal components through turnover in the control embryos while 40 μm of labeled axoneme is being added to 20 μm of existing, unlabeled axoneme in the induced embryos. In addition, the labeling of regenerated cilia obtained from these two embryo populations is likewise identical, suggesting proportionately labeled total pools (Fig. 11.7). That ciliary proteins are constantly being replenished in a proportionate manner is also consistent with the observation that steady-state levels of tektin-A mRNA are proportional to the length of cilia that an embryo has constructed (Norrander et al., 1988). Taken together, these data suggest that the embryo precisely controls the amounts of each protein needed to make (or replace) the required length of cilia and does so in the proper proportion with respect to the pool of that protein, regardless of how long the

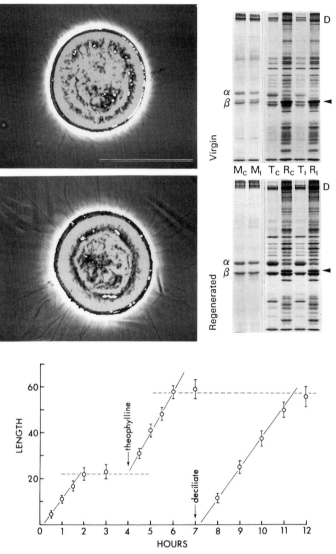

FIGURE 11.4 *(top left)*. Zinc-arrested blastula of *Tripneustes gratilla*. Phase-contrast video image of a swimming embryo, processed with a two-step gray scale to suppress the brightness of the blastomeres and enhance the contrast of the cilia (scale bar = 100 μm). FIGURE 11.5 *(middle left)*. Theophylline-induced hyperciliation of zinc-arrested blastula of *Tripneustes gratilla*. Imaging and magnification as in Figure 11.4. FIGURE 11.6 *(bottom)*. Regeneration kinetics of cilia from zinc-arrested blastulae and elongation kinetics of cilia from theophylline-induced blastulae of *Tripneustes gratilla*. Each point is the mean ± SD of more than twenty measurements; length in micrometers. FIGURE 11.7 *(top right)*. Incorporation of labeled proteins into ciliary axonemes from zinc-animalized *Tripneustes gratilla* embryos labeled during steady-state or theophylline-induced growth. One hour pulse and 1-h chase; 24°C. Upper set: Fluorograms of membrane/matrix, tubulin, and remnant fractions of virgin cilia from control (M$_C$, T$_C$, R$_C$) and induced (M$_I$, T$_I$, R$_I$) embryos. Lower set: Same for regenerated cilia. Equivalent protein loadings. Designations as in Fig. 11.2.

cilia are destined eventually to become. Put another way, the entire set of axonemal proteins is regulated on a per-unit-length of axoneme basis.

The protein synthetic subroutine for cilia formation in the form of elongation had to have been carried out in order to obtain these proportionate labeling results. This raised the question of whether theophylline induction of hyperlong cilia was accompanied by the production of new mRNA for tektin-A, not at all an unreasonable expectation in light of the *de novo* synthesis of tektin-A mRNA that accompanies ciliogenesis and regeneration (Norrander et al., 1988). Surprisingly, the answer was no. In collaboration with Drs. Jan Norrander and Richard Linck of the University of Minnesota, preliminary Northern blot analysis has revealed that a high level of tektin-A mRNA already exists in the zinc-arrested yet normally ciliated embryos and that this mRNA rapidly decreases coincident with ciliary elongation. The proportionate labeling of most other proteins would suggest similar behavior for the mRNAs encoding these proteins as well.

In *Tripneustes* embryos, zinc arrest is apparently accompanied by the accumulation of stable ciliary mRNAs, and the addition of theophylline somehow triggers their translation and degradation, an interpretation consistent with the more rapid growth kinetics and the lack of a lag period during elongation in comparison with normal regeneration. These embryos evidently have already run the subroutine for hyperciliation in terms of proportionate and quantal mRNA production, but they are blocked in terms of translational regulation. This situation is reminiscent of dormancy in brine shrimp embryos, where there is abundant mRNA but translation is blocked at initiation (Hoffmann and Hand, 1992). These observations for ciliary architectural proteins further reinforce the importance of posttranscriptional regulation in zinc-animalized sea urchin embryos, which, as noted above, also show an increased accumulation of tubulin mRNA (Harlow and Nemer, 1987b).

Questions for Future Study

These examples raise a number of intriguing questions concerning the regulation of ciliogenesis and ciliary length while also providing model systems with which to begin the search for answers.

Deciliation immediately triggers the production of new ciliary mRNAs and subsequently axonemal proteins. Increased transcription is in response to the loss of cilia, not to the hypertonic salt treatment, and can occur even when regeneration is prevented (Gong and Brandhorst, 1987). What is the primary signal that informs the nucleus that deciliation has taken place?

In the biflagellate alga *Chlamydomonas*, partial amputation of one flagellum triggers the partial resorption of its mate, back to an equivalent length, and then both regenerate together (Rosenbaum et al., 1969). As noted above, this cell has very limited pools of most of its flagellar components. Does a similar mechanism exist in metazoan embryos wherein the cell can somehow sense either the length of its cilium or the size of its precursor pool? Is the observation of high protein turnover in fully motile cilia (Stephens, 1991b) indicative of a feedback loop? In sea urchin or surf clam embryos, what happens to transcription and translation if partially regrown cilia are removed?

The induction work discussed above, plus evidence from normal and zinc-animalized embryos (Norrander et al., 1988), indicates that the amount of tektin-A mRNA, like the protein itself, is quantal with respect to the anticipated length of cilia. How are the precise amounts of mRNAs for quantal architectural components such as tektin-A regulated? What feedback mechanism inhibits transcription of the highly insoluble tektins? Since autoregulation involves the binding of excess product to a nascent polypeptide chain on the polysome, destabilizing the mRNA (Gong and Brandhorst, 1988), one would predict a soluble tektin-A pool and posttranslational modification before incorporation into the axoneme.

The architectural proteins and the major building blocks, tubulin and dynein, fill pools of quite different capacities in sea urchin embryos but very similar capacities in surf clam embryos. How are the synthesis of limited architectural components and the synthesis of nonlimiting tubulin and dynein differentially regulated in sea urchin embryos, and why are all of these proteins coordinately regulated in surf clam embryos? Although these organisms are on separate evolutionary lines, these differences may simply reflect larval strategies, and hence an investigation of lecithotrophic sea urchin larvae would now seem in order.

Positional information on the embryo dictates the length of the cilium that each blastomere produces, yet this can be overridden by a variety of animalizing agents. At what level (cell surface signaling, pool sequestering, transcriptional, translational, post-translational control, animal-vegetal gradient, etc.) do the various animalizing agents act?

Finally, with respect to the mechanism of ciliary assembly, how are newly synthesized proteins such as tektin-A targeted for quantitative assembly into the axoneme without first appearing in detectable amounts in the membrane plus periaxonemal matrix compartment while, at the same time, that compartment is dominated by tubulin destined for axonemal growth?

To be able to replace cilia efficiently while carrying on the program of development is of obvious survival value for actively swimming and feeding larvae. One thing is very clear; sea urchin embryos and larvae are exquisitely designed to cope with their environment in this respect. This unique ability thus provides an excellent means to approach a number of fundamental questions of signaling, coordinate gene expression, and autoregulation.

Acknowledgments

This work was supported by U.S. Public Health Service Grant GM 20,644 from the National Institute of General Medical Sciences. I thank Dr. Robert E. Kane and the staff of the Kewalo Marine Laboratory, University of Hawaii, for their help and hospitality over the years and Dr. Robert E. Palazzo for his assistance with surf clam reproductive biology.

Literature Cited

Alexandraki, D., and J. V. Ruderman. 1985. Expression of α- and β-tubulin genes during development of sea urchin embryos. Dev. Biol. 109: 436–451.

Amemiya, S. 1971. Relationship between cilia formation and cell association in sea urchin embryos. Exp. Cell Res. 64: 227–230.

Auclair, W., and B. W. Siegel. 1966. Cilia regeneration in the sea urchin embryo: Evidence for a pool of ciliary proteins. Science 154: 913–915.

Bibring, T., and J. Baxandall. 1981. Tubulin synthesis in sea urchin embryos. II. Ciliary A tubulin derives from the unfertilized egg. Dev. Biol. 83: 122–126.

Burns, R. G. 1973. Kinetics of regeneration of sea urchin cilia. J. Cell Sci. 13: 55–67.

———1979. Kinetics of regeneration of sea urchin cilia. II. Regeneration of animalized cilia. J. Cell Sci. 37: 205–215.

Foltz, K. R., and D. J. Asai. 1990. Molecular cloning and expression of sea urchin embryonic ciliary dynein β heavy chain. Cell Motil. Cytoskel. 16: 33–46.

Gong, Z., and B. P. Brandhorst. 1987. Stimulation of tubulin gene transcription by deciliation of sea urchin embryos. Molec. Cell. Biol. 7: 4238–4246.

———1988. Autogenous regulation of tubulin synthesis via RNA stability during sea urchin embryogenesis. Development 102: 31–43.

Harlow, P., and M. Nemer. 1987a. Developmental and tissue specific regulation of β-tubulin gene expression in the embryo of the sea urchin *Strongylocentrotus purpuratus*. Genes Dev. 1: 147–160.

———1987b. Coordinate and selective β-tubulin gene expression associated with cilium formation in sea urchin embryos. Genes Dev. 1: 1293–1304.

Hoffmann, G. E., and S. C. Hand. 1992. Comparison of messenger RNA pools in active and dormant *Artemia franciscana* embryos: Evidence for translational control. J. Exp. Biol. 164: 103–116.

Lallier, R. 1959. Recherches sur l'animalisation de l'oeuf d'Oursin par les ions Zinc. J. Exp. Morph. 7: 540–548.

————1975. Animalization and vegetalization. *In* G. Czihak, ed. The Sea Urchin Embryo—Biochemistry and Morphogenesis. Springer-Verlag, New York, pp. 473–509.

Lepage, T., and C. Gache. 1990. Early expression of a collagenase-like hatching enzyme gene in the sea urchin embryo. EMBO J. 9: 3003–3012.

Linck, R. W. and G. L. Langevin. 1982. Structure and chemical composition of insoluble filamentous components of sperm flagellar microtubules. J. Cell Sci. 58: 1–22.

Linck, R. W., and R. E. Stephens. 1987. Biochemical characterization of tektins from sperm flagellar doublet microtubules. J. Cell Biol. 104: 1069–1075.

Masuda, M. 1979. Species specific pattern of ciliogenesis in developing sea urchin embryos. Dev. Growth Differ. 21: 545–552.

Masuda, M., and H. Sato. 1984. Reversible resorption of cilia and the centriole cycle in dividing cells of sea urchin blastulae. Zool. Sci. 1: 445–462.

Norrander, J. M., R. E. Stephens, and R. W. Linck. 1988. Levels of ciliary 55kD tektin mRNA during sea urchin development. J. Cell Biol. 107: 20a.

Palazzo, R. E., E. Vaisberg, R. W. Cole, and C. L. Rieder. 1992. Centriole duplication in lysates of *Spisula solidissima* oocytes. Science 256: 219–221.

Remillard, S. P., and G. B. Witman. 1982. Synthesis, transport, and utilization of specific flagellar proteins during flagellar regeneration in *Chlamydomonas*. J. Cell Biol. 93: 615–631.

Riederer-Henderson, M.A. 1986. Effects of theophylline on expression of the long cilia phenotype in sand dollar blastulae. J. Exp. Zool. 246: 17–22.

Rosenbaum, J. L., J. E. Moulder, and D. L. Ringo. 1969. Flagellar shortening and elongation in *Chlamydomonas*. J. Cell Biol. 41: 600–619.

Spiegel, E., and L. Howard. 1983. Development of cell junctions in sea-urchin embryos. J. Cell Sci. 62: 27–48.

Stephens, R. E. 1972. Studies on the development of the sea urchin *Strongylocentrotus droebachiensis*. III. Embryonic synthesis of ciliary proteins. Biol. Bull. 142: 489–504.

————1977. Differential protein synthesis and utilization during cilia formation in sea urchin embryos. Dev. Biol. 61: 311–329.

————1986. Isolation of embryonic cilia and sperm flagella. Meth. Cell Biol. 27: 217–227.

————1989a. Quantal tektin synthesis and ciliary length in sea urchin embryos. J. Cell Sci. 92: 403–413.

————1989b. Additive induction of long cilia phenotypes in zinc-animalized sea urchin embryos by theophylline. J. Cell Biol. 109: 176a.

————1991a. Tubulin in sea urchin embryonic cilia: Characterization of the membrane-periaxonemal matrix. J. Cell Sci. 100: 521–531.

————1991b. Sea urchin embryonic cilia exist in a dynamic equilibrium. J. Cell Biol. 115: 343a.

————1992. Tubulin in sea urchin embryonic cilia: Post-translational modifications during regeneration. J. Cell Sci. 101: 836–845.

Stephens, R. E., and G. Prior. 1991. Tektins from *Spisula solidissima* cilia. Biol. Bull. 181: 339–340.

Stephens, R. E., S. Oleszko-Szuts, and R. W. Linck. 1989. Retention of ciliary nine-fold structure after removal of microtubules. J. Cell Sci. 92: 391–402.

Part III

Larval Morphology

and Evolution

12 Functional Consequences of Simple Cilia in the Mitraria of Oweniids (an Anomalous Larva of an Anomalous Polychaete) and Comparisons with Other Larvae

Richard B. Emlet and Richard R. Strathmann

ABSTRACT The mitraria larvae of oweniid polychaetes have simple cilia in both the prototroch and metatroch, in contrast to other larvae with an opposed-band feeding apparatus. Observations of the paths of particles revealed no qualitative differences between capture of particles by the mitraria and capture of particles by other larvae with the opposed-band apparatus. In contrast, the simple prototrochal cilia of the mitraria are shorter than the compound prototrochal cilia of annelid trochophores and molluscan veligers of comparable or even much smaller body size than the mitraria. The simple prototrochal cilia of the mitraria produce a slower current than the longer, compound prototrochal cilia of other larval forms. The short simple cilia of the mitraria thus limit the rate at which a unit length of ciliary band clears food particles from suspension. This limitation may account for several peculiar features of the mitraria. Development of a long and sinuous ciliary band may have evolved in the mitraria because, in the absence of longer and faster prototrochal cilia, high maximum clearance rates depend largely on the length of the ciliary band. The mitraria's very thin body wall and large body cavity provide a large surface for distribution of a long and sinuous ciliary band at a low cost in ephemeral larval structures that are destroyed upon metamorphosis. These compensations for the restrictions associated with simple cilia provide a starting point for Wilson's (1932) hypothesis on the evolution of the cataclysmic metamorphosis of the mitraria. The thin body wall and sinuous ciliary band of the mitraria superficially resemble the body wall and ciliary band in the auricularia and bipinnaria, echinoderm larval forms, which are subject to similar functional restrictions associated with simple cilia but possess a very different feeding mechanism from that used by the mitraria.

Introduction

The mitraria larvae of oweniid polychaetes are unusual in form, metamorphosis (Wilson, 1932), and nephridia (Smith et al., 1987). Another unusual feature is a prototroch composed of simple cilia (Figs. 12.1a,d, 12.2, and 12.3). Simple prototrochal cilia occur in mitraria larvae in the genus *Owenia* (Wilson, 1932; Smith et al., 1987) and in the unidentified mitraria larvae that we have obtained from the plankton. The reason that these larvae have simple rather than compound prototrochal cilia is unclear. Worms of the genus *Owenia* have only monociliated cells, a condi-

Richard B. Emlet, Department of Biological Sciences, University of Southern California, Los Angeles, CA 90089–0371; present address: Oregon Institute of Marine Biology, Charleston, OR 97420.
Richard R. Strathmann, Friday Harbor Laboratories and Department of Zoology, University of Washington, 620 University Road, Friday Harbor, WA 98250.

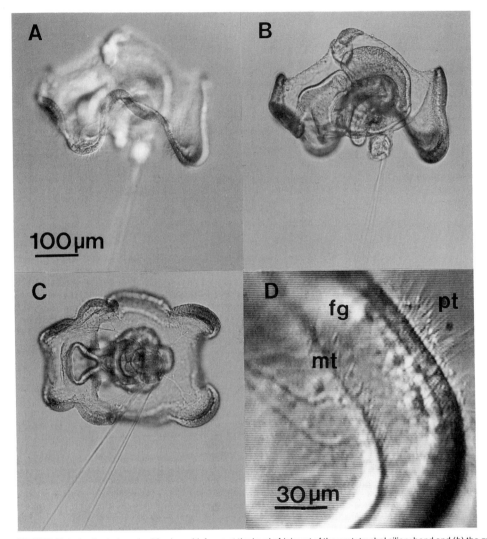

FIGURE 12.1. A mitraria larva in side view with focus at the level of (a) part of the prototrochal ciliary band and (b) the gut, and (c) a posterior view that shows the prototrochal ridge and less prominent metatrochal ridge separated by the food groove. a, b, and c are to the same scale. A video image at higher magnification (d) shows the prototroch (pt), metatroch (mt), and ciliated food groove (fg).

tion apparently unique among polychaetes (Gardiner, 1978). It is tempting to attribute the simple prototrochal cilia to a restriction in oweniids to monociliated cells; however, not all oweniid adults or mitraria larvae possess only monociliated cells (Smith et al., 1987; S. L. Gardiner, unpub. obs.). Also, compound cilia might be formed from the cilia of monociliated cells, as in the posterior locomotory ring of cilia of the actinotroch larva of phoronids (Nielsen, 1987).

Whatever the cause of simple prototrochal cilia in the mitraria, the condition is predicted to limit the amount of water that passes a unit length of ciliary band. Simple prototrochal cilia are more constrained than compound cilia in the lengths and velocities that can be attained (Sleigh, 1962). A limitation on water moved per unit length of ciliary band may account for other unusual features of the larva.

To assess the functional implications of simple prototrochal cilia for the mitraria, we have

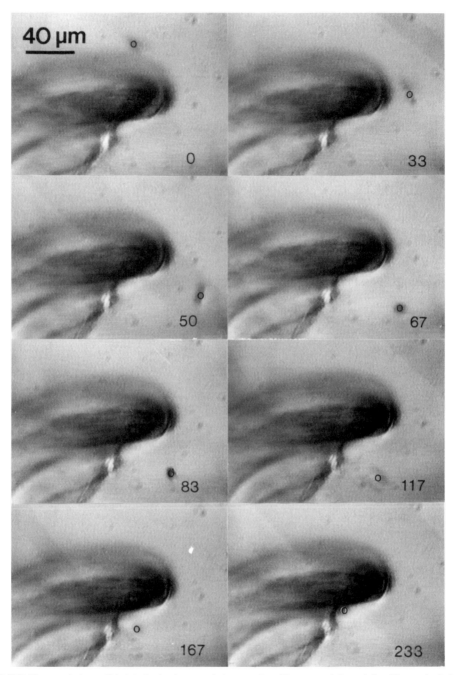

FIGURE 12.2. Movement of a particle into the food groove during a capture. Time on each frame is in milliseconds. A circle marks the particle.

measured ciliary length in the prototroch and movement of suspended particles by the prototroch. We compared the values with quantitative observations for (1) larvae with opposed-band feeding mechanisms (like the mitraria) but with compound cilia and (2) larvae with upstream capture of food particles (unlike the mitraria) and with simple cilia. We also recorded captures of particles by a mitraria for comparison with the more usual opposed-band feeding apparatus.

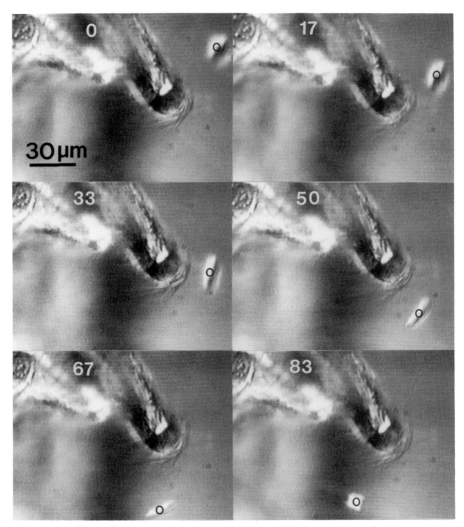

FIGURE 12.3. Movement of a particle through the prototroch without a capture. Time on each frame is in milliseconds. A circle marks the particle.

These observations and comparisons are the basis for a discussion of the implications of simple prototrochal cilia as a constraint that may have influenced evolution of other unusual features of the mitraria: the form of the band, the form of the larva, and the cataclysmic metamorphosis. This discussion provides a plausible explanation for the unusual form of the mitraria and thus fills a gap that Wilson (1932) recognized in his reasoning about the evolution of the cataclysmic metamorphosis of the mitraria. We also note convergent features shared by the mitraria and some other larval forms with simple cilia.

Some background information is necessary for an appreciation of the comparison of ciliary bands of larvae and the unusual features of the mitraria. The requirement for particulate food places functional constraints on the body plans of suspension-feeding invertebrate larvae (Strathmann, 1978, 1987; Emlet, 1991). Concentrations of food are often low enough to limit rates of growth (Strathmann, 1987; Olson and Olson, 1989). These observations suggest that high maximum clearance rates (the volume of water cleared of food particles per unit time) are often needed

Table 12.1. Types and lengths of cilia in ciliary bands that produce feeding currents for upstream or opposed-band mechanisms of particle capture

Taxon	Location of cilia	Simple or compound	Length (μm)	Reference[a]
Echinoderm larvae				
Dendraster excentricus	Feeding band	s	20–25	1
Lytechinus anamesus	Feeding band	s	20–25	2
Enteropneust larvae				
Ptychodera flava	Feeding band	s	20–25	3
Bryozoan larvae				
Membranipora sp.	Lateral feeding cilia	s	20	4
Polychaete larvae				
Mitraria larva	Prototroch	s	34	6
Galeolaria caespitosa	Prototroch	c	60	5
Spirobranchus spinosus	Prototroch	c	30–40	2
Molluscs, gastropod larvae				
Charonia tritonis	Prototroch	c	100	2
Crepidula fornicata	Prototroch	c	43	7
Ilyanassa obsoleta	Prototroch	c	70	8
Jorunna sp.	Prototroch	c	46	9
Lacuna sp.	Prototroch	c	150	10
Philine aperta (early stage)	Prototroch	c	10–11	11
Philine aperta (late stage)	Prototroch	c	55–60	11
Tritonia diomedia	Prototroch	c	40	8
Molluscs, bivalve larvae				
Crassotrea gigas	Prototroch	c	30	8
C. gigas (late stage)	Prototroch	c	91	12
Mercenaria mercenaria	Prototroch	c	48	13
Adult rotifer				
Philodina sp.	Prototroch	c	15	2

[a]1, McEdward (1984); 2, Strathmann et al. (1972); 3, Strathmann and Bonar (1976); 4, McEdward and Strathmann (1987); 5, Marsden and Anderson (1981); 6, this study; 7, Werner (1955); 8, Strathmann and Leise (1979); 9, Sleigh (1968); 10, Emlet (unpublished); 11, Hansen and Ockelman (1991); 12, Emlet (1990); 13, Gallager (1988).

to achieve high growth rates. One way to achieve higher clearance rates is to move more water. In the evolution of larval forms, there have been several ways that greater movement of water has been achieved. For ciliated larvae, placement of bands of cilia on ridges on the body increases the volume of fluid moved per stroke (Emlet, 1991). Other ways of achieving greater movement of water are development of more cilia or longer cilia (Strathmann et al., 1972). Mechanical limits to velocities and stiffness are encountered as a single cilium is increased in length, and an especially effective way to circumvent these limits is the bundling of simple cilia into compound cilia (Sleigh, 1962). A longer ciliary band, rather than longer cilia, is yet another way to increase clearance rates, and in the absence of compound cilia it may be the only way to obtain substantial increases.

Different mechanisms of capturing particles are found in the various invertebrate phyla. The opposed-band apparatus is found in larvae of several spiralian phyla (molluscan veligers, echiuran

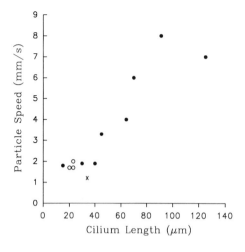

FIGURE 12.4. Comparison of reported maximum speeds of particles (mm/s) through bands of cilia with cilia of different lengths (μm). All data are for ciliary bands that produce feeding currents: × = prototroch of the mitraria larva, filled circles = prototrochs with compound cilia (and trochus of a rotifer), open circles = bands of simple cilia with upstream particle capture. Data from Strathmann et al. (1972), Strathmann and Leise (1979), Emlet (1983), McEdward and Strathmann (1987), Gallager (1988), Emlet (1990), unpublished observations of Emlet on *Ilyanassa obsoleta* and *Lacuna* sp., and Figures 12.7 and 12.8.

and entoproct trochophores, and many of the annelid trochophores) as well as in many rotifers (Strathmann, 1978; Nielsen, 1987). The opposed-band apparatus consists of (1) a band of prototrochal cilia that beat posteriorly and create a current for feeding and swimming and (2) a band of shorter metatrochal cilia that beat anteriorly toward the prototroch. Between these parallel ciliary bands lies a food groove, with cilia that beat toward the mouth and orthogonal to the cilia of prototroch and metatroch (Strathmann, 1987). The mouth is positioned on the ventral midline within the food groove. The term metatroch is applied here solely to the band of cilia that lies posterior to the food groove and beats towards the prototroch. This definition contrasts with the definition of Wilson (1932) that also included the cilia of the food groove in this band. With the opposed-band feeding mechanism, particles are retained between the prototroch and metatroch and passed along the food groove to the mouth. In larvae that have this feeding apparatus (with the exception of the mitraria), the cilia of the prototroch are compound cilia, and the cilia may reach great lengths. Lengths of compound prototrochal cilia range from about 10 μm to more than 100 μm (Table 12.1) and increase during development of an individual larva (Strathmann and Leise, 1979; Nielsen 1987; Gallager, 1988; Hansen and Ockelman, 1991). Longer prototrochal cilia are associated with greater angular velocities, faster movement of water, or greater clearance of particles in the few cases in which one or more of these parameters have been measured (Strathmann and Leise, 1979; Gallager, 1988; Hansen and Ockelman, 1991; Fig. 12.4) Thus clearance rates with the opposed-band apparatus can be enhanced by increases in the length and velocity of the prototrochal cilia as well as by increases in the length of the prototrochal band (Strathmann et al., 1972).

In contrast, particles are captured upstream from a band of simple cilia in the larvae of enteropneust hemichordates, echinoderms, and many lophophorates (Strathmann, 1978; Nielsen, 1987). In all of these cases, the band that produces the feeding current is composed of simple cilia, usually about 20 μm in length and never much longer (Strathmann, 1987). The number of cells per unit length of ciliary band and clearance rates per unit length of band do vary among these larval forms (M. Hart, University of Washington, unpub. obs.). However, major increases during development in the rate at which particles can be cleared from suspension are achieved only by increases in the length of the ciliary band (Strathmann, 1971; Hart, 1991).

The mitraria, with an opposed-band feeding apparatus but simple cilia in the prototroch, shares some features of two widespread larval feeding mechanisms. Therefore the mitraria can be

Table 12.2. Dimensions for
mitraria larvae

Larva	Body length (μm)	Prototrochal cilium length (μm)
1	326	33
2	396	33
3[a]	511	35
4	588	32
5	686	36
6	784	33

[a]Subject of videotape.

used to test predictions about the different evolutionary constraints and opportunities that are usually associated with these two feeding mechanisms. If, as predicted, the type of cilium limits length and velocity of the cilia, then velocity of feeding currents of the mitraria and lengths of its prototrochal cilia should resemble the currents and cilia of larvae with upstream particle capture. The mitraria would be expected to increase the length of the ciliary band during development rather than to increase the length of cilia. In contrast, if ciliary lengths and ciliary velocities of the mitraria were inconsistent with these predicted limits, then possession of simple cilia could be rejected as a reason for peculiarities of the form and length of the mitraria's ciliary band. Moreover, convergent similarities between the mitraria and larvae with upstream capture could not be attributed to the shared trait of simple cilia.

Materials and Methods

Mitraria larvae 1 to 3 in Table 12.2 were collected from San Juan Channel in the San Juan Islands of Washington State by oblique and horizontal tows of a 125-μm mesh net in May and August of 1991. Additional mitraria larvae (4 to 6 in Table 12.2) were obtained in May of 1992 in an oblique tow near Catalina Island, California. Mitraria larvae sorted from the live plankton were placed in glass bowls with sea water that had been filtered through a bag filter of less than 10 μm mesh. The bowls were cooled by surrounding aquarium water (12° to 14°C at San Juan Island and 17°C at Catalina Island) and held several days before observation.

Velocities near the ciliary bands and movements of cilia were determined for larva 3 of Table 12.2 in a chamber constructed from a circular ring of about 1 cm diameter and 0.5 cm height glued to a microscope slide. The chamber was filled with sea water at room temperature (about 17°C). Currents were indicated by plastic spheres of 10 μm diameter, cells of about 12 μm length by 5 μm width of a *Rhodomonas* species, and particles naturally occurring in the sea water. A coverslip was added, and air pockets were excluded to create a stationary viewing surface.

Observations were on untethered larvae that swam near the side wall or the coverslip of the chamber. Drag associated with walls could have reduced current velocities, but we concluded that wall effects were insufficient to affect our comparisons among larval forms for the following three reasons:. (1) The differences in speeds of currents passing through ciliary bands of different larvae are very great. (2) The published studies include data from larvae observed under diverse conditions: for example, in smaller chambers than we used here, on tethers but farther from walls, on tethers and in a flow approximating the swimming speed. Yet, in all studies, particle speeds through the band are related to length and type of cilia in a regular pattern (Fig. 12.4). (3) Speeds of particles passing through the ciliary band are less affected by tethering than speeds of particles beyond the reach of cilia (Emlet, 1990); we expect the same to be true for the effects of the chamber's wall on our observations.

Motion of particles and cilia was recorded for a larva with a 60-Hz, charge-coupled device (CCD) videocamera attached to the phototube of a compound microscope fitted with optics for differential interference contrast (DIC). Recordings were made in two sessions lasting 26 min and 16 min. Videotapes were analyzed with a two-dimensional motion analyzer (Motion Analysis Corp.) that automatically tracked particles past the ciliary band. Distances of particles from the body surface were measured for particles that passed through the region swept by prototrochal cilia and were captured. Velocities and distances from the base of the prototrochal cilia were measured for particles that passed in the plane of the beat of prototrochal cilia but were not captured. Velocities of particles transported along the food groove were also measured. The figures of video images are photographs of the video monitor and contain distortions of scale not included in the data from the videotapes. Thus, the scale lines for these figures are necessarily approximations.

Comparisons between the mitraria and published data for other larvae were augmented with observations on laboratory-reared larvae of the echinoid *Dendraster excentricus*. Plutei of *D. excentricus* were tethered by a suction micropipet in an observation chamber at 17°C. Motion of 2-μm plastic spheres was recorded with a high speed (200-Hz) videocamera (NAC Corp.). Velocity profiles were constructed with the same motion analysis system and additional software.

Lengths of two to four prototrochal cilia and body length were measured for each of five additional mitraria larvae (Table 12.2). Larvae 1 and 2 were from San Juan Channel, and larvae 4 to 6 from the waters near Catalina Island. The body sizes were measured from the ventral edge of the prototroch to the dorsal edge of the prototroch, the same designation of dorso-ventral axis (Fig. 12.1c) that Wilson (1932) used.

Results

The simple cilia of the mitraria's opposed bands moved suspended particles into the food groove (Figs. 12.2, 12.3, 12.5, and 12.6). Prototrochal cilia showed no response to particles passing within the region swept by the effective strokes. Some particles that passed through the region swept by the prototrochal cilia were carried into the food groove (Figs. 12.2 and 12.5) whereas others that passed through the same region at about the same distance from the larva were not conveyed to the food groove (Fig. 12.3). The paths of particles thus resembled paths of particles for other suspension feeding larvae with opposed prototrochal and metatrochal bands of cilia.

The lengths of the prototrochal cilia of the mitraria did not appear to change during the growth of the stages in Table 12.2. Mean lengths of prototrochal cilia were 32 to 36 μm for six different mitraria larvae, ranging in length from 326 μm to 784 μm (Table 12.2). Only two to four cilia were measured per larva, but no great differences in length were evident between larvae. These observations of short cilia and little change in length of cilia during development agree with the predictions for simple cilia.

For larva 3 of Table 12.2, the height of the prototrochal recovery stroke height was 9 μm, metatrochal ciliary length was 22 μm, and the width of the food groove was 33–40 μm. A similar estimate of 8 μm for prototrochal recovery stroke height was obtained for larva 6.

Peak velocity was less than 1250 μm/s for particles passing through the prototrochal cilia at a distance of about two-thirds of a cilium length from the base of the cilia (Fig. 12.7). Particles that

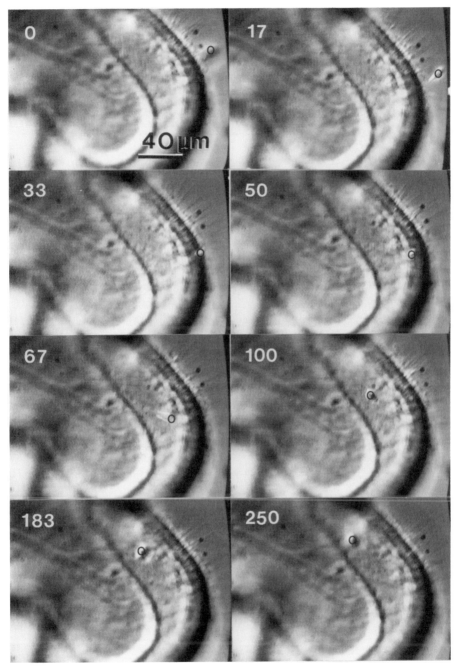

FIGURE 12.5. Movement of a particle into the food groove during a capture. Time on each frame is in milliseconds. A circle marks the particle.

were captured passed through the prototroch at distances of 12 to 30 μm from the base of the cilia (Fig. 12.7).

We confirmed that particle velocities were similar to those observed for larvae that capture particles upstream from bands of simple cilia by new observations at the same temperature on particle velocities through the ciliary band on the vibratile lobes of a pluteus (Fig. 12.8).

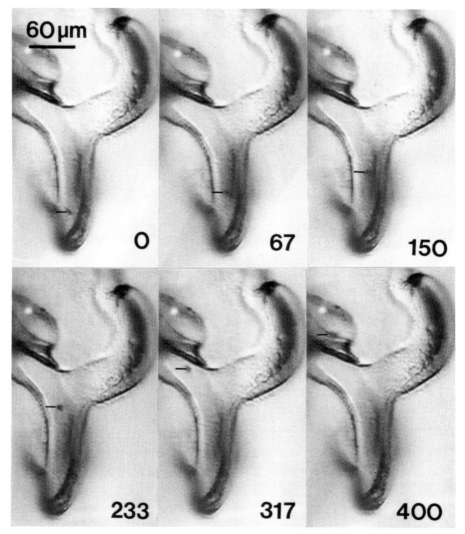

FIGURE 12.6. Transport of a particle along the food groove and into the mouth. Time on each frame is in milliseconds. A horizontal line to the left of the particle marks its position. Two food grooves lead to the mouth on each side of the larva; only the proximal parts of food grooves on one side of the larva are shown.

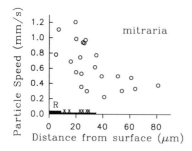

FIGURE 12.7. Velocity profile for the prototrochal ciliary band of a mitraria. Open circles are particles moving by the proto-trochal cilia in the plane of beat. X's are distances from the base of the cilia that particles passed through the ciliary band and were later caught. R identifies the height of the recovery stroke.

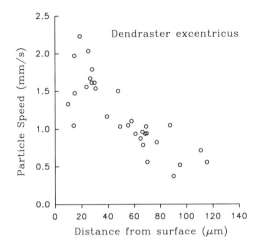

FIGURE 12.8. Velocity profile for a pluteus of *Dendraster excentricus*. Open circles are particles moving past a vibratile lobe located on the ventral surface near the base of a postoral arm.

Particles were transported along the food groove to the mouth as in other larvae with the opposed-band ciliary apparatus (Fig. 12.6). Particles in the food groove moved toward the mouth at speeds of 100 to 250 μm/s ($N = 3$) in the sinuous regions of the band distant from the mouth and increased speed to 250 to 400 μm/s ($N = 4$) in parts of the food groove near the mouth.

In contrast to most other polychaete larvae with the opposed-band feeding mechanism, as the mitraria grows (1) the larval body expands much more in the transverse than in the anterior-posterior axis and (2) the parallel bands of the prototroch, food groove, and metatroch form sinuous curves instead of a simple equatorial ring (Fig. 12.1a,c) (Wilson, 1932). Both of these features of larval growth produce ciliary bands that are long relative to the rest of the body wall. In the lateral projection of the prototroch in Fig. 12.1a, the sinuous band is 2.1 times the dorso-ventral axis (greatest body length) because of departures from the equatorial plane. The bends add more than 0.92 mm to the total length of the prototroch of a larva that is 0.42 mm long.

Discussion

The paths of particles captured by the mitraria are similar to those observed for polychaete and mollusc larvae with compound cilia (Strathmann et al., 1972; Strathmann and Leise, 1979; Gallager, 1988). It is possible that similar particle paths can result from quite different physical mechanisms of particle capture (Strathmann and McEdward, 1986). Also, Gallager and Langdon's (1988) experiments with a bivalve veliger indicate capture of particles by more than one physical mechanism. With these qualifications, we can say that the unique occurrence of simple cilia in the mitraria does not appear to change the mechanisms of feeding with opposed bands of cilia.

The simple prototrochal cilia of the mitraria were predicted to be constrained in length of cilium and in speed of particles passing through the band (and therefore constrained in maximum clearance rate per unit length of band). These predictions were confirmed.

The prototrochal cilia of the mitraria are shorter than the prototrochal cilia of most animals with the opposed-band feeding mechanism (Table 12.1), as was predicted from the greater limitations on simple cilia than compound cilia (Sleigh, 1962). The contrast becomes even more striking when one considers that the body sizes of most of the listed larvae of polychaetes and molluscs are less than the size of the mitraria. The few animals listed with compound prototrochal cilia that are shorter than the simple prototrochal cilia of the mitraria are much smaller in body size than the mitraria. These are the rotifers and the small early-stage veligers of bivalves and opisthobranchs (Table 12.1).

The constant length of the simple prototrochal cilia during growth of the mitraria (Table 12.2) contrasts with the increases in length of compound prototrochal cilia that have been noted during growth of other larvae with an opposed-band feeding mechanism (Gallager, 1988; Hansen, 1991; unpublished observations of R. B. Emlet on a *Lacuna* sp.; B. Rivest and R. R. Strathmann on *Ilyanassa obsoleta*). Developmental increases in lengths of prototrochal cilia are best documented for molluscan veligers, however. Data for polychaete larvae are scarce. Wilson's (1929) figures of *Sabellaria spinosa* suggest that lengths of prototrochal cilia increase during growth of some polychaete larvae with the opposed-band mechanism, but little increase is indicated by his figures of prototrochal cilia of *Sabellaria alveolata*. It is possible that Wilson may not have drawn the cilia to exact scale in his semidiagrammatic figures.

The simple cilia of the prototroch of the mitraria fall within the range of lengths of simple cilia in ciliary bands of other larval forms. They are longer than the simple cilia of about 20 to 25 μm length that are used in upstream capture of particles in the lophophorates, enteropneusts, and echinoderms (Table 12.1) but are not longer than simple cilia with a solely locomotory role in these other phyla, such as the 35-μm-long cilia of the epaulettes of echinoplutei (Strathmann, 1971) or the locomotory coronal cilia of about 55 μm of the advanced cyphonautes larvae of bryozoans (McEdward and Strathmann, 1987). These comparisons suggest that upstream capture of particles may involve constraints on ciliary lengths beyond the general constraints on simple cilia. The rapid reversal of beat required for upstream capture (Strathmann et al., 1972) may impose more severe constraints on lengths of cilia used in upstream capture , but there have been no theoretical calculations on such constraints and no experimental tests of hypotheses.

The maximum velocities of particles passing through the prototroch of a mitraria larva was indeed lower than those reported for particles passing through prototrochs composed of compound cilia in other larvae and in rotifers that have the opposed-band feeding apparatus (Fig. 12.4). One complication in a comparison of maximum particle speeds through prototrochs is that particles passing through the prototroch nearer to the tips of the cilia may not be caught. The capture zone varies among species and perhaps with ciliary length. Nevertheless, longer compound prototrochal cilia are associated with higher clearance rates per unit length of ciliary band (Strathmann and Leise, 1979; Gallager, 1988). The maximum particle speeds for the mitraria are similar to those observed for larvae with simple bands of cilia but with upstream capture of particles (Fig. 12.4).

Thus the quantitative comparisons confirm the predicted constraints on performance of simple prototrochal cilia. Possible consequences of these constraints for other features of development of the mitraria can now be examined. Wilson (1932) attempted to explain in functional terms the cataclysmic metamorphosis of the mitraria, in which most of the larval body rapidly disintegrates and is ingested by the juvenile worm. The large and flat episphere and hyposphere of the larval body is awkward baggage for a juvenile worm, and Wilson (1932) attributed the cataclysmic metamorphosis to the advantages of a rapid transition to benthic life. He lacked an explanation of the large larval body but had considerable insight into the requirements for a functional explanation: "Let us assume that for some unknown reason the larva was induced to increase the size of its prototroch. One is tempted to suggest that a premium was put on increased swimming powers until one reflects that such an increase would be more effectively brought about by an advance in power of the cilia (those of *Owenia* are very weak) and that an enlargement of the prototroch adds to water resistance. However, for some obscure reason the prototroch was enlarged, while the head and trunk remained about the same size." Functional restrictions associated with simple prototrochal cilia may be the reason that was obscure for Wilson. If a higher clearance rate per unit length of ciliary band cannot be achieved by increasing the length of the cilia or speed of currents through the band, then the remaining option is to increase the length of the ciliary band. A longer ciliary band requires a larger larval body, which is then rapidly diminished at metamorphosis. The simple cilia in the prototroch may provide the starting point that Wilson lacked. We would now change Wilson's explanation of body form and metamorphosis to begin with the statement "For some obscure reason the prototroch of the mitraria is composed of simple cilia. To increase the maximum clearance rate, the prototroch was therefore elongated with concomitant expansion of the larval body . . . " and continue with Wilson's argument from there. The reasons for the simple cilia are still unclear.

The mitraria, with a sinuous band, thin epidermis, and spacious body cavity, superficially resembles the auricularia or bipinnaria of echinoderms. The resemblance is especially striking in the posterior view (Fig. 12.1c). The anteroventral extension of prototroch, metatroch, and food groove adds to the similarity. This topological peculiarity of the mitraria raises questions about the lineage, movement, and fates of cells that form the anteroventral extension of the ciliary bands. The similarity between the mitraria and the auricularia or bipinnaria could have a common functional basis. The bands that produce the feeding currents are composed of simple cilia in both forms. Because the simple cilia limit the rate at which water moves through the ciliary band, high clearance rates are achieved by deployment of a longer ciliary band, which requires a large larval body. Investment in a transitory supporting larval body wall is minimized by development of a sinuous band and by development of a large body cavity bounded by a thin epidermis.

There are other ways to increase the length of ciliary bands with little investment in temporary larval structures. In some larvae loops of the band are found on narrow projections of the body. For example, many of the larger larvae of prosobranch gastropods have long and narrow velar lobes. The rostraria, probably the larva of amphinomid polychaetes (Schroeder and Hermans, 1975), also deploys its prototroch on long narrow lobes (Jägersten, 1972). Among the echinoderm larvae the echinoplutei, the ophioplutei, and some advanced-stage bipinnariae also extend the ciliary band on narrow arms or lobes. Narrow projecting lobes of the body are one solution to the requirement for an extended band length, and a sinuous band that loops around a more compactly shaped body is another. Reasons why the latter evolved in the mitraria, the auricularia, and early stages of the bipinnaria are as yet unclear. Convergence on this means of achieving high clearance rates may be an accident of history. For different reasons, the two kinds of larvae were limited to simple cilia. This common limitation resulted in evolution of long ciliary bands for high clearance rates for similar functional reasons in both kinds of larvae. In both kinds of larvae a long band was achieved by sinuous curves distributed on an inflated larval body rather than by narrow loops on thin projecting lobes.

Acknowledgments

This research was supported by National Science Foundation (NSF) grant BSR9058139 and equipment funds from the University of Southern California to R. Emlet, NSF grant OCE8922659 to R. Strathmann, and facilities at the Friday Harbor Laboratories. We are grateful for discussions of metamorphosis with C. G. Reed and D. P. Wilson and for editorial comments from C. O. Hermans, W. H. Wilson, Jr., and an anonymous reviewer.

Literature Cited

Emlet, R. B. 1983. Locomotion, drag, and the rigid skeleton of larval echinoderms. Biol. Bull. 164: 433–445.

———1990. Flow fields around ciliated larvae: The effects of natural and artificial tethers. Mar. Ecol. Prog. Ser. 63: 211–225.

———1991. Functional constraints on the evolution of larval forms of marine invertebrate larvae: Experimental and comparative evidence. Am. Zool. 31: 707–725.

Gallager, S. M. 1988. Visual observations of particle manipulation during feeding in larvae of a bivalve mollusc. Bull. Mar. Sci. 43: 344–365.

Gallager, S. M., and C. J. Langdon 1988. High-speed video analysis of particle capture by a ciliated suspension feeder. Eos 69: 1086. (Abstract)

Gardiner, S. L. 1978. Fine structure of the ciliated epidermis on the tentacles of *Owenia fusiformis* (Polychaeta: Oweniidae). Zoomorph. 91: 37–48.

Hansen, B. 1991. Feeding behaviour in larvae of the opisthobranch *Philine aperta*. II. Food size spectra and particle selectivity in relation to larval behaviour and morphology of the velar structures. Mar. Biol. 111: 263–270.

Hansen, B., and K. W. Ockelman. 1991. Feeding behaviour in larvae of the opisthobranch *Philine aperta*. I. Growth and functional response at different developmental stages. Mar. Biol. 111: 255–261.

Hart, M. W. 1991. Particle captures and the method of suspension feeding by echinoderm larvae. Biol. Bull. 180: 12–27.

Jägersten, G. 1972. Evolution of the Metazoan Life Cycle. Academic Press, New York.

Marsden, J. R., and D. T. Anderson. 1981. Larval development and metamorphosis of the serpulid polychaete *Galeolaria caespitosa* Lamarck. Aust. J. Mar. Freshwat. Res. 32: 667–680.

McEdward, L. R. 1984. Morphometric and metabolic analysis of the growth and form of an echinopluteus. J. Exp. Mar. Biol. Ecol. 82: 259–287.

McEdward, L. R., and R. R. Strathmann. 1987. The body plan of the cyphonautes larva of bryozoans prevents high clearance rates: Comparisons with the pluteus and a growth model. Biol. Bull. 172: 30–45.

Nielsen, C. 1987. Structure and function of metazoan ciliary bands and their phylogenetic significance. Acta Zool. 68: 205–262.

Olson, R. R., and M. H. Olson. 1989. Food limitation of planktotrophic marine invertebrate larvae: Does it control recruitment success? Ann. Rev. Ecol. Syst. 20: 225–247.

Schroeder, P. C., and C. O. Hermans. 1975. Annelida: Polychaeta. *In* A. C. Giese and J. S. Pearse, eds. Reproduction of Marine Invertebrates, Vol. 3. Annelids and Echiurans. Academic Press, New York, pp. 465–550.

Sleigh, M. A. 1962. The Biology of Cilia and Flagella. Pergamon Press, New York.

———1968. Patterns of ciliary beating. Symp. Soc. Exp. Biol. 22: 131–150.

Smith, P. R., E. E. Ruppert, and S. L. Gardiner. 1987. A deuterostome-like nephridium in the mitraria larva of *Owenia fusiformis* (Polychaeta, Annelida). Biol. Bull. 172: 315–323.

Strathmann, R. R. 1971. The feeding behavior of planktotrophic echinoderm larvae: mechanisms, regulation and rates of suspension-feeding. J. Exp. Mar. Biol. Ecol. 6: 109–160.

———1978. The evolution and loss of feeding larval stages of marine invertebrates. Evolution 32: 894–906.

———1987. Larval feeding. *In* A. C. Giese, J. S. Pearse, and V. B. Pearse, eds. Reproduction of Marine Invertebrates, Vol. 9. General Aspects: Seeking Unity in Diversity. Blackwell Scientific, Palo Alto, Calif., pp. 465–550.

Strathmann, R. R., and D. Bonar 1976. Ciliary feeding of tornaria larvae of *Ptychodera flava* (Hemichordata: Enteropneusta). Marine Biology 34: 317–324.

Strathmann, R. R., T. L. Jahn, and J. R. C. Fonseca. 1972. Suspension feeding by marine invertebrate larvae: clearance of particles by ciliated bands of a rotifer, pluteus, and trochophore. Biol. Bull. 142: 505–519.

Strathmann, R. R., and E. Leise. 1979. On feeding mechanisms and clearance rates of molluscan veligers. Biol. Bull. 157: 524–535.

Strathmann, R. R., and L. R. McEdward. 1986. Cyphonautes' ciliary sieve breaks a biological rule of inference. Biol. Bull. 171: 694–700.

Werner, Von B. 1955. Über die Anatomie, die Entwicklung und Biologie des Veligers und der Veliconcha von *Crepidula fornicata* L. (Gastropoda Prosobranchia). Helgo. Meeres. 5: 169–217.

Wilson, D. P. 1929. The larvae of the British sabellarians. J. Mar. Biol. Assoc. 16: 221–268.

———1932. On the mitraria larva of *Owenia fusiformis* Delle Chiaje. Philos. Trans. R. Soc. London B 221: 231–334.

13 Biology and Morphology of Living *Meiopriapulus fijiensis* Morse (Priapulida)

Gustav Paulay and Bern V. Holthuis

ABSTRACT Of the four known lineages of priapulids, the small, meiobenthic *Meiopriapulus fijiensis* is the most recently described, and its biology remains poorly known. Field and laboratory observations indicate that the species may be gregarious and that, in routine movements, it does not extend its mouth cone. Many individuals in a population have double cuticles, indicating that the old cuticle is retained for a long period prior to molting. Unusual, vacuolelike spaces are evident in the epidermis. A portion of the protonephridial duct of males is expanded to form a bladder that is packed with immotile, apparently mature spermatozoa; the bladder thus functions as a seminal vesicle. At present, evidence is equivocal about the mode of fertilization and development in these enigmatic worms.

Introduction

The discovery of three lineages of meiobenthic priapulids in the past twenty-five years has greatly expanded the known functional diversity of this small animal phylum, previously thought to comprise only six morphologically and ecologically similar, carnivorous, macrobenthic species (Land, 1968, 1970; Por and Bromley, 1974; Morse, 1981). Recent morphological, especially ultrastructural, work on both meio- and macrobenthic species has led to the discovery of several novel anatomical structures, often restricted to one or two genera, indicating that the four major lineages known at present have had a long and independent evolutionary history (e.g., Storch et al., 1989a,b; Higgins and Storch, 1989). This hypothesis is corroborated by the diversity of body plans already evident among Middle Cambrian priapulids (Conway Morris, 1977).

Meiopriapulus fijiensis, the most recently discovered lineage of meiobenthic priapulids, was described from coarse intertidal coral sands at Korolevu, Fiji (Morse, 1981). Although in Fiji it is known only from the type locality, the report of very similar specimens from the Andaman Islands (Westheide, 1990) indicates a wider distribution. Recent studies on the morphology of *Meiopriapulus*, based largely on preserved materials, have revealed several structures unique within the Priapulida (Morse, 1981; Storch et al., 1989a,b; Higgins and Storch, 1991). Below we discuss observations on living specimens that shed additional light on the morphology and mode of life of this distinct lineage.

Materials and Methods

Meiopriapulus specimens were collected at the type locality, Korolevu Beach (Viti Levu Island), Fiji, during August 1982 and July–August 1985. Water temperature at the locality varied from 24°C (3 August 1985) to 27°C (14 August 1985). Samples of subtidal sand were placed in plastic

Gustav Paulay and Bern V. Holthuis, Marine Laboratory, University of Guam, Mangilao, GU 96923.

bags with ample water, shaken, and decanted. The decanted water was searched for priapulids. Additional sand samples were searched with the aid of a dissecting microscope. Worms were kept in petri dishes with a thin layer of sand and ample sea water. In 1982 and 1985, they were held in a 24°C temperature-controlled room at the University of the South Pacific. In 1985, several hundred live specimens were transported to the University of Washington and maintained at 20°–22°C for over a month. Animals from this latter collection also formed the basis for some of the studies of Storch et al. (1989a,b). All anatomical observations below were made on live, whole-mounted, or freshly dissected animals.

Results and Discussion

Morse (1981: Fig. 25) found *Meiopriapulus* to extend from 0.2 m below to 1.8 m above spring low tide level at Korolevu Beach, in an area partially lacking reef protection and exposed to wave action through a wide reef channel. These sands are particularly rich in meiofauna (Morse, 1981, 1987). We found *M. fijiensis* to be most common at and slightly below the 0.0-m tide level on the western edge of this beach. The distribution of worms even within this area was highly patchy, with densities ranging from zero to over 150 worms per liter of sand collected from a one square meter area. *Meiopriapulus* kept in petri dishes display a similarly clumped distribution, with most individuals within one or two body lengths of each other and large areas of sand remaining unoccupied. These observations suggest that *M. fijiensis* may be actively gregarious.

Meiopriapulus is considerably larger than indigenous sediment grains and is effectively a burrowing rather than interstitial animal. Animals kept in shallow sand are constantly active if left undisturbed, everting and retracting the introvert at an average rate of 3.9 ± 0.5 s/cycle ($N = 10$ cycles each on thirteen individuals, 24°C). The introvert expands laterally upon eversion (Fig. 13.1). The introvert eversion cycle is accompanied by direct peristaltic contractions of the body, as has been described for *Priapulus* (e.g., Hunter et al., 1983). The mouth cone is usually not everted in these putative burrowing cycles; it was, however, everted several times sequentially by a single observed animal, an activity which was preceded and followed by cycles in which only the introvert was everted. It may be that the mouth cone is everted only during feeding, while the rest of the movements are locomotory. Storch et al. (1989a) described a unique coelomlike hydrostatic skeleton that presumably supports the mouth cone during feeding.

Worms kept without sand exhibit the same locomotory behavior as described above, but gradually become lethargic and die in one or two days, perhaps due to starvation. In contrast, most worms kept in shallow sand but otherwise under identical conditions survived for over a month, until they were preserved.

Molting

In a high percentage of the animals, apparently preparing to molt, a narrow (ca. 5–10-μm wide, wider at posterior end), fluid-filled space separated two layers of cuticle (Fig. 13.2). Such a double cuticle was also noted by Storch et al. (1989b) for one specimen of *Meiopriapulus*; it is apparently common in *Priapulus*, in which Land (1970) described it as a separation of the outer and inner layers of cuticle, occurring especially frequently among larvae. The common occurrence of a double cuticle indicates that this stage in the molting cycle may be of long duration. In the single molting event observed, the animal slowly emerged head first from the outer cuticle, which was separated from the inner one by a much wider space than that described above (Fig. 13.3).

The cuticular surface bears abundant, characteristic tubercles, about 1 μm in diameter (Morse, 1981; Storch et al., 1989b, see also Fig. 13.4), which appear to be responsible for

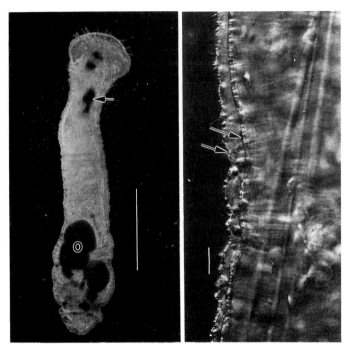

FIGURE 13.1 *(left)*. Mature female. Note expanded introvert, polythyridium (arrow), and oocytes (O) (scale bar = 0.5 mm). FIGURE 13.2 *(right)*. Double cuticle (arrows). Circular muscle layer and three introvert retractor muscles also visible. (scale bar = 10 μm).

differences in color among individuals. Newly molted individuals, whose tubercles are not obvious, are cream colored. As tubercles grow in size and become more apparent, the animals become darker: tan or brown. Tubercle density remains constant. Storch et al. (1989b) described a tubercle as consisting of a cap of flocculent material around a cuticular thickening. Perhaps the flocculent material accumulates gradually after molting and is the source of the color.

Flat, roughly circular, vacuolelike structures (10–25 μm diam) are scattered within the body wall between the cuticular surface and the outer, circular muscle layer (Fig. 13.4). These match Storch et al.'s (1989b) description of an intraepidermal channel system in approximate diameter and location, but are circumscribed vacuoles rather than capillarylike channels. Further work is needed to clarify this difference.

Internal anatomy

The spacious body cavity of *M. fijiensis* contains abundant free cells of two types. The more common type is a larger (ca. 10–15-μm-diam cell bodies), flattened, roughly discoid cell with sparse, thin, pseudopodial extensions and a centrally located, bulging nucleus (Fig. 13.5); it appears to correspond in shape (except for the presence of sparse pseudopodia) and size to the erythrocytes of Storch et al. (1989a). These larger cells were clumped in one or a few masses within the body cavities of some live animals. Whether this clumping was natural or an artifact of handling is not clear. Schreiber et al. (1991) described a similar tendency for in vitro aggregation of amoebocytes in *Halicryptus spinulosus*. The less common, smaller cell type is round, with a granular cytoplasm under the light microscope.

The most conspicuous part of the alimentary tract in *M. fijiensis* is the polythyridium, at the junction of the fore- and midguts. It is more elongated than that of *Tubiluchus* (cf. Kirsteuer,

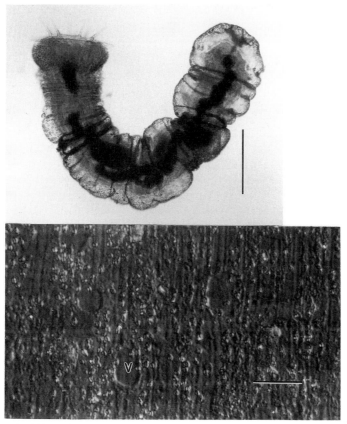

FIGURE 13.3 *(top)*. Molting individual. Introvert has emerged, while rest of body remains covered by old cuticle; dark granules on old cuticle are tubercles (scale bar = 250 μm). FIGURE 13.4 *(bottom)*. Body wall surface, showing three putative epidermal vacuoles (V) beneath cuticular surface and above circular muscles. Abundant shiny granules are tubercles on surface of cuticle (Scale bar = 20 μm).

1976), the only other priapulid genus known to have such a structure. It comprises eight elongate cuticular ridges that come together centrally to form a valvelike convergence (Fig. 13.6). Food material is most commonly seen in the region of the polythyridium, which may function as a valve (Morse, 1981) or in food maceration (Storch, 1989a). Kirsteuer and Land (1970) likewise hypothesized that the function of the polythyridium in *Tubiluchus corallicola* is food maceration. We did not see any movement within the polythyridium in any of the live animal preparations.

Urogenital System and Reproduction

A pair of protonephridial clusters are clearly visible through the body wall in both sexes. Each unit within these clusters presumably corresponds to the mass of about nine terminal cells described by Storch et al. (1989a). The activity of the protonephridial flagella is obvious in live preparations. In all mature males, each protonephridial cluster drains directly into a bladderlike expansion of the protonephridial duct (Figs. 13.7 and 13.8). In all males examined, the bladder contains numerous, immotile, apparently mature spermatozoa, and thus functions as a seminal vesicle. The spermatozoa lie in a dense packet in the basal half of the bladder, with a thin, diminishing trail of spermatozoa extending from this mass toward the protonephridium. The distal half of the bladder is filled with clear fluid, presumably the product of protonephridial filtration.

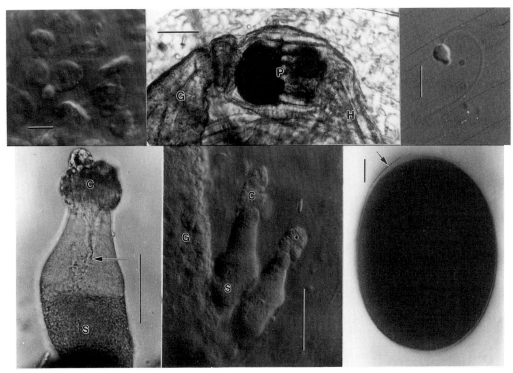

FIGURE 13.5 *(top left)*. Body cavity's large, discoidal cells bearing sparse, thin pseudopodia (scale bar = 10 μm). FIGURE 13.6 *(top middle)*. Polythyridium (P) with dark food bolus and cuticular ridges, between pharynx (H) and midgut (G) (scale bar = 50 μm). FIGURE 13.7 *(bottom middle)*. Pair of protonephridial bladders, basally filled with spermatozoa (S) and with protonephridial clusters (C) apically; gut (G) to left; in situ (scale bar = 50 μm). FIGURE 13.8 *(bottom left)*. Protonephridial cluster (C) with beating flagella; protonephridial bladder filled with immotile spermatozoa (S) at base, with trail of spermatozoa (arrow) extending toward protonephridia; dissected lab preparation (scale bar = 50 μm). FIGURE 13.9 *(top right)*. Spermatozoon dissected from seminal vesicle (scale bar = 5 μm). FIGURE 13.10 *(bottom right)*. Large oocyte dissected from female. Note oocyte membrane (arrow) (scale bar = 20 μm).

The bladder is drained by a long, wide, apparently muscular protonephridial duct, the inner end of which is regulated by what appears to be a thick, muscular sphincter. The protonephridial duct contracts frequently and vigorously in intact worms on slide preparations, jerking the protonephridia posteriorly. We were unable to trace the protonephridial ducts to their external openings.

Testicular tissue, as evidenced by a variety of developing spermatids in squash preparations, is visible as a mass near the posterior end of the hindgut. It is unclear whether the testis is single or paired. Although both Morse (1981) and Storch et al. (1989a) reported paired testes in *Meiopriapulus*, it is not clear whether they were referring to the actual testes or to the paired, sperm-filled protonephridial bladders. Females have only a single ovary, located in the posterior body cavity.

Organs functioning as seminal vesicles evidently do not exist in other known priapulids, although Land (1982) inappropriately labeled as such sperm-filled seminal receptacles in female *Tubiluchus remanei*. In some macrobenthic priapulids, the protonephridia empty into a bladderlike expansion of the urogenital canal, although spermatozoa have not been observed in these expansions (Land, 1970; Storch, 1991).

In *Meiopriapulus*, the route by which sperm enter the protonephridial bladders is unclear. In other priapulids whose reproductive systems have been described, testes and protonephridia empty into a common urogenital duct which carries both sperm and protonephridial filtrate (Hyman, 1951; Land, 1970; Storch, 1991). It seems probable, then, that in *Meiopriapulus* the testis empties into the protonephridial bladder or the duct draining the bladder. Alternatively, it is possible that sperm are released from the testis into the body cavity, entering the protonephridial bladder via (1) the nephridial clusters, (2) a coelomostome, or (3) the wall of the bladder or duct. The first seems unlikely: the presence of nephridial holes large enough for the passage of spermatozoa would presumably defeat the process of ultrafiltration by the protonephridia. The second case is likewise improbable: we found no evidence for a coelomostome associated with the protonephridia (as in the "protonephromixia" of phyllodocid polychaetes (Goodrich, 1945)) although sperm could feasibly enter through a transient coelomostomelike structure. In the third case, sperm may enter the protonephridial bladder or duct via permanent or transient pores, or by direct penetration.

The functional morphology of the seminal vesicle/protonephridia apparatus in *Meiopriapulus* is puzzling; the dense pack of spermatozoa lie directly in the path of the protonephridial filtrate being propelled by the activity of the flagella toward the protonephridial duct. More work is needed to clarify the plumbing and operation of this enigmatic system.

Spermatozoa released from dissected seminal vesicles move about lethargically for several minutes. The spermatozoa are of the primitive type, with a round, about $2.75 \mu m \times 3.5 \mu m$ head bearing a cap-shaped acrosome (Fig. 13.9); fine structure is illustrated by Storch et al. (1989a). The flagellum appears to arise at an angle relative to the acrosome (Fig. 13.9), and is about $32 \mu m$ long, thinning out abruptly about $5 \mu m$ from its end.

In females, the protonephridial clusters empty via a duct that lacks a bladder and apparently lacks the sphincter present in males. The single ovary contains a few white, yolky, ovoid oocytes of a range of sizes. The largest seen were $192 \mu m \times 144 \mu m$ in live preparations; Higgins and Storch (1991) report maximum oocyte diameters of $250 \mu m$ in preserved specimens. Germinal vesicles are clearly visible. Large ($>150\text{-}\mu m$) oocytes dissected out of animals are surrounded by a thin membrane, $0\text{--}10 \mu m$ off the egg surface (Fig. 13.10).

The embryonic development of *M. fijiensis* is as yet unknown. We were unable to find developing embryos in the petri dishes in which adult animals were maintained. The largest eggs dissected out of the ovaries, when combined with sperm dissected out of the seminal vesicles, failed to develop on numerous trials.

Higgins and Storch (1991) presented evidence indicating that this species lacks the characteristic priapulid larval stage. The smallest juveniles encountered were of comparable size to the unusually large oocytes of this species, and one such juvenile was found apparently partially protruding from a female. We failed to find any larval stages among hundreds of worms collected. Efforts to locate larvae by searching through numerous fresh sand samples were also unsuccessful. The smallest juveniles encountered, which lacked developed gonads (i.e., having neither visible oocytes nor sperm in the protonephridial bladders), measured $700 \mu m$ in length.

Higgins and Storch's (1991) observation of a juvenile apparently protruding from a female raises the possibility that this species may be viviparous and may have internal fertilization. If so, one would expect to find developing embryos within female worms, although Higgins and Storch (1991) do not describe any such embryos. We likewise found no evidence for them among the numerous worms examined. However, our specimens were collected during austral winter in both years; reproduction may occur seasonally at another time. The largest oocytes we observed (192 μm) were considerably smaller than the largest (250 μm) noted by Higgins and Storch (1991, albeit the latter were preserved), and may have been immature. The mature spermatozoa seen may last for long periods in the protonephridial bladder.

Tubiluchus, the only priapulid known to have internal fertilization (see Land [1982] and Alberti and Storch [1988] for descriptions of sperm within females of two species), has highly specialized spermatozoa (Alberti and Storch, 1983; Storch and Higgins, 1989). In contrast, the primitive-type sperm of *Meiopriapulus*, quite similar to that of *Priapulus* (Afzelius and Ferraguti, 1978), is suggestive of external fertilization (Storch et al., 1989a). The existence of a seminal vesicle in *Meiopriapulus* should not necessarily be taken as evidence of internal fertilization, as sperm can be stored by free spawners (e.g., echiurans, Gould-Somero, 1975) as well as species with internal fertilization. The seminal vesicle may be an adaptation for maximizing the probability of fertilization in such a small, meiobenthic species; the observed gregarious behavior of the worm may be a similar adaptation.

Storch et al. (1989a) and Higgins (pers. comm., 1992) noted a paucity of males in their samples. In August 1985, the sex ratio was not significantly different from 1:1 ($N = 19$, chi-square test, $p > .5$).

Acknowledgments

We thank Bill Kenchington, Alison Haynes, and the late John Gibbons (University of the South Pacific) for access to microscopes in Fiji, Dennis Willows (director) for access to facilities at the Friday Harbor Laboratories, Bob Higgins for discussion, and George Shinn for critical comments. This work was supported by a Lerner-Gray Fund for Marine Research to G. P. and a National Science Foundation grant to A. J. Kohn. This paper is contribution 334 from the University of Guam Marine Laboratory. We dedicate this study to the memory of Chris Reed, who shared our fascination with invertebrate morphology and practical jokes.

Literature Cited

Afzelius, B. A., and M. Ferraguti. 1978. The spermatozoa of *Priapulus caudatus* Lamarck. J. Submicr. Cytol. 10: 71–80.

Alberti, G., and V. Storch. 1983. Structure of the developing and mature spermatozoa in *Tubiluchus* (Priapulida, Tubiluchidae). Zoomorph. 103: 219–227.

———1988. Internal fertilization in a meiobenthic priapulid worm: *Tubiluchus philippinensis* (Tubiluchidae, Priapulida). Protoplasma 143: 193–196.

Conway Morris, S. 1977. Fossil priapulid worms. Spec. Pap. Palaeont. 20: 1–95.

Goodrich, E. S. 1945. The study of nephridia and genital ducts since 1895. Q. J. Microsc. Sci. 342/343: 113–301.

Gould-Somero, M. 1975. Echiura. *In* A. C. Giese and J. S. Pearse, eds. Reproduction of Marine Invertebrates, Vol. 3. Annelids and Echiurans. Academic Press, New York, pp. 277–311.

Higgins, R. P., and V. Storch. 1989. Ultrastructural observations of the larva of *Tubiluchus corallicola* (Priapulida). Helgo. Meeres. 43: 1–11.

———1991. Evidence for direct development in *Meiopriapulus fijiensis* (Priapulida). Trans. Am. Microsc. Soc. 110: 37–46.

Hunter, R. D., V. A. Moss, and H. Y. Elder. 1983. Image analysis of the burrowing mechanisms of *Polyphysia crassa* (Annelida: Polychaeta) and *Priapulus caudatus* (Priapulida). J. Zool. 199: 305–323.

Hyman, L. H. 1951. The Invertebrates, Vol. 3: Acanthocephala, Aschelminthes, and Entoprocta: The Pseudocoelomate Bilateria. McGraw-Hill, New York.

Kirsteuer, E. 1976. Notes on adult morphology and larval development of *Tubiluchus corallicola* (Priapulida), based on *in vivo* and scanning electron microscopic examinations of specimens from Bermuda. Zool. Scripta 5: 239–255.

Kirsteuer, E., and J. van der Land. 1970. Some notes on *Tubiluchus corallicola* (Priapulida) from Barbados, West Indies. Mar. Biol. 7: 230–238.

Land, J. van der. 1968. A new aschelminth, probably related to the Priapulida. Zool. Meded. 42: 237–250.

———1970. Systematics, zoogeography and ecology of the Priapulida. Zool. Verh. 112: 1–118.

———1982. A new species of *Tubiluchus* (Priapulida) from the Red Sea. Netherlands J. Zool. 32: 324–335.

Morse, M. P. 1981. *Meiopriapulus fijiensis* n. gen., n. sp.: An interstitial priapulid from coarse sand in Fiji. Trans. Am. Microsc. Soc. 100: 239–252.

———1987. Distribution and ecological adaptations of interstitial molluscs in Fiji. Am. Malac. Bull. 5: 281–286.

Por, F. D., and H. J. Bromley. 1974. Morphology and anatomy of *Maccabeus tentaculatus* (Priapulida; Seticoronaria). J. Zool. 173: 173–197.

Schreiber, A., V. Storch, M. Powilleit, and R. P. Higgins. 1991. The blood of *Halicryptus spinulosus* (Priapulida). Can. J. Zool. 69: 201–207.

Storch, V. 1991. Priapulida. *In* F. W. Harrison and E. E. Ruppert, eds. Microscopic Anatomy of Invertebrates, Vol. 4. Aschelminthes. Wiley-Liss, New York, pp. 333–350.

Storch, V., and R. P. Higgins. 1989. Ultrastructure of developing and mature spermatozoa of *Tubiluchus corallicola* (Priapulida). Trans. Am. Microsc. Soc. 108: 45–50.

Storch, V., R. P. Higgins, and M. P. Morse. 1989a. Internal anatomy of *Meiopriapulus fijiensis* (Priapulida). Trans. Am. Microsc. Soc. 108: 245–261.

———1989b. Ultrastructure of the body wall of *Meiopriapulus fijiensis* (Priapulida). Trans. Am. Microsc. Soc. 108: 319–331.

Westheide, W. 1990. *Meiopriapulus fijiensis* Morse (Priapulida) from South Andaman, another example of large-scale geographic distribution of interstitial marine meiofauna taxa. Proc. Biol. Soc. Wash. 103: 784–788.

14 Uptake of Protein by an Independently Evolved Transitory Cell Complex in Encapsulated Embryos of Neritoidean Gastropods

Brian R. Rivest and Richard R. Strathmann

ABSTRACT During intracapsular development in *Nerita picea*, an unusual group of cells forms at the tip of the foot. This pedal cell complex is dominated by a vacuolated cell which may reach 100 μm in diameter due to its single, fluid-filled vacuole. Five to ten basal cells develop around the proximal, ventral margin of the vacuolated cell. The complex begins to differentiate during the trochophore stage of development and reaches its maximum size during the early veliger stage. However, it is fully resorbed by the time the planktotrophic veligers are ready to hatch. Experimental evidence suggests that the cells of the complex take up albumen proteins, but further work is needed to determine the function of the large vacuole. The pedal cell complex appears to be limited to the genus *Nerita*; it was not found in confamilial genera *Neritina*, *Septaria*, *Smaragdia*, and *Clithon*, although in *Clithon neglectus* there are cells that may be comparable to the basal cells of *Nerita*. The position of the pedal cell complexes in the encapsulated embryos of neritoideans and the phylogenetic position of these snails suggest that these embryonic structures have evolved independently of other transitory cell complexes associated with encapsulated development in gastropods.

Introduction

Encapsulation of embryos is found in diverse taxa (Thorson, 1946; Pechenik, 1979). The original function of encapsulation may have been protection of embryos or retention of embryos in a favorable habitat (see review by Pechenik, 1979), but encapsulation can produce a new environment for embryos that sometimes leads to special adaptations for intracapsular development. At least within the gastropods, transitory structures that occur only in encapsulated stages have evolved many times. One hypothesis regarding the influence of encapsulation on the evolution of development is that encapsulation leads to an environment that is richer in extraembryonic nutritive materials. Subsequently, natural selection leads to modifications of embryos that enhance their uptake of these materials (Fioroni, 1966). However, other functions of transitory structures of encapsulated embryos are possible. Unfortunately, most comparative information on structure and function of transitory embryonic structures is limited to observations of gross morphology.

Prosobranch gastropods enclose their eggs in capsules in which development occurs until either a veliger larva or juvenile snail emerges. During their intracapsular development, many species show nearly unmodified patterns of development to the veliger stage (Fretter and Graham, 1962; Strathmann, 1987), but some show developmental modifications. Species that hatch as

Brian R. Rivest, Department of Biological Sciences, State University of New York at Cortland, Cortland, NY 13045.
Richard R. Strathmann, Friday Harbor Laboratories, University of Washington, Friday Harbor, WA 98250.

juvenile snails may develop only small velar lobes (e.g., Fretter and Graham, 1962; Rivest, 1983). Those that include nurse eggs in their capsules may have embryos with morphological specializations that assist in the handling of this nutritive material (Lyons and Spight, 1973; Hadfield and Iaea, 1989). Some encapsulated embryos have large structures that are absent in planktonic embryos. These appear to be adaptations for intracapsular life. Some of these cells are capable of endocytotic uptake of materials from the extraembryonic fluid within egg capsules. For example, in some prosobranch gastropods, cells associated with the kidney complex are specialized for the uptake of capsular albumen (Fioroni, 1985; Rivest, 1992).

In embryos of snails belonging to the genus *Nerita*, a group of large cells occurs at the tip of the embryonic foot and composes a substantial part of the body volume of the embryo. This pedal cell complex is a transitory structure that develops during early embryogenesis but disappears before the embryos hatch as planktotrophic veligers. Its presence has been noted in several species of *Nerita* (Risbec, 1932; Anderson, 1962; Kolipinski, 1964), but a detailed description of the cells composing this structure and experimental work on the function of the complex are lacking. In this report, we describe this unusual cluster of cells in *Nerita picea* Récluz, 1841, test its capability to take up protein from surrounding fluids, and examine the distribution of this structure and other structures of possibly similar function in embryos of neritoidean gastropods.

Materials and Methods

Egg capsules of *Nerita picea* used in this study came from two sources. Some were collected on coral rock during the winter months on Oahu, Hawaii, at Wailupe Beach Park and Kaneohe Bay. The egg capsules were identified as being produced by *N. picea*, because capsules of neritids are easily recognized (Andrews, 1935; Houston, 1990) and *N. picea* was the only neritid found at these localities. Additional egg capsules were obtained from adult snails that were maintained in the laboratory. Though *N. picea* normally occur in much warmer waters, adult snails laid capsules in the laboratory at temperatures as low as 8°–11°C. Embryos removed from the capsules were examined using light microscopy as well as scanning and transmission electron microscopy.

Other neritoideans were also examined. Egg capsules of *Smaragdia bryanae* (Pilsbry, 1918) were collected from leaves of the sea grass *Halophila* at Paiko Point and Kaneohe Bay, Oahu, Hawaii. Capsules of *Clithon neglectus* (Pease, 1860) were collected at Wailupe Beach Park on Oahu on coral rocks. Embryos of these species were examined with light microscopy.

Egg capsules of *Nerita tessellata* Gmelin, 1791, and *N. versicolor* Gmelin, 1791, were collected on coral rock in the rocky intertidal zone near the Hofstra University Marine Laboratory in St. Ann's Bay, Jamaica. Capsule identification was initially established by examining capsules collected on isolated boulders on which only one species of *Nerita* was found. Capsules of *N. albicilla* Linné, 1758 were collected on the Great Barrier Reef at Lizard Island. Preserved egg capsules of the confamilial freshwater *Septaria porcellana* (Linné, 1758) and *Neritina turtoni* Récluz, 1841 from Fiji were provided by B. Holthuis. The embryos from all these capsules were examined using light microscopy.

Embryos prepared for scanning electron microscopy (SEM) were fixed in 2 percent osmium tetroxide in either filtered sea water or 1.25 percent sodium bicarbonate adjusted to pH 7.2 with HCl. The embryos were then rinsed in distilled water, dehydrated in ethanol and 2,2-dimethoxypropane acidified with HCl, critical point dried, mounted and coated with gold. The coated specimens were examined with a JEOL JSM-35 scanning electron microscope.

Attaining satisfactory fixation of encapsulated embryos for transmission electron microscopy (TEM) can be problematic. Previous TEM studies on the encapsulated embryos of *Searlesia dira* and *Nucella canaliculata* showed that some embryonic cells may show fixation artifacts while adjacent ones do not (Rivest, 1992). This was the case in *Nerita picea*, but an additional difficulty was encountered in preserving the membrane surrounding the large central vacuole of the largest cell in the pedal cell complex. This membrane was not well preserved with any of the fixation recipes used, even though nearby vesicular membranes and the plasmalemma were still present. However, the existence of the vacuolar membrane could be confirmed in living embryos examined by light microscopy.

Although satisfactory fixation for *N. picea* embryos was not achieved, the following details are provided to assist in the development of better fixation protocols in the future. The best preservation of *N. picea* embryos was achieved using a modified version of the Cavey and Cloney (1972) primary fixative that contained cacodylate-buffered glutaraldehyde with ruthenium red. For these fixations, embryos removed from their capsules and rinsed with filtered sea water were placed in 2 percent glutaraldehyde in a 0.2 M solution of sodium cacodylate adjusted to 1000–1100 mOsM with sucrose and containing 0.05 percent ruthenium red and 0.002 M calcium chloride. Final pH was adjusted to 7.4. A comparable fixation was achieved by first placing the embryos in a solution consisting of 0.05 percent osmium tetroxide in 3 percent glutaraldehyde in 0.1 M (pH 7.35) phosphate buffer with the osmolarity raised to 990 mOsM with sucrose. After 10 min, the embryos were transferred to a solution containing the same ingredients but lacking the osmium tetroxide. After an hour in either of the primary fixatives, an equal volume of 10 percent EDTA was added to decalcify the developing shell. An hour or two later the embryos were postfixed in two changes of 2 percent osmium tetroxide in freshly mixed 1.25 percent sodium bicarbonate (pH 7.2) for 1 h at room temperature. The embryos were then dehydrated and embedded in Epon (Luft, 1961). One-micrometer-thick sections were stained with a mixture of Azure II and methylene blue (Richardson et al., 1960) and examined by light microscopy. Thin sections were stained with uranyl acetate and lead citrate (Reynolds, 1963) and examined with a Philips EM-300 electron microscope.

To test for the absorptive capabilities of the cells of the pedal complex, some embryos were exposed to a solution that demonstrates receptor-mediated endocytosis. A particularly useful solution for quickly revealing the rapid uptake of exogenous proteins is a 2–10-mg/ml isosmotic solution of fluoresceinisothiocyanate-labeled bovine serum albumin (FITC-BSA, Sigma Chemical Company) (Rivest, 1992). Embryos of *Nerita picea*, *Smaragdia bryanae*, and *Clithon neglectus* were soaked in such a solution for periods ranging from 10 min to several hours before being examined by epifluorescence microscopy. To confirm receptor-mediated endocytosis with TEM, some embryos of *N. picea* were first soaked for 30–60 min in a sea water solution of ferritin prior to fixation and subsequent processing.

Results

Observations on *Nerita picea*

The capsules examined in this study were planoconvex to doubly convex, depending on the surface topography of the rock to which they were affixed. They were usually oblong, measuring approximately 1.5 mm × 1.8 mm. Embedded in the upper surface of the capsule were numerous spherical calcareous granules as described by Andrews (1935). We confirmed that the calcareous granules dissolve in hydrochloric acid, leaving the organic matrix intact, and that the organic matrix is removed by a 4 percent solution of sodium hypochlorite, leaving the granules intact. Each capsule contained 40 to 90 eggs. The eggs were 160 μm in diameter and were surrounded by a watery fluid.

The pedal cell complex found in *Nerita picea* becomes evident early in embryogenesis. A swelling develops on the posterio-ventral corner of the developing trochophores, before the foot or velar lobe rudiments begin to protrude from the embryonic body. The swelling continues to enlarge during subsequent development, becoming spherical. It initially appears to consist of a single transparent cell with a large, refractile central vacuole. This cell is located at the posterior tip of the embryo's foot and reaches its largest size during the early veliger phase of development (Fig. 14.1). The vacuolated cell can attain a diameter of 70 to 100 µm, resulting in a cell volume of up to 5×10^{-4} µl. Observations on live embryos and sectioned material show that most of the cell volume is due to a large central vacuole (Fig. 14.2). The vacuolar contents always appear light and electron-lucent, with only an occasional bit of flocculent material seen in TEMs. Also, in embryos examined with the SEM which had their vacuolated cells teased open the vacuoles appeared empty. However, when examining live embryos on a microscope slide while a solution of 10 percent trichloracetic acid was drawn under the coverslip, a small amount of a fine precipitate formed within the vacuole the moment the acid front arrived. This suggests that some, albeit little, protein was present in the vacuole. The absence of schlieren (lines of refraction) when the contents of a punctured vacuole spill into sea water also indicates that the vacuole does not contain a concentrated protein solution.

The rim of cytoplasm that surrounds the lateral and distal sides of the vacuole may be only 1.5 to 3 µm thick and contains scattered mitochondria as well as some rough endoplasmic reticulum. The nucleus also is commonly found within this rim, usually dorsally (Fig. 14.2).

By the time the vacuolated cell has reached its maximum size, 5 to 10 other cells have become enlarged around its base (Figs. 14.1 and 14.3). These basal cells are centered around the midline along the junction between the vacuolated cell and the ventral surface of the foot. In embryos with relatively few basal cells, the cells are restricted to the ventral surface of the foot (Fig. 14.3). However, in embryos where the cells are more numerous, basal cells also may be found laterally. The basal cells vary in size, but may attain a greatest dimension of 40 µm, although most are closer to half that size. The basal cells appear colorless and do not contain a vacuole. No specializations other than junctional complexes were seen along the border between the adjacent pedal cells and the basal cells or vacuolated cell.

The outer surfaces of the vacuolated cell and basal cells possess microvilli and are active in receptor-mediated endocytosis (Figs. 14.4–14.6). In embryos incubated in a ferritin solution, the reddish ferritin was taken up in endocytotic vesicles of coated membrane that apparently fused into larger heterophagosomes within the cytoplasm of both the vacuolated and basal cells. Ferritin within these heterophagosomes was sufficiently concentrated, and the heterophagosomes were large and numerous enough that these cells took on a light-reddish color. This color was more evident in the basal cells where the heterophagosomes were more concentrated than in the vacuolated cell where the heterophagosomes were more dispersed within the thin rim of cytoplasm surrounding the central vacuole. No ferritin was seen in the central vacuole of the vacuolated cell. However, the embryos were exposed to ferritin for no more than 60 min before they were fixed.

FITC-BSA was taken up and concentrated within the vacuolated cell and basal cells via a mechanism apparently identical to that observed in the ferritin studies (Figs. 14.7 and 14.8). Uptake of FITC-BSA was evident in the vacuolated cell as early as the trochophore stage, when the cell was just beginning to bulge from the outline of the embryo. As the basal cells differentiated, they too absorbed the protein. After short exposure times to FITC-BSA, numerous small heterophagosomes located just under the plasmalemma fluoresced under ultraviolet epi-

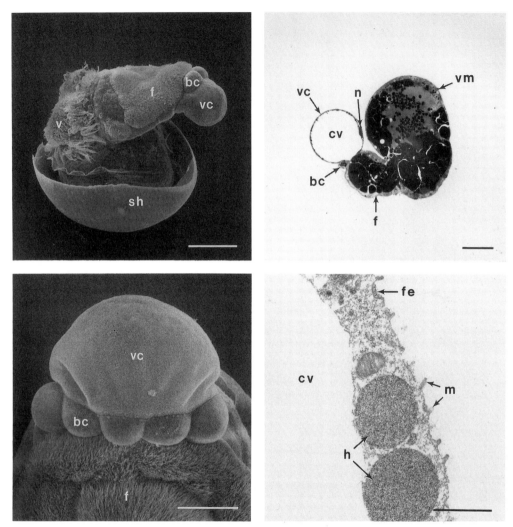

FIGURE 14.1 *(top left)*. SEM of a *Nerita picea* early veliger, showing the pedal cell complex protruding from the tip of the foot. The pedal cell complex consists of a vacuolated cell and several basal cells. (Scale bar = 50 μm; bc, basal cell; cv, central vacuole; f, foot; fe, ferritin adhering to endocytotic impocketing; h, heterophagosomes filled with ferritin; m, microvilli; n, nucleus of vacuolated cell; sh, shell; v, velum; vc, vacuolated cell; vm, visceral mass.). FIGURE 14.2 *(top right)*. One-micrometer-thick midsagittal section of an *N. picea* early veliger, illustrating how the volume of the vacuolated cell is principally due to a large central vacuole. At this stage of development the pedal cell complex has reached its maximum size. (Scale bar = 50 μm; for label designations, see Figure 14.1.) FIGURE 14.3 *(bottom left)*. SEM of the ventral surface of the pedal cell complex in *N. picea* , showing five basal cells around the proximal margin of the vacuolated cell. Microvilli on the surface of the vacuolated and basal cells give them a slightly rough appearance. (Scale bar = 20 μm; for label designations, see Figure 14.1.) FIGURE 14.4 *(bottom right)*. TEM of a section through the edge of the vacuolated cell in *N. picea*. The membrane of the central vacuole was not preserved. The embryo had been exposed for 30 min prior to fixation to ferritin, which fills the two heterophagosomes visible here. (Scale bar = 1 μm; for label designations, see Figure 14.1.)

illumination. With increased exposure times, larger, more intensely fluorescing heterophagosomes were evident deeper inside the basal cells and the proximal cytoplasm of the vacuolated cell. Possibly because of space constraints, heterophagosomes within the thin rim of cytoplasm surrounding the vacuole never achieved the size of those in the base of the vacuolated cell or in the basal cells. Although the membrane of the central vacuole was not preserved in preparations for TEM (Figs. 14.4 and 14.5), the membrane was clearly seen within some of the enlarged vacuo-

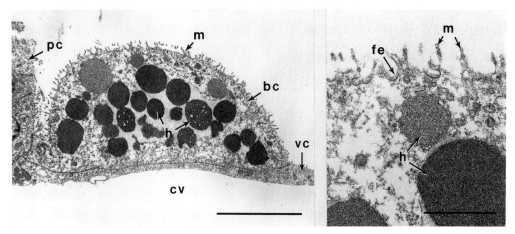

FIGURE 14.5 *(left).* TEM through a basal cell in a *Nerita picea* early veliger that had been exposed to ferritin, which filled the heterophagosomes. The clear area around the central heterophagosomes appears to be a fixation artifact. (Scale bar = 5 μm; bc, basal cell; cv, central vacuole; fe, ferritin adhering to coated membrane; h, heterophagosomes filled with ferritin; m, microvilli; pc, pedal cells; vc, vacuolated cell.) FIGURE 14.6 *(right).* Enlarged view of the external surface of the basal cell shown in Figure 14.5, showing ferritin adhering to coated membrane which subsequently would have formed endocytotic vesicles that transport the ferritin to heterophagosomes. (Scale bar = 1 μm; for label designations, see Figure 14.5.)

lated cells of FITC-BSA-treated embryos (Fig. 14.9). It should be noted that no structures other than the heterophagosomes in the vacuolated and basal cells visibly fluoresced except for some heterophagosomes in the digestive gland, presumably from ingested and absorbed FITC-BSA. Some proteins absorbed by the vacuolated cell can end up in the central vacuole. Figure 14.10 shows the vacuolated cell of an early veliger that had been soaked for 6 h in FITC-BSA. After another hour on a microscope slide, the plasma membrane of the vacuolated cell broke, pulling the rim of cytoplasm back toward the basal cells and revealing the fluorescing vacuole.

As indicated above, the vacuolated cell reaches its maximum size during the early veliger stage of development. As the velar lobes continue to enlarge, the size of the vacuolated cell diminishes. The basal cells also decrease in size. Soon, none of the cells protrudes from the tip of the foot. Even so, these cells continue to be endocytotically active for a while, as they still can rapidly take up FITC-BSA. However, there are no endocytotically active cells present by the time the veligers are ready to hatch.

Observations on Other Species in the Family Neritidae

Based on light microscopic observations, early veligers of *Nerita tessellata* and *N. versicolor* collected in Jamaica resembled the early veligers of *N. picea* in that they possessed a pedal cell complex consisting of a large vacuolated cell and several basal cells. *Nerita albicilla* embryos collected at Lizard Island had at least the large vacuolated cell. In contrast, no such enlarged vacuolated cell was seen in the preserved embryos of the freshwater *Septaria porcellana* and *Neritina turtoni* from Fiji. Similarly, no vacuolated cell occurred in live embryos of *Smaragdia bryanae* and *Clithon neglectus* from Hawaii. However, in early to mid-veligers of *C. neglectus* exposed to FITC-BSA, several somewhat enlarged but nonvacuolated cells fluoresced at the tip of the foot. Such fluorescing pedal cells were not found in embryos of *Smaragdia bryanae*, but in contrast to the other species examined with FITC-BSA, cells of the mantle margin in early and mid-veligers took up FITC-BSA into heterophagosomes.

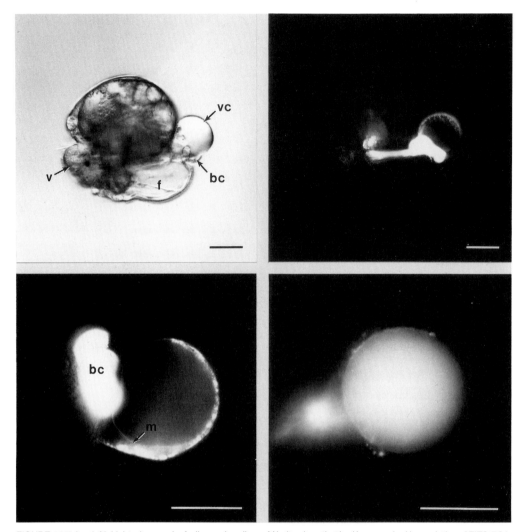

FIGURE 14.7 *(top left)*. Light micrograph of a live early veliger of *Nerita picea* that had been soaked in a solution of FITC-BSA. (Scale bar = 50 μm; bc, basal cells; f, foot; m, membrane of the central vacuole; o, operculum; sh, shell; v, velum; vc, vacuolated cell.) FIGURE 14.8 *(top right)*. An epifluorescence micrograph of the same embryo pictured in Figure 14.7. Individual fluorescing heterophagosomes can be seen in the rim of cytoplasm around the central vacuole of the vacuolated cell. However, because of their density and intensity, the larger heterophagosomes in the basal cells cannot be distinguished. The bright horizontal line extending to the left of the basal cells is the operculum, which is not fluorescing but is refracting light emitted from the pedal cell complex. (Scale bar = 50 μm; for label designations, see Figure 14.7.) FIGURE 14.9 *(bottom left)*. An epifluorescence micrograph of the pedal cell complex of an early *N. picea* veliger soaked in FITC-BSA. The membrane of the central vacuole, which was not preserved in other embryos examined with TEM, is clearly seen here. (Scale bar = 50 μm; for label designations, see figure 14.7.) FIGURE 14.10 *(bottom right)*. An epifluorescence micrograph of the central vacuole of the vacuolated cell of an early *N. picea* veliger soaked in FITC-BSA for 6 h. The plasmalemma of the vacuolated cell had ruptured, and most of the thin rim of cytoplasm around the vacuole retreated toward the basal cells, revealing that labeled protein had entered the central vacuole. (Scale bar = 50 μm; for label designations, see Figure 14.7.)

Discussion

The pedal cell complex of *Nerita picea* is a transitory structure consisting of a large vacuolated cell and several basal cells located at the tip of the foot of encapsulated embryos. It becomes evident as a swelling on trochophores, but becomes quite large on early veligers, where it may be one-third to one-half the length of the developing foot. It then diminishes in size until it is no longer evident in fully developed veligers prior to hatching.

The pedal cell complex appears to be unique to the genus *Nerita*. Because of its size, it would be difficult to miss in other groups if appropriate stages of encapsulated embryos were examined. We have described the complex in *N. picea*, *N. versicolor*, and *N. tessellata*. We also saw the large vacuolated cell in *N. albicilla*. Kolipinski described what he called a "transparent sac" (1964: 61) on the intracapsular trochophores of *N. fulgurans* Gmelin, 1791, in Florida. According to Baker (1923) and Starmuhlner (1988), *N. fulgurans* may be synonymous with *N. tessellata*. Risbec (1932) briefly described swollen structures on the foot of the embryos of the New Caledonian *N. reticulata* Karsten, 1789 (as well as *N. albicilla*), that may be homologous to the pedal cell complex of *N. picea*. Anderson (1962: 66) illustrated a "refractile vesicle" on the tip of the foot of an early veliger of *N. atramentosa* Reeve, 1855 (= *Melanerita melanotraga* [Smith, 1855]) from New South Wales. He also showed that veligers ready to hatch lacked the vesicle. Finally, a pedal cell complex is suggested in *N. plicata* Linné, 1758, from Guam, because Smalley (1981) described its trochophores as consisting of two unequal lobes, with the smaller one being unpigmented. Conversely, no such pedal cell complex was seen in the confamilial *Septaria porcellana*, *Neritina turtoni*, *Smaragdia bryanae*, and *Clithon neglectus* examined by us, or in the intertidal *Puperita pupa* (Linné, 1767) or the riverine *Fluvinerita tenebricosa* (Adams, 1851) (=*Nerita alticola* Pilsbry, 1932) examined by B. Holthuis in Jamaica (pers. comm.). The confamilial *Theodoxus fluviatilis* (Linné, 1758) also does not appear to have a pedal cell complex (Claparède, 1858). Such a complex has not been reported in the literature for other gastropod species.

The functional significance of the pedal cell complex in *Nerita* is enigmatic. Its transitory nature suggests that it is a specialization for intracapsular development. Its precocious development and ability to absorb and concentrate FITC-BSA from the external milieu further suggest that the complex normally functions to take up intracapsular albumen prior to the functional differentiation of the digestive tract, which probably does not occur until the embryos are developing the veliger morphology. In this way the pedal cell complex is similar to that of the absorptive cells of the larval kidney complex of many other prosobranch gastropods (Rivest, 1992). In both cases, the cells begin to swell with accumulated albumen during the trochophore stage of development but are resorbed prior to, or shortly after, hatching. The fate of the protein absorbed by either structure has not been elucidated, but presumably it is translocated into the body of the embryo and used for nutritional purposes. The amount of protein absorbed by the absorptive cells of the larval kidneys can be considerable, judging from the fact that these cells can compose up to 20 percent of the tissue volume of the embryo (Rivest, 1992). By comparison, the volume of the pedal cell complex in *Nerita*, ignoring the central vacuole, is considerably less, indicating that the pedal cell complex may contain less absorbed albumen than do the larval kidneys. Volume comparisons can be misleading, however, for translocation of proteins absorbed by the pedal cell complex into the body of the embryo may be steady throughout development whereas in larval kidneys, translocation may not occur until later in development. It should be noted that the protein content of the intracapsular fluid of *Nerita picea* is unknown and may be small. In contrast to the viscous nature of the intracapsular fluid reported for *N. fulgurans* (Kolipinski, 1964), the fluid of newly laid *N. picea* capsules does not appear more viscous than sea water nor does it create schlieren when spilled into sea water.

The pedal cell complex may function in part as an albumen absorptive structure, but the functional significance of the large clear vacuole in the vacuolated cell is unknown. Although the evidence from experimental exposure to FITC-BSA indicates that proteins absorbed by the vacuolated cell may end up in the vacuole, it is clear that in untreated embryos the vacuole contains little protein. The vacuole is thus not a large heterophagosome storing materials absorbed from the capsular fluid by receptor-mediated endocytosis as is found in the absorptive cell of the larval kidneys of some other gastropods (Rivest, 1992). It is also unlikely that the vacuolated cell of the pedal cell complex is heavily involved in the regulation of ion concentrations, for it lacks the numerous mitochondria that are characteristic of osmoregulatory cells (Lawn, 1960; Copeland, 1967; Sardet et al., 1979). The pedal cell complex in *Nerita* also does not appear to be homologous to the podocyst of terrestrial pulmonate embryos, a structure consisting of a hollow ball of ciliated epidermal cells that pumps hemolymph through the embryo (Cather and Tompa, 1972). There are also functional differences between the pedal cell complex and podocysts, since although the podocyst cells do take up some proteins endocytotically, their principal function is assumed to be associated with the respiratory activity of the embryo and possibly with dissolution of egg shell calcium (Cather and Tompa, 1972; Tompa, 1979).

Vera Fretter suggested to Milton Kolipinski that the so-called transparent sacs in *Nerita* embryos might function as a water reserve (Kolipinski, 1964). This is an attractive hypothesis, for the vacuolar contents appear to be basically water, and so far this vacuolated cell has been reported only a genus of tropical to subtropical gastropods that lay their egg capsules on rocks in the intertidal zone. However, it remains to be demonstrated that the vacuolated cell functions principally as a water reserve, for the advantages of having a store of water isolated at the tip of the foot instead of elsewhere within the body are unclear.

The neritoideans appear to be an isolated offshoot within the prosobranchs. Fretter (1965, 1966) and Haszprunar (1988) inferred early separation of the clade, not only from the higher prosobranchs but also from many prosobranchs of the archaeogastropod grade. This isolated phylogenetic position assigned to the neritoidean gastropods within the prosobranchs suggests that the transitory embryonic pedal cell complex arose independently of other transitory structures that take up encapsulated nutritive material in other gastropods. Such an interpretation is supported by the location of the pedal cell complex in *Nerita* being different from that of albumen-absorbing cells in other prosobranch embryos. We therefore conclude that the albumen-absorbing cells in embryos of *Nerita* are not homologous with those in other prosobranch lineages.

Acknowledgments

We wish to thank M. Hadfield, C. Unabia, and B. Holthuis for advice and for collecting some of the neritoidean embryos and adults. A. Belloma, B. Holthuis, and two anonymous reviewers provided useful comments on the manuscript. Facilities were provided by the Friday Harbor Laboratories, the State University of New York at Cortland, the Kewalo Marine Laboratory, the Hawaii Institute of Marine Biology, the Hofstra University Marine Laboratory, and the Lizard Island Research Station. This study was supported in part by National Science Foundation grant OCE8922659.

Literature Cited

Anderson, D. T. 1962. The reproduction and early life histories of the gastropods *Bembicium auratum* (Quoy and Gaimard) (Fam. Littorinidae), *Cellana tramoserica* (Sower.) (Fam. Patellidae) and *Melanerita melanotragus* (Smith) (Fam. Neritidae). Proc. Linn. Soc. N. S. W. 87: 62–68.

Andrews, E. A. 1935. The egg capsules of certain Neritidae. J. Morph. 57: 31–59.

Baker, H. B. 1923. Notes on the radula of the Neritidae. Proc. Acad. Nat. Sci. 75: 117–178.

Cather, J., and A. Tompa. 1972. The podocyst in pulmonate evolution. Malacol. Rev. 5: 1–3.

Cavey, M., and R. Cloney. 1972. Fine structure and differentiation of ascidian muscle. I. Differentiated caudal musculature of *Distaplia occidentalis* tadpoles. J. Morph. 138: 349–374.

Claparède, E. 1858. Anatomie und Entiwicklungsgeschichte der *Neritina fluviatilis*. Mueller's Arch. f. Anat. Phys. u. wiss. Med. 23: 109–248, taf. IV-VIII.

Copeland, D. E. 1967. A study of salt secreting cells in the brine shrimp (*Artemia salina*). Protoplasma 63: 363–384.

Fioroni, P. 1966. Zur Morphologie und Embryogenese des Darmtraktes und der transitorischen Organe bei Prosobranchiern (Mollusca, Gastropoda). Rev. Suisse Zool. 73: 621–876.

———1985. Struktur und Funktion der larvalen Zellen des Cephalopodiums bei jungen intracapsulären Larven von *Nucella lapillus* (Gastropoda, Prosobranchia). Zool. Beitr. N. F. 29: 103–117.

Fretter, V. 1965. Functional studies of the anatomy of some neritid prosobranchs. J. Zool. 147: 46–74.

———1966. Some observations on neritids. Malacologia 5: 79–80.

Fretter, V., and A. Graham. 1962. British Prosobranch Molluscs. Ray Society, London.

Hadfield, M. G., and D. K. Iaea. 1989. Velum of encapsulated veligers of *Petaloconchus* (Gastropoda). Bull. Mar. Sci. 45: 377–386.

Haszprunar, G. 1988. On the origin and evolution of major gastropod groups, with special reference to the Streptoneura. J. Moll. Stud. 54: 367–441.

Houston, R. 1990. Reproductive systems of neritimorph archaeogastropods from the Eastern Pacific, with special reference to *Nerita funiculata* Menke, 1851. Veliger 33: 103–110.

Kolipinski, M. C. 1964. The life history, growth, and ecology of four intertidal gastropods (Genus *Nerita*) of southeast Florida. Ph.D. dissertation, University of Miami, Miami, Fl.

Lawn, A. M. 1960. Observations on the fine structure of the gastric parietal cells of the rat. J. Biophys. Biochem. Cytol. 7: 161–166.

Luft, J. H. 1961. Improvements in epoxy resin embedding methods. J. Biophys. Biochem. Cytol. 9: 409–414.

Lyons, A., and T.M. Spight. 1973. Diversity of feeding mechanisms among embryos of Pacific Northwest *Thais*. Veliger 16: 189–194.

Pechenik, J. 1979. Role of encapsulation in invertebrate life histories. Am. Nat. 114: 859–870.

Reynolds, E. S. 1963. The use of lead citrate at high pH as an electron-opaque stain in electron microscopy. J. Cell Biol. 17: 208–211.

Richardson, K. C., L. Jarrett, and E. H. Finke. 1960. Embedding in epoxy resins for ultrathin sectioning in electron microscopy. Stain Technol. 35: 313–323.

Risbec, J. 1932. Notes sur la ponte et le développement de mollusques gasteropodes de Nouvelle-Caledonie. Bull. Soc. Zool. Fr. 57: 358–374.

Rivest, B. R. 1983. Development and the influence of nurse egg allotment on hatching size in *Searlesia dira* (Reeve, 1846) Prosobranchia: Buccinidae). J. Exp. Mar. Biol. Ecol. 69: 217–241.

———1992. Studies on the structure and function of the larval kidney complex of prosobranch gastropods. Biol. Bull. 182: 305–323.

Sardet, C., M. Pisam, and J. Maetz. 1979. The surface epithelium of teleostean fish gills. Cellular and junctional adaptations of the chloride cell in relation to salt adaptation. J. Cell Biol. 80: 96–117.

Smalley, T. L. 1981. The distribution and abundance of *Nerita plicata* in a tropical, rocky intertidal habitat. Master's thesis. University of Guam, Mangilao.

Starmuhlner, F. 1988. Results of the Austrian-French Hydrobiological Mission 1979 to Guadeloupe, Dominica and Martinique (Lesser Antilles). II. Contributions to the knowledge of the freshwater and brackish water mollusks of Guadeloupe, Dominica and Martinique. Ann. Naturhist. Mus. Wien. Ser. B. Bot. Zool. 90(0): 221–340.

Strathmann, M. F. 1987. Reproduction and Development of Marine Invertebrates of the Northern Pacific Coast. University of Washington Press, Seattle.

Thorson, G. 1946. Reproduction and larval development of Danish marine bottom invertebrates. Medd. Dan. Fisk-Havunders. 4: 1–523.

Tompa, A. S. 1979. Localized egg shell dissolution during development in *Stenotrema leai* (Pulmonata: Polygyridae). Nautilus 94: 136–137.

15 Temperature Sensitivity, Rate of Development, and Time to Maturity: Geographic Variation in Laboratory-Reared *Nucella* and a Cross-Phyletic Overview

A. Richard Palmer

ABSTRACT Unlike most other physiological processes, the rate of development generally shows little evidence of temperature compensation—the tendency, for example, for physiological processes at the same temperature to proceed more rapidly in cold-adapted than warm-adapted organisms. As a consequence, a substantial fraction of the variation in rate of development can be explained on the basis of taxonomic affinity and temperature. These patterns suggest rather strong taxon-dependent constraints on the duration of the prehatching period. I report here that the rate of development in the laboratory of a rocky-shore thaidine gastropod, "northern" *Nucella emarginata*, lies very near that predicted from other muricacean gastropods. In spite of this close agreement, however, at temperatures less than 10°C, *N. emarginata* from southeast Alaska hatched in significantly less time (up to 15 percent less at 8°C) than snails from Barkley Sound, British Columbia. Over a rather narrow temperature range (8°–11°C), seasonal variation in laying frequency was more pronounced in Alaskan snails. On the other hand, rate of development (1/days to hatch) was less sensitive to temperature in Alaskan than British Columbian snails (Q_{10} values of 2.63 versus 4.66, respectively). These data thus provide evidence for intraspecific latitudinal temperature compensation in the average rate of development, as well as evidence for geographic variation in the temperature sensitivity of development rate. A review of the literature suggests that, in spite of considerable variation in reproductive characteristics, direct-developing muricacean gastropods not only develop more slowly than many other marine organisms but also exhibit a more precise relationship between temperature and rate of development. This slower rate of development, coupled with more precise temperature dependence, suggests that encapsulation during the entire prejuvenile period may impose significant constraints on the rate of development.

Introduction

Temperature has a profound impact on the rates of many physiological processes (Cossins and Bowler, 1987). In addition, many species exhibit the phenomenon of temperature compensation, where at a given temperature the rate of a physiological process in cold-adapted organisms is higher than that of warm-adapted ones. In contrast to most other physiological processes, however, rate of development appears to show little if any temperature compensation (Patel and Crisp, 1960; Emlet et al., 1987; Johnston, 1990; but see McLaren et al., 1969). So pervasive is the effect

A. Richard Palmer, Department of Zoology, University of Alberta, Edmonton, Alberta T6G 2E9, and Bamfield Marine Station, Bamfield, British Columbia V0R lB0.

177

of temperature on development that, to a reasonable approximation, one can predict the development time of species of meso- and neogastropods simply by knowing their taxonomic affinities and their temperature of development (Spight, 1975).

Many marine invertebrates lay their eggs as gelatinous masses or enclosed in egg capsules attached to the substratum (Strathmann, 1987). Unlike solitary pelagic eggs, eggs in masses or in capsules experience limited rates of diffusion of metabolites, respiratory gases, and ions because of the medium by which they are surrounded and because of their proximity to other developing embryos (Strathmann and Strathmann, 1989). Encapsulation of embryos is particularly common among meso- and neogastropods (Strathmann, 1987), yet surprisingly little is known about its costs in terms of reduced fecundity or rate of development (Grahame and Branch, 1985; but see Perron, 1981b).

Species with limited gene flow should be more likely to exhibit geographic differentiation in response to different climatic regimes than species with greater dispersal potential. Prior studies on temperature compensation of development rate have examined species with moderate to high dispersal potential, including barnacles (Patel and Crisp, 1960), copepods (McLaren et al., 1969), echinoderms (Emlet et al., 1987), and fishes (Johnston, 1990). To assess the extent of latitudinal physiological adaptation of development rate in a species with limited dispersal capability, I examined data from the rocky-shore thaidine gastropod "northern" *Nucella emarginata* that were collected as part of a long-term study of the genetic basis of shell variation (Palmer, 1984a, 1985a,b). Successive generations of snails were reared in the laboratory for over ten years, and sea water temperature was recorded throughout this time. Because *N. emarginata* spawn throughout the year, and because ambient sea water temperature varied seasonally in the laboratory, the effect of temperature on rate of development and time to maturity could be assessed. In addition, because genetic studies were being conducted on snails from geographically widespread populations, I could test for temperature compensation in these traits. Finally, because snails were reared from laying to maturity in the laboratory, the prejuvenile period as a fraction of total time to maturity could be computed. To place these values in context, I compared development rate, the temperature sensitivity of development rate, and the prejuvenile period as a fraction of the total time to maturity, to published values for a wide variety of other organisms.

Methods

Taxonomic Concerns

The taxonomic status of "*Nucella emarginata*" (Deshayes, 1839) on the Pacific Coast of North America remains unsettled (Palmer et al., 1990). I have inspected Deshayes' syntypes, and they are clearly of the "southern" species. Hence the "northern" species, the focus of this chapter, requires a new name. Several varietal names are eligible (Dall, 1915), but shells alone are ambiguous and locality data are notoriously unreliable. More important, I have been unable to inspect one of Middendorf's specimens synonymized with *emarginata* by Dall (1915). Until these taxonomic problems can be resolved, I shall follow earlier recommendations (Palmer et al., 1990) and refer to the species studied here as the *northern* species of *N. emarginata*. For convenience, I have omitted *northern* from the frequent references to this species in the text, but in all instances I mean to refer only to the northern species, as the southern species is not found north of San Francisco Bay (Palmer et al., 1990).

Table 15.1. Approximate schedule for monitoring various aspects of reproduction and growth in laboratory-reared northern *Nucella emarginata*

Aspect of rearing procedure	Period of study		
	1982–84	1985–86	1987–92
Check for egg capsules produced by parents	1 week	2 weeks	3 weeks
Duration of capsules in mesh bags	80 days	70 days	70 days
Check cages for hatching and replenish small barnacles for hatchling snails	2 weeks	3 weeks	4 weeks
Size at which hatchlings were transferred to cages with larger mesh and larger barnacles	3–8 mm	3–10 mm	3–15 mm
Replenish barnacles for juvenile snails	3 weeks	4 weeks	6 weeks
Size at which male and female juveniles were isolated into separate cages	15–22 mm	15–22 mm	15–22 mm
Replenish barnacles for adult snails and check for production of capsules	3 weeks	4 weeks	6 weeks

Sources of Snails and Rearing Procedure

Snails were collected from several localities on the Pacific coast of North America, but extensive data were only available for two: (1) Torch Bay, southeast Alaska (58°19′41″ N, 136°47′56″ W), in July 1980 and July 1982, and (2) Barkley Sound, Vancouver Island, British Columbia (two sites: Wizard Island 48°51′30″ N, 125°09′33″ W; Cape Beale 48°47′05″ N, 125°12′54″ W), in August 1980 and July 1982. All snails from subsequent years were reared as successive generations from these initial collections.

Snails were reared following the general protocol outlined in Palmer (1985b). In brief, snails were sexed based on relative penis size, and single pairs of snails were placed in individual cages and provided with their preferred prey (Palmer, 1984b), the barnacle *Balanus glandula*, on small stones. Food was replenished and cages were monitored for egg-capsules at regular intervals (Table 15.1). Capsules were removed from the rock with fine forceps and placed in 3 cm × 3 cm plastic screen bags (2-mm mesh) with Velcro closures and suspended from fishing line in gently running sea water until shortly before hatching (Table 15.1). Before they began to hatch, capsules were transferred to freezer containers (10 × 10 × 15 cm) with 500-μm Nitex mesh sides that contained stones with small barnacles for food (*B. glandula* and *Chthamalus dalli*, 3–5-mm basal diam). After snails had begun to hatch, barnacles were replenished on a regular basis (Table 15.1). When juveniles reached an adequate size, they were transferred to larger freezer container cages with larger mesh and provided with larger barnacles on a regular basis (Table 15.1). Prior to maturity, male and female offspring were assigned to separate cages based on a visual inspection of relative penis size. Mature progeny were monitored for the production of egg capsules and provided with fresh barnacles at regular intervals (Table 15.1). To avoid introducing unwanted snails from the field, barnacle stones were collected only from quiet-water shores where *N. emarginata* were absent.

Cages and egg capsules were held continuously immersed in running sea water and under a natural photoperiod maintained by fluorescent lighting in the main aquarium room at the Bamfield Marine Station, British Columbia.

Sea Water Temperature

Sea water temperature varied seasonally and annually. All cages were held under ambient conditions in flow-through aquaria. Temperature was monitored periodically with a mercury thermometer ($\pm 0.1°C$) in the aquaria where the snails were held, usually every 2–4 days but sometimes less frequently. At least one temperature measurement was available for all but 14 of the 120 months of detailed records. Because temperature sometimes varied by up to 1°C over the course of each day and more so throughout the month, and because some months were sampled more intensively than others, the raw temperature data were rather noisy (Fig. 15.1). I attempted to remove the effects of short-term variability and the effects of gaps in the data by fitting a smooth curve via polynomial regression and moving averages. The final "best fit" to the regular seasonal variation was determined by eye. This best fit curve was then used to generate the expected daily temperature for each day of each month over the full 10 year period (Fig. 15.1). These best fit estimates of average temperature for a given day were probably accurate to within 0.1°C.

The effect of temperature on development time was determined by computing the average temperature over the roughly 3-month period from laying to hatching for each individual clutch based on the data from the best fit curve through the relevant time interval. Because snails spawned throughout the year (see Fig. 15.2), development time could be monitored over a natural range of temperature (8°–11°C).

Duration of Life History Stages

Although key life history events (laying, hatching, and maturity) were monitored on a regular basis, the uncertainty associated with each varied from inspection to inspection, and over the duration of the study (Table 15.1). The estimated time and the associated uncertainty of each event were determined in the same fashion for each clutch (Fig. 15.3). For example, laying date was estimated as the average of the date capsules were first observed and the previous date on which that cage had been checked. The maximum possible error in the estimated laying date for that clutch was thus the number of days between those two dates (e.g., $E1$, Fig. 15.3). The actual uncertainty associated with a given laying date would, of course, be half the value of $E1$ in days. Dates of hatching (first hatchlings observed in a cage) and maturity (first capsules produced by any mature offspring, whether in holding cages or as individual snails used in subsequent crosses) and their uncertainties were estimated in the same manner (Fig. 15.3). The mean ($\pm SE$) maximum uncertainty associated with each life history event over the entire period of study was: laying, 13.0 days (± 0.23, $N = 591$), hatching, 28.7 days (± 0.56, $N = 607$), and maturity, 23.7 days (± 1.19, $N = 279$).

By chance the uncertainties for certain stages for some clutches were large, and for others small. To avoid introducing unnecessary errors, some observations with large uncertainties were excluded from the analyses. All analyses were restricted to clutches whose uncertainties were less than a predefined threshold: time to hatch ≤ 50 days (i.e., $E1 + E2$, Fig. 15.3), juvenile period ≤ 60 days ($E2 + E3$); and total time to maturity ≤ 50 days ($E1 + E3$). The average maximum uncertainties after removing observations above these threshold levels were 30.2 days ($N = 357$), 39.0 days ($N = 166$), and 24.6 days ($N = 169$) for time to hatch, juvenile period, and total time to maturity, respectively. Note that these thresholds yielded maximum possible uncertainties and that actual uncertainties would be half of these values (see preceding paragraph). Sample sizes varied for the estimates of different life history events and stages because reliable dates for all three events (laying, hatching, and maturity) were not available for all clutches.

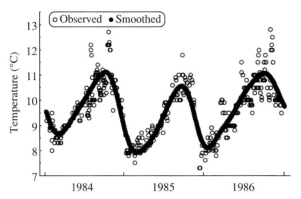

FIGURE 15.1. Observed and smoothed (best fit) daily sea water temperatures at the Bamfield Marine Station main aquarium room for the period 1 January 1984 to 31 December 1986.

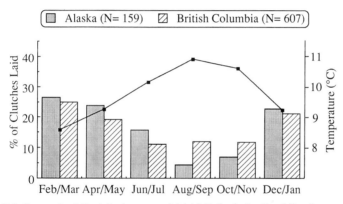

FIGURE 15.2. Seasonal variation in the frequency of clutch initiation for "northern" *Nucella emarginata* from two geographically distant populations held continuously immersed in the laboratory: Torch Bay, Southeast Alaska, and Barkley Sound, British Columbia. Temperatures are mean monthly sea water temperature over the duration of the study (1982–1992).

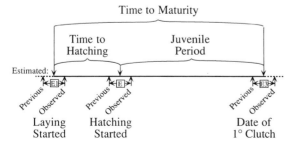

FIGURE 15.3. The procedure for estimating the duration and uncertainty of various life history stages. The horizontal line indicates time moving from left to right for an individual clutch. Vertical arrows below the line indicate actual dates, whereas those above it indicate estimated dates. E1, E2, E3 the maximum error or uncertainty associated with estimated dates used in the analyses.

Time to hatch as a fraction of total time to maturity was computed separately for each clutch for which both time to hatch and juvenile period were available. The average value was computed as the average for all clutches from each of the two geographic regions.

Comparisons with Other Species

To place the results obtained for *Nucella emarginata* in context, similar data were obtained from the literature for other species. Where data on time to hatch or juvenile period were expressed as a range of values at a given temperature, I used only the minimum for each. If data were given for a range of temperatures, I estimated time to maturity only for one temperature near the "normal" temperature range experienced by the species. Where graphical data were available describing the dependence of time to hatch on temperature, these data were digitized from the published figures and used to compute Q_{10} values.

Q_{10} Computations

Q_{10} values were computed from the slope (A) of the least-squares linear regression: X = temperature (°C), Y = log(1/days to hatch). Hence, for a given species $Q_{10} = 10^{(10 \cdot A)}$. Least-squares linear regression seemed more appropriate here than reduced major axis regression (LaBarbera, 1989) because errors in the estimate of time to hatch were probably considerably larger than those for temperature.

With two exceptions, Q_{10} values for time to hatch were estimated using all the available data published for a species. First, where data departed significantly from a linear relationship of log(1/days to hatch) versus temperature at extreme high temperature, observations near the stressful limits were not included. Second, for one species of snail upon which several studies had been conducted (*Urosalpinx cinerea*), one author's results differed from all the others and were excluded (see Spight, 1975).

Statistical analyses

Data manipulation, editing, and computation of uncertainties were conducted with Excel (Ver. 3.0a, Microsoft). Descriptive statistics, least-squares linear regressions, contingency table analyses, and analysis of variance (ANOVA) were computed with Statview II (Ver. 1.03, Abacus Concepts). Analysis of covariance (ANCOVA) was conducted with SuperAnova (Ver. 1.1, Abacus Concepts).

Results

Although both British Columbian and southeast Alaskan populations of *Nucella emarginata* produced egg capsules throughout the year, both spawned more frequently during the winter months (December–May) when laboratory water temperatures were lowest (Fig. 15.2). Alaskan snails, however, were significantly more sensitive to seasonal cues and spawned less frequently than those from British Columbia during the late summer and fall (August–November; $p = .04$, contingency table analysis).

Over the temperature range examined (8°–11°C), time to hatch decreased significantly with increasing temperature for capsules from both British Columbia and southeast Alaska (Fig. 15.4). The effect of temperature on hatching time was significantly more pronounced for capsules from the more southern region (see Fig. 15.4 legend). For British Columbian capsules, the Q_{10} for development rate (1/time to hatch) was 4.93 (± 1.05 SE) whereas for those from southeast Alaska the Q_{10} was 2.63 (± 0.69, see Table 15.2). Based on the slopes of the regressions in Figure 15.4,

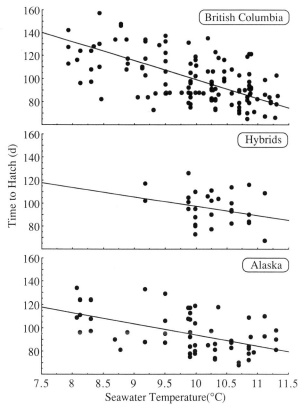

FIGURE 15.4. Time to hatch as a function of seawater temperature in laboratory-reared "northern" *Nucella emarginata* from two geographically distant populations, and in hybrid crosses between them. Only development times with uncertainties ≤ 50 days were included (see "Duration of Life History Stages" in Methods). Lines correspond to least-squares linear regression equations (slope ± S.E., intercept ± S.E.): BC, −16.3 ± 1.40, 262.5 ± 1.16 ($N = 203$, $r = 0.634$); hybrids, −8.2 ± 5.26, 179.0 ± 2.55 ($N = 32$, $r = 0.272$); AK, −9.5 ± 1.91, 188.5 ± 1.62 ($N = 81$, $r = 0.488$). The slopes for BC and AK differed significantly from each other ($p = .013$, ANCOVA). No further analysis was attempted for hybrids because the data spanned too narrow a range of temperature to be useful and because the slope for this group by itself did not differ significantly from zero. The data are included here only for completeness.

time to hatch increased 16.3 days for each degree drop in temperature for capsules from British Columbia and 9.5 days per degree for capsules from southeast Alaska. These would correspond to an impact on development time per degree of approximately 16 percent and 10 percent, respectively, of the total time to hatch at 10°C. No analyses were conducted for hybrid snails since the small sample size and limited temperature range yielded no significant association between development time and temperature.

Because of the different slopes for these two regions, capsules from British Columbia took significantly longer to hatch than those from southeast Alaska only at temperatures below about 10°C. For example, at 8°C British Columbian capsules took 17 percent longer to hatch than Alaskan capsules (140 versus 120 days), whereas at 10°C they differed by only about 5 percent (99 versus 94 days). The regression lines crossed at approximately 10.7°C. Because of the generally inverse exponential dependence of hatching time on temperature (Johnston, 1990), these linear regressions cannot be extrapolated reliably to higher temperatures.

Average juvenile period did not differ significantly between British Columbian and southeast Alaskan populations ($p = .80$, one-way ANOVA). For both, the juvenile period in the laboratory was approximately 430 days (Table 15.3). Similarly, the average time to maturity, measured as the

Table 15.2. Q_{10} values of rate of development (1/time to hatch) for selected fish and invertebrates

Species	N^a	Temperatures tested (°C) Min.	Max.	Mean	Q_{10}^b	Source
Teleost fishes						
Clupea harengus (Atl. herring)	9	5.3	14.2	9.0	4.18	Johnston, 1990
Cyprinodon macularis (pupfish)^c	19	15.0	28.4	22.6	3.72	Kinne, 1963
Enchelyopus cimbrius (rockling)	12	5.3	23.1	13.4	2.80	Johnston, 1990
Gadus macrocephalus (Pac. cod)	11	2.5	13.2	7.4	4.18	Johnston, 1990
G. morhua (Atl. cod)	9	−1.0	14	6.8	2.84	Kinne, 1963
Harpagifer spp.	8	−0.4	0.6	0.1	3.89	Johnston, 1990
Morone saxatilis (striped bass)	8	15.3	22.5	18.8	3.61	Johnston, 1990
Mugil cephalus (grey mullet)	11	10.4	29.1	19.4	2.15	Johnston, 1990
Osmerus eperlanus (smelt)	11	4.9	17.8	10.3	5.77	Johnston, 1990
Pleuronectes platessa (plaice)	11	2.4	14.3	7.7	3.06	Johnston, 1990
Scomber scombrus (mackerel)	10	10.3	21	15.5	3.77	Johnston, 1990
Prosobranch gastropods						
Eupleura caudata	13	21.1	26.6	24.9	2.89	MacKenzie, 1961
Ilyanassa obsoleta	7	11.0	28.2	20.0	2.81	Scheltema, 1967
Urosalpinx cinerea	12	10.2	29.7	19.1	1.71	Spight, 1975
Nucella emarginata, AK	81	8.1	11.3	10.0	2.63	This study
N. emarginata, BC	203	7.9	11.3	9.9	4.93	This study
Opisthobranch gastropod						
Onchidoris bilamellata	7	5.0	15.0	10.7	2.17	Todd, 1991
Cephalopod						
Octopus vulgaris	13	15.8	24.7	20.2	2.23	Spight, 1975
Marine copepods						
Acartia clausi	3	0	10.8	5.2	4.73	McLaren et al., 1969
A. tonsa	5	10	25.4	15.5	4.13	McLaren et al., 1969
Calanus finmarchicus	4	0	20	9.9	2.41	McLaren et al., 1969
C. glacialis	4	0	8	4	2.94	McLaren et al., 1969
C. hyperboreas	3	0	7	3.4	3.06	McLaren et al., 1969
Centropages furcatus	5	16	30	21.8	3.24	McLaren et al., 1969
Eurytemora hurundoides	3	0	12	4.9	3.50	McLaren et al., 1969
Metridia longa	5	0	12	5.3	3.08	McLaren et al., 1969
Pseudocalanus minutus	5	0	12	5.9	2.86	McLaren et al., 1969
Temora longicornis	4	0	12	5.6	3.64	McLaren et al., 1969
Tortanus discaudatus	3	0	10.8	5.2	4.34	McLaren et al., 1969

time from the date of laying to the date upon which the first capsules were produced by progeny of a particular cross (see Fig. 15.3), did not differ between these populations either $p = .83$, one-way ANOVA), and was approximately 540 days for both (Table 15.3).

Discussion

Temperature Compensation

That temperature should have an effect on the development rate of northern *Nucella emarginata* is not terribly surprising in light of the extensive effects of temperature on the rates of physiological

Table 15.2. Continued

Species	N^a	Temperatures tested (°C)			Q_{10}^b	Source
		Min.	Max.	Mean		
Lepadomorph barnacle						
Lepas anatifera	2	19.7	24.8	22.2	3.05	Patel and Crisp, 1960
Balanomorph barnacles						
Balanus amphitrite[c]	5	17.6	31.7	24.2	1.91	Patel and Crisp, 1960
B. balanus	4	3.1	13.1	7.8	3.15	Patel and Crisp, 1960
B. crenatus[c]	5	3.1	18.1	10.3	2.92	Patel and Crisp, 1960
B. perforatus	7	9.1	27.4	19.9	2.56	Patel and Crisp, 1960
Chthamalus stellatus[c]	5	9.1	27.8	18.6	2.17	Patel and Crisp, 1960
Elminius modestus[c]	5	3.1	18.0	10.3	4.13	Patel and Crisp, 1960
Semibalanus balanoides	5	3.0	13.9	8.9	1.31	Patel and Crisp, 1960
Verruca stroemia[c]	6	3.0	18.1	10.8	2.99	Patel and Crisp, 1960
Echinoid echinoderms						
Sterechinus neumayeri	4	−1.9	−0.1	−0.8	9.58	Bosch et al., 1987
Strongylocentrotus droebachiensis	4	0.0	9.4	5.4	3.12	Bosch et al., 1987
S. franciscanus	2	10.2	12.7	11.5	2.13	Bosch et al., 1987

[a]N sample size.
[b]$Q_{10} = 10^{(10 \times A)}$, where A is the slope of log(1/days to hatch) as a function of temperature (°C).
[c]Data at stressful high temperature excluded.

Table 15.3. Juvenile period and time to maturity for laboratory-reared *Nucella emarginata* from two geographically distant populations

(No observations of juvenile period were obtained for hybrid clutches.)

	Juvenile period (d)			Time to Maturity (d)		
	N	Mean	SE	N	Mean	SE
Barkley Sound, British Columbia	126	429	7.8	126	541	8.7
Torch Bay, Alaska	30	434	18.1	32	545	17.7

processes (Kinne, 1963; Johnston, 1990). However, the discovery of latitudinal temperature compensation in *N. emarginata* is somewhat surprising given the apparently minimal or undetectable temperature compensation reported either within or among other species, including barnacles (Patel and Crisp, 1960), echinoderms (Emlet et al., 1987), and fishes (Johnston, 1990). Although temperature compensation has been reported in the short-term larval growth rates of several gastropods (Dehnel, 1955), Patel and Crisp (1960) note that short-term rates of size increase in larvae may not accurately reflect the impact of temperature on the entire process of growth and differentiation, a sentiment echoed by Johnston (1990).

  N. emarginata may exhibit temperature compensation among distant populations because of the limited gene flow associated with direct development. The other species for which temperature compensation has been noted as lacking (barnacles, echinoids, and fishes) all have pelagic larval stages.

Questions persist about the proper analysis of temperature dependence of physiological processes (see the appendix in McLaren, 1963). For example, McLaren reports evidence of temperature compensation in rates of development of copepods both within (McLaren, 1966) and among species (McLaren et al., 1969) when using a different analytical approach that attempts to circumvent some of the inadequacies of conventional Q_{10} analyses. Hence the apparent lack of temperature compensation in rates of development reported by some may in part reflect a poor choice of analysis.

Q_{10} Values for Rate of Development in Northern *Nucella emarginata* and Other Taxa

Not only did Alaskan populations of *N. emarginata* develop more rapidly at temperatures less than 10°C (Fig. 15.4), but their rate of development was less sensitive to temperature variation than British Columbian snails over the temperature range examined (8°–11°C, Table 15.2). This difference in temperature sensitivity may have been due to the relative differences between lab and field sea water temperatures for these two populations. For British Columbia snails, although the range of monthly mean temperatures experienced in the laboratory was less than half that of surface sea water at nearby Cape Beale for the period November 1982 to August 1988 (laboratory range = 3.4°C [min. = 8.0, max. = 11.4] versus Beale range = 7.6°C [min. = 6.6, max. = 14.2]), the mean temperature experienced in the laboratory was 0.6°C less than that of surface sea water at Cape Beale over the same time period (9.9°C ± 0.11 SE versus 10.5°C ± 0.26, $N = 70$). For Alaskan snails, the mean laboratory sea water temperature was likely higher than those experienced in the field. Unfortunately, because I did not manipulate temperatures experimentally to try to mimic those experienced in air (i.e., up to 25°C during the summer months), I cannot be certain whether the rates of development of British Columbian snails would be faster than (as implied by the regressions of Fig. 15.4), or the same as, those for Alaskan snails at temperatures above 11°C.

The Q_{10} values reported here for rate of development (1/time to hatch) were comparable to those determined from a casual survey of published values for other marine taxa (Table 15.2, Fig. 15.5). At a Q_{10} of 2.6, rates of development of Alaskan *N. emarginata* were close to those of the majority of other species, which exhibited Q_{10} values in the range of 2 to 3. The Q_{10} of 4.9 for British Columbian snails seemed rather high in comparison since only two other species, one fish and one echinoid (Table 15.2), exhibited higher Q_{10} values for rate of development. Although Q_{10} values have been noted to increase at lower temperature in other taxa (Bosch et al., 1987), this is in part an artifact of Q_{10} analyses. Even for chemical reactions that behave precisely according to Arrhenius' function, Q_{10} is expected to increase with decreasing temperature (Schmidt-Nielsen, 1979: 537).

Juvenile Period and Time to Maturity in Northern *Nucella emarginata* and Other Taxa

In spite of significant differences in the rate of development reported above at temperatures less than 10°C, differences were not apparent in either juvenile period or total time to maturity between British Columbian and Alaskan snails (Table 15.3). Hence, differences in time to hatch over the majority of temperatures experienced in the laboratory did not translate into a longer overall time to maturity as, for example, reported for the nudibranch *Phestilla sibogae* with an increase in the larval period (Miller and Hadfield, 1990). In part this may be due to the cumulative effects of many small uncertainties affecting laboratory growth during the relatively long juvenile period for *N.*

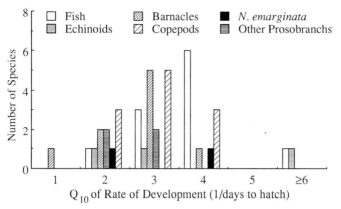

FIGURE 15.5. Temperature sensitivity of rate of development for a variety of marine organisms (see Table 15.2 for detailed data). The two values for *N. emarginata* correspond to separate estimates for the British Columbian and Alaskan populations.

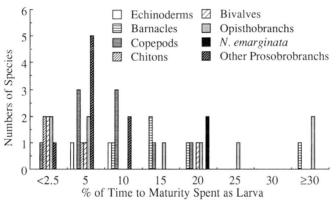

FIGURE 15.6. Percentage of time to maturity (see Fig. 15.3 for definition) spent as larvae or "prejuvenile" for a variety of marine organisms (see Table 15.4 for detailed data). The two values for *N. emarginata* correspond to separate estimates for the British Columbian and Alaskan populations.

emarginata compared to *P. sibogae* (430 versus 21 d). In further contrast to these results for *N emarginata*, Huntley and Lopez (1992) report dramatic effects of temperature on generation time (fertilization to maturity) in a broad survey of copepod growth: more than 90 percent of the variance in growth rate of thirty-three species over a temperature range from −1.7° to 30.7°C could be accounted for by temperature alone.

When compared to a selection of published values for other marine invertebrates, the time to hatch as a fraction of the total time to maturity for *N. emarginata* seemed high (Table 15.4, Fig. 15.6). This difference was apparent even though the prejuvenile period included time in the plankton for species with a pelagic stage. First, only six of thirty-one other species spent a longer fraction of their time to maturity as a larvae, and of these three were short-lived opisthobranch gastropods (time to maturity less than two months) and one was a short-lived balanomorph barnacle (time to maturity approximately three months). Second, of the eight species of meso- and neogastropods for which I could find comparable data, *N. emarginata* spent at least half again as long in the prejuvenile period compared to the next most similar species, *Strombus costatus* (17.7

Table 15.4. Time to juvenile, juvenile period, and time to maturity for selected marine invertebrates

Taxon	Time to juvenile					Juvenile period		Time to juvenile as % of time to maturity
	Lay to hatch (d)	Hatch to juvenile (d)	Lay to juvenile (d)	Temp. (°C)	Ref.[a]	(days)	Ref.[b]	
Annelida								
Polychaeta								
Pseudopolydora kempi japonica	4	15	19	?	(13)	21	(13)	47.5
Mollusca								
Gastropoda								
Prosobranchia								
Mesogastropoda								
Polinices lewisi	42	1	43	?	(1)	915	(1)	4.5
Strombus costatus	6	36	42	27	(2)	300	(2)	12.3
S. gigas	4.5	27	31.5	28	(2)	1095	(2)	2.8
Neogastropoda								
Conus pennaceus	25	0	25	25	(9)	730	(10)	3.3
C. quercinus	15.5	30	45.5	25	(9)	2555	(10)	1.7
Eupleura caudata	18	0	18	25	(6)	710	(6)	2.5
Nucella canaliculata	90	0	90	?	(12)	915	(11)	9.0
N. emarginata "northern", BC	100.5	0	100.5	9.9	(15)	429	(15)	19.0
N. emarginata "northern", AK	93.7	0	93.7	9.9	(15)	434	(15)	17.7
N. lamellosa	67	0	67	?	(13)	1460	(11)	4.4
Opisthobranchia								
Nudibranchia								
Archidoris pseudoargus	?	?	37	10	(14)	730	(14)	4.8
Doridella obscura	?	?	9	25	(14)	26	(14)	25.7
D. steinbergae	?	?	25	14	(14)	24.5	(14)	50.5
Onchidoris bilamellata	?	?	53	10	(14)	270	(14)	16.4

O. muricata	17.7	(14)	270	(14)	10	58	?	?
Phestilla melanobranchia	5.6	(14)	135	(14)	23	8	?	?
P. sibogae	36.4	(8)	21	(8)	23	12	7	5
Tritonia hombergi	0.2	(14)	730	(14)	9	1.5	?	?
Adalaria proxima	0.7	(14)	300	(14)	9	2	?	?
Polyplacophora								
Katharina tunicata	0.9	(13)	730	(13)	14	7	?	?
Lepidochitona fernaldi	7.1	(13)	182	(13)	10	14	2	12
Mopalia muscosa	1.6	(13)	730	(13)	15	11.5	10.5	1
Bivalvia								
Heterodonta								
Bankia setacea	18.9	(13)	120	(13)	14	28	?	?
Panope abrupta	1.4	(13)	1275	(13)	17	18	?	?
Tapes philippinarum	6.2	(13)	365	(13)	15	24	?	?
Tresus capax	2.3	(13)	1095	(13)	10	26	?	?
Arthropoda								
Crustacea								
Copepoda								
Acartia clausi	20.9	(4)@10°C	31	(7)	5.2	8.2	8.2	
A. tonsa	17.1	(4)@16°C	16	(7)	16	3.3	3.3	
Calanus finmarchicus	5.1	(4)@10°C	45.2	(7)	9.9	2.4	2.4	
C. glacialis	2.3	(4)@1.5°C	180	(7)	2.6	4.3	4.3	
C. hyperboreas	3.5	(4)@0°C	185	(7)	0.0	6.64	6.64	
Centropages furcatus	6.8	(4)@15°C	19.3[b]	(7)	22	1.4	1.4	
Eurytemora hurundoides	9.4	(4)@2°C	75.1[c]	(7)	3	7.81	7.81	
Pseudocalanus minutus	10.9	(4)@6.6°C	47.5	(7)	5.9	5.8	5.8	
Temora longicornis	8.1	(4)@12°C	26.3	(7)	12	2.31	2.31	
Cirripedia								
Lepadomorpha								
Pollicipes polymerus	13.2	(13)	365	(13)	14	55.4	25.4	30

Table 15.4. Continued

Taxon	Time to juvenile					Juvenile period		Time to juvenile as % of time to maturity
	Lay to hatch (d)	Hatch to juvenile (d)	Lay to juvenile (d)	Temp. (°C)	Ref.[a]	(days)	Ref.[b]	
Balanomorpha								
Balanus glandula	27	21	48	12	(3)	182	(3)	20.9
B. pacificus	8	14	22	15	(5)	105	(5)	17.3
Chthamalus fissus	14	21	35	12	(3)[d]	60	(3)	36.8
Tetraclita squamosa	40	30	70	12	(3)[d]	730	(3)	8.8
Echinodermata								
Asteroidea								
Pisaster ochraceus	1.25	76	77.3	12	(13)	1725	(13)	4.3
Echinoidea								
Strongylocentrotus purpuratus	1	74	75	10	(13)	730	(13)	9.3

[a](1) Bernard, 1967; (2) Brownell, 1977; (3) Hines, 1978; (4) Huntley and Lopez, 1992; (5) Hurley, 1973; (6) MacKenzie 1961; (7) McLaren, et al., 1969; (8) Miller and Hadfield, 1990; (9) Perron, 1981b; (10) Perron, 1986; (11) Spight, 1975; (12) Spight, 1976; (13) Strathmann, 1987; (14) Todd, 1991; (15) this study.
[b]For *Centropages hamatus*.
[c]For *Eurytemora herdmanni*.
[d]Estimated from settlement data.

percent versus 12.3 percent of total time to maturity; Table 15.4). Some of this difference may have to do with the relatively short juvenile period of *N. emarginata* in the laboratory. But even if this period were two years (Spight and Emlen, 1976), the fraction of time spent in the prejuvenile period would still be approximately 14 percent.

These data suggest that although natural selection has been able to reduce the time from hatching to maturity in *N. emarginata* compared to other meso- and neogastropods, it has had little impact on the time spent in the egg capsule. This implies that substantial constraints may limit the rate of intracapsular development, as suggested for gelatinous egg masses (Strathmann and Strathmann, 1989). Whether these constraints are related to rates of gas exchange, or metabolite or ion diffusion, across the capsule wall remains to be seen.

Some caution should be exercised when interpreting Figure 15.6 since apparent differences among taxa may have arisen due to a nonrandom selection of modes of development within particular taxa. A more systematic survey would be required to reject this possibility.

Is Development More Tightly Constrained in Muricacean Gastropods with Direct Development?

Spight (1975) was the first to observe that hatching time for meso- and neogastropods could be predicted with reasonable accuracy simply by knowing taxonomic affinity and temperature. An expanded search of the literature suggests this is largely true across an even broader range of taxa (Fig. 15.7). The results of this expanded search further suggest that direct-developing muricacean gastropods are unusual in two respects. First, relative to the total time to maturity, they take longer to hatch at a given temperature than many other marine organisms. Second, they exhibit an unusually precise dependence of time to hatch on temperature.

When compared at a given temperature (Fig. 15.7a), not only do direct-developing muricacean gastropods take among the longest times to develop of prosobranch gastropods, but prosobranch gastropods generally take longer to hatch than opisthobranch gastropods (Fig. 15.7b), and molluscs generally take longer than barnacles, echinoid echinoderms, and teleost fishes (Fig. 15.7c). All of these taxa exhibit a pronounced effect of temperature on rate of development, but what is striking about direct-developing muricaceans is the precision of this temperature dependence across species. Because this precision is even greater than that for muricaceans with a pelagic stage following their intracapsular period (compare solid versus open circles in Fig. 15.7a), constraints on the rate of development in direct-developing species would seem to be acting during the later stages of development, during the transformation from a veliger into a hatchling snail.

Alternatively, Richard Strathmann has suggested to me that the seemingly more precise temperature dependence of development rate in direct-developing muricaceans may have less to do with constraints imposed by encapsulation and more to do with variation in the stage, and hence time of development, at which species with pelagic veligers hatch. I cannot rule out this possibility for muricaceans. However, other prosobranchs with encapsulated direct development exhibit much less precise temperature dependence of development than do muricaceans (compare filled triangles to filled circles in Fig. 15.7a), which suggests that encapsulation per se is not the only constraint.

Taken together, the unusually long time to hatch, coupled with the rather precise temperature dependence of development rates, suggest that the rate of development is constrained at some very fundamental level in direct-developing muricacean gastropods, even though egg size, number of

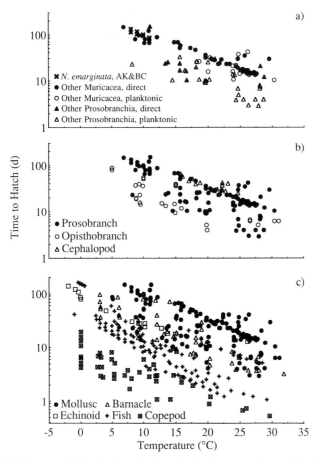

FIGURE 15.7. Time to hatch (in days) as a function of water temperature for a variety of marine organisms: (a) prosobranch gastropods, (b) various molluscs, (c) higher marine taxa. For some species, development time was measured over a range of temperatures; for others data are for only a single temperature. Direct: entire prejuvenile period spent within the egg capsule; planktonic: some fraction of the prejuvenile period spent in the plankton. Compiled from several sources: "northern" *Nucella emarginata* (this study); prosobranch gastropods (Fourteen species [MacKenzie, 1961; Scheltema, 1967; Spight, 1975]); opisthobranch gastropods (Twenty-three species [Spight, 1975; Todd, 1991]), cephalopods (one species [Spight, 1975]), barnacles (Twelve species [Patel and Crisp, 1960; Hines, 1978]), echinoid echinoderms (Seven species [Bosch et al., 1987]), fish (Eleven species [Kinne, 1963; Johnston, 1990]), and copepods (Eleven species [McLaren et al., 1969]).

embryos per capsule, nurse egg to embryo ratio, and egg capsule morphology all vary substantially among these species.

Acknowledgments

My warmest thanks go to George Shinn, Steve Stricker, and Herb Wilson for having the inspiration and energy to initiate and coordinate this conference in honor of Chris Reed. I am also deeply indebted to Robin Boal, Barbara Bunting, and Jeanne Ferris for their care in monitoring and maintaining the laboratory crosses. Thanks also to Dick Strathmann, Herb Wilson, and an anonymous reviewer for comments on the manuscript. As always, I thank the staff at the Bamfield Marine Station for their patience and assistance over this prolonged study. This research was supported by National Sciences and Engineering Research Council of Canada operating grant A7245.

Literature Cited

Bernard, F. R. 1967. Studies on the biology of the naticid clam drill *Polinices lewisi* (Gould). Fisheries Research Board of Canada, Tech. Rpt. 42: 1–41.

Bosch, I., K. A. Beauchamp, M. E. Steele, and J. S. Pearse. 1987. Development, metamorphosis, and seasonal abundance of embryos and larvae of the Antarctic sea urchin *Sterechinus neumayeri*. Biol. Bull. 173: 126–135.

Brownell, W. N. 1977. Reproduction, laboratory culture, and growth of *Strombus gigas*, *S. costatus* and *S. pugilus* in Los Roques, Venezuela. Bull. Mar. Sci. 27: 668–680.

Cossins, A. R., and K. Bowler. 1987. Temperature biology of animals. Chapman & Hall, London.

Dall, W. H. 1915. Notes on the species of the molluscan subgenus *Nucella* inhabiting the northwest coast of America and adjacent regions. Proc. U.S. Nat. Mus. 49: 557–572.

Dehnel, P. A. 1955. Rates of growth of gastropods as a function of latitude. Physiol. Zool. 28: 115–143.

Emlet, R. B., L. R. McEdward, and R. R. Strathmann. 1987. Echinoderm larval ecology viewed from the egg. Echinoderm Stud. 2: 55–136.

Grahame, J., and G. M. Branch. 1985. Reproductive patterns of marine invertebrates. Oceanogr. Mar. Biol. Ann. Rev. 23: 373–398.

Hines, A. H. 1978. Reproduction in three species of intertidal barnacles from central California. Biol. Bull. 154: 262–281.

Huntley, M. E., and M. D. G. Lopez. 1992. Temperature-dependent production of marine copepods: a global synthesis. Am. Nat. 140: 201–242.

Hurley, A. C. 1973. Fecundity of the acorn barnacle *Balanus pacificus* Pilsbry: A fugitive species. Limnol. Oceanogr. 18: 386–393.

Johnston, I. A. 1990. Cold adaptation in marine organisms. Philos. Trans. Roy. Soc. Lond. B 326: 655–667.

Kinne, O. 1963. The effects of temperature and salinity on marine and brackish water animals. Oceanogr. Mar. Biol. Ann. Rev. 1: 301–340.

LaBarbera, M. L. 1989. Analyzing body size as a factor in ecology and evolution. Ann. Rev. Ecol. Syst. 20: 97–117.

MacKenzie, C. L. J. 1961. Growth and reproduction of the oyster drill *Eupleura caudata* in the York River, Virginia. Ecology 42: 317–338.

McLaren, I. A. 1963. Effects of temperature on growth of zooplankton, and the adaptive value of vertical migration. Can. J. Fish. Aquat. Sci. 20: 685–727.

————1966. Predicting development rate of copepod eggs. Biol. Bull. 131: 457–469.

McLaren, I. A., C. J. Corkett, and E. J. Zillioux. 1969. Temperature adaptations of copepod eggs from the arctic to the tropics. Biol. Bull. 137: 486–493.

Miller, S. E., and M. G. Hadfield. 1990. Developmental arrest during larval life and life-span extension in a marine mollusc. Science 248: 356–358.

Palmer, A. R. 1984a. Species cohesiveness and genetic control of shell color and form in *Thais emarginata* (Prosobranchia, Muricacea): Preliminary results. Malacologia 25: 477–491.

————1984b. Prey selection by thaidid gastropods: Some observational and experimental field tests of foraging models. Oecologia 62: 162–172.

————1985a. Genetic basis of shell variation in *Thais emarginata* (Prosobranchia, Muricacea). I. Banding in populations from Vancouver Island. Biol. Bull. 169: 638–651.

————1985b. Quantum changes in gastropod shell morphology need not reflect speciation. Evolution 39: 699–705.

Palmer, A. R., S. C. Gayron, and D. S. Woodruff. 1990. Reproductive, morphological, and genetic evidence for two cryptic species of northeastern Pacific *Nucella*. Veliger 33: 325–338.

Patel, B., and D. J. Crisp. 1960. Rates of development of the embryos of several species of barnacles. Physiol. Zool. 33: 104–119.

Perron, F. E. 1981a. The partitioning of reproductive energy between ova and protective capsules in marine gastropods of the genus *Conus*. Am. Nat. 118: 110–118.

———1981b. Larval biology of six species of the genus *Conus* (Gastropoda: Toxoglossa) in Hawaii, USA. Mar. Biol. 61: 215–220.

———1986. Life history consequences of difference in developmental mode among gastropods of the genus *Conus*. Bull. Mar. Sci. 39: 485–497.

Scheltema, R. S. 1967. The relationship of temperature to the larval development of *Nassarius obsoletus* (Gastropoda). Biol. Bull. 132: 253–265.

Schmidt-Nielsen, K. 1979. Animal Physiology. Cambridge University Press, New York.

Spight, T. M. 1975. Factors extending gastropod embryonic development and their selective cost. Oecologia 21: 1–16.

———1976. Ecology of hatching size for marine snails. Oecologia 24: 283–294.

Spight, T. M., and J. M. Emlen. 1976. Clutch sizes of two marine snails with a changing food supply. Ecology 57: 1162–1178.

Strathmann, M. F. 1987. Reproduction and Development of Marine Invertebrates of the Northern Pacific Coast, University of Washington Press, Seattle, 670 pp.

Strathmann, R. R., and M. F. Strathmann. 1989. Evolutionary opportunities and constraints demonstrated by artificial gelatinous egg masses. *In* J. S. Ryland and P. A. Tyler, ed. Reproduction, Genetics and Distributions of Marine Organisms. Olsen & Olsen, Fredensborg, Denmark, pp. 201–209.

Todd, C. D. 1991. Larval strategies of nudibranch molluscs: Similar means to the same end? Malacologia 32: 273–289.

16 Current Knowledge of Reproductive Biology in Two Taxa of Interstitial Molluscs (Class Gastropoda: Order Acochlidiacea and Class Aplacophora: Order Neomeniomorpha)

M. Patricia Morse

ABSTRACT Two taxa are important members of the molluscan assemblage in the interstitial environment, the well-represented gastropod order Acochlidiacea and the lesser known aplacophoran family, Meiomeniidae. Adaptations of the reproductive systems, discussed for representatives of these taxa, are consistent with a general reduction in the organ systems associated with minute species of molluscs. In the acochlidiaceans, four out of five families are interstitial; two of the families are composed of monoecious species and one of dioecious species. Spermatophores have evolved for sperm transfer with a subsequent reduction in accessory copulatory organs. In mature females, only a few large vitellogenic eggs are present at any one time, and where known, egg capsules with few developing embryos enclosed are reported. The aplacophorans, represented by the family Meiomeniidae, are monoecious, with small numbers of vitellogenic eggs and copulatory spicules.

Introduction

Adaptations have evolved for interstitial molluscs, ranging from 0.5 mm to 3.0 mm in length, to live in the pore spaces of sediments. Such adaptations are evident among representatives of the gastropod order Acochlidiacea and the aplacophoran order Neomeniomorpha, which constitute a major part of the molluscan assemblage characteristic of coarse sand habitats (Morse, 1987). These molluscs are cosmopolitan in distribution, occurring in subtidal and low intertidal habitats in both temperate and tropical waters. The reproductive biology of these interstitial species is poorly known. Swedmark (1959, 1964) was the first to consider adaptations of organisms living in the interstitial environment. He investigated the biological adaptations of interstitial molluscs and noted that they possess a small number of reproductive cells, reduced organ systems, and a tendency for encapsulated development of the larvae (Swedmark, 1968a).

Acochlidiaceans are shell-less opisthobranch gastropods, arranged in five families, Microhedylidae (*Unela, Microhedyle*), Asperspinidae (*Asperspina*), Ganitidae (*Ganitis, Paraganitis*), Hedylopsidae, (*Hedylopsis, Pseudunela*), and Acochlidiidae (*Acochlidium* and *Strubellia*) (Arnaud et al., 1986). The first four families of acochlidiaceans are minute (0.5 mm to 3 mm in length) and are known only from the interstitial environment. Members of the family Acochlidiidae are larger (30 mm in length) and are found mainly on Pacific Islands in freshwater streams where they live under rocks (Rankin, 1979). An interstitial freshwater form from the Caribbean

M. Patricia Morse, Marine Science Center and Biology Department, Northeastern University, Nahant, MA 01908.

was described by Rankin (1979) although no information on reproduction in the species was provided.

Aspects of the reproductive biology of acochlidiaceans have been described by Kowalevsky (1901), Challis (1968, 1970), Swedmark (1968a,b), and Morse (1976). In a monograph on the Acochlidiacea, Rankin (1979) reviewed various organ systems, including the reproductive system. However, her revisions were based solely on the literature, thereby creating new inconsistencies.

Within the class Aplacophora, only one family, the Meiomeniidae (Salvini-Plawen, 1985), is interstitial. The type species, *Meiomenia swedmarki* Morse, 1979, first found by Dr. Bertil Swedmark in sand from 58 fathoms at Reid Rock off San Juan Island, Washington, was recollected at the type locality and described by Morse (1979). A second species, *Meiomenia arenicola*, found off the coast of North Carolina, was scantily described by Salvini-Plawen (1985). Subsequently, large numbers were found in sands off the coast of Fort Pierce, Florida, and this species was redescribed (Morse and Norenburg, 1992). General information on reproduction in the aplacophorans can be found in Hyman (1967), Hadfield (1979), and, more recently, in Scheltema and Tscherkassky (in press).

It is the purpose of this chapter to review information on reproduction of representative interstitial molluscs and to discuss their adaptations for the interstitial environment. Some new observations and illustrations with light and electron microscopy on the reproductive anatomy are included.

Materials and Methods

Specimens of *Asperspina* sp. were collected from interstitial sediments taken from the base of Reid Rock, off San Juan Island. Sand was collected by the *RV Nugget* (Friday Harbor Laboratories) at 58 m with an anchor dredge outfitted with a canvas bag in the mouth of the dredge. *Unela nahantensis* was collected from low intertidal sand at Reid State Park, Maine. *Meiomenia arenicola* was collected from sand taken 6 miles off Fort Pierce, Florida, with a sled dredge in 16 m of water (Morse and Norenburg, 1992). Organisms were separated from the sediments by agitating sand in fresh sea water, collecting the suspended organisms on a 150-μm nylon screen, and resuspending them in petri dishes of sea water. For light microscopy, specimens were relaxed in 7 percent magnesium chloride in distilled water, fixed in Hollande's fixative, embedded in polyester wax, sectioned at 7 μm, and stained with Heidenhain's iron alum hematoxylin and eosin or Heidenhain's Azan. For transmission (TEM) and scanning (SEM) electron microscopy, organisms were routinely fixed in 3 percent glutaraldehyde in 2 M cacodylate buffer and postfixed in 1 percent osmium tetroxide. For SEM, organisms were critical point dried in ethanol and coated with gold-palladium. For TEM, organisms were embedded in LX-12 (Ladd Co.) and sectioned on a Porter-Blum 2B ultramicrotome. Thick sections were stained with 1 percent toluidine blue in borax. Thin sections were stained with uranyl acetate and lead citrate and viewed under a Philips 300 transmission electron microscope.

Reproduction in the Interstitial Acochlidiaceans

Within the four interstitial families of acochlidiaceans, members of the Microhedylidae and the Ganitidae are dioecious while those of the Asperspinidae and the Hedylopsidae are monoecious. The exact location of the minute freshwater form from the Caribbean, *Tantulum elegans* Rankin, 1979, is not known, and no information on the reproductive system was included in Rankin's description. Reproduction in dioecious species of the family Microhedylidae has been described by Marcus (1953), Marcus and Marcus (1954), Doe (1974), and Westheide and Wawra (1974). In all specimens examined, the thin-walled gonad has an undivided glandular and ciliated duct that leads to a gonopore at the junction of the head-foot and visceral mass on the right side of the body. Spermatophores have been reported, and it has been assumed that the eggs are fertilized within the body cavity (Wawra, 1992). Spermatophores have been reported in other members of the Microhedylidae, that is, in three species in the genus *Microhedyle* (Hertling, 1930; Swedmark, 1968b; Westheide and Wawra, 1974); and in one species of *Unela* (Doe, 1974).

Unela nahantensis Doe, 1974, has been found in intertidal coarse habitats at Nahant, Massachusetts, and Reid State Park, Maine, during most of the year. As described by Doe (1974), males and females can be readily differentiated in squash preparations of the adults. Externally, the females have a groove along the right side of the animal from the genital pore at the junction of the head-foot and the visceral mass to the base of the right tentacle. The male canal is not obvious externally; it lies within the epidermis of the right lateral body wall, extending from the genital pore anteriorly to a position under the base of the right rhinophore (Robinson and Morse, 1976). Internally, the female reproductive system consists of a posterior saclike ovary that runs adjacent to, and fills, the area central to the dorsal tubular digestive gland (Fig. 16.1a). The oviduct courses anteriorly from the ovary to the genital opening. Several large vitellogenic eggs and a series of smaller eggs with less yolk can be seen within the ovary. This species differs from other species of acochlidiaceans (as seen in *Ganitis* sp., Fig. 16.2e) in the scarcity of vitellogenic eggs, as mature specimens of *U. nahantensis* seldom have more than one to three eggs with a full complement of yolk at any one time. The testis of *U. nahantensis* is a posterior sac that, like the ovary, parallels the dorsal tubular digestive gland (Fig. 16.1b). The testis empties mature sperm in packets into a saccular seminal receptacle that connects with a ciliated glandular duct leading to the gonopore at the junction of the head-foot and visceral mass on the right side of the animal. I have observed numerous *U. nahantensis* in finger bowls on a water table and have noted that the males may attach more than one spermatophore to another animal. Figure 16.1b shows a spermatophore attached to the right of the head-foot of *U. nahantensis,* although it could not be determined whether it was released from the gonopore or was attached to the body by a male. Spermatophores were observed associated with various surfaces of the molluscs in different stages of emptying their contents into the body cavity. Also, sperm were seen within the hemocoel of the recipients. After the spermatophores are attached to the body surface, the tips fuse with the epidermis and lyse the underlying epidermal cells. During a 2-h period, sperm empty into the hemocoel, and the emptied spermatophore sac eventually falls off. However, how fertilization and development of the eggs and deposition of the egg capsules occur in these species is not known.

In the Ganitidae two genera, *Ganitus* (Marcus, 1953) (Fig. 16.2) and *Paraganitus* (Challis 1968) are dioecious species. As described by Challis (1968) and observed by me in 1979 in animals from Guadalcanal, the gonads in *Paraganitis ellynnae* are tubular and closely associated with the digestive gland. In the female, the ovary releases eggs into a ciliated oviduct which leads through albumen, membrane, and mucous glands to discharge on the right side near the kidney opening at the region of the junction of the head and foot. The testis empties into a seminal vesicle that connects to a ciliated ejaculatory duct opening near the kidney pore. *Paraganitis ellynnae* form spermatophores. Spermatophores also were found attached to *Ganitus sp.* that were collected in North Carolina, observed, and photographed by Professor W. Westheide (Fig. 16.2). Within the thin-walled spermatophores, the sperm are tightly packed (Figs. 16.2c,d).

Members of the Asperspinidae, *Asperspina riseri* (Morse, 1976) and a very similar, undescribed *Asperspina* sp. from the San Juan Islands, Washington, are simultaneous hermaphrodites.

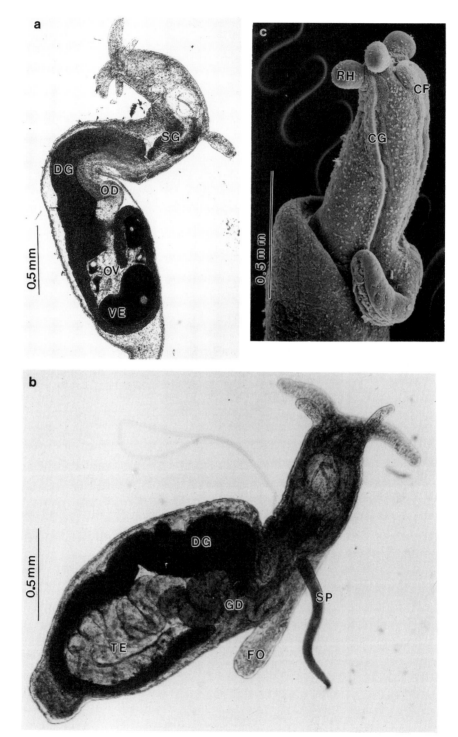

FIGURE 16.1. *a. Unela nahantensis* Doe, 1974—female with a large vitellogenic egg (VE) (DG, digestive gland; OD, oviduct; OV, ovary; SG, salivary gland). *b. Unela nahantensis.*—Male with spermatophore (SP) in close association with gonopore region on right side of body near the junction of head foot (note the protruding foot) and visceral mass (DG, digestive gland; GD, genital duct; FO, foot; TE, testis). *c. Asperspina* sp—scanning electron micrograph of anterior end showing external ciliated groove (CG) (CF, ciliated foot; RH, rhinophore).

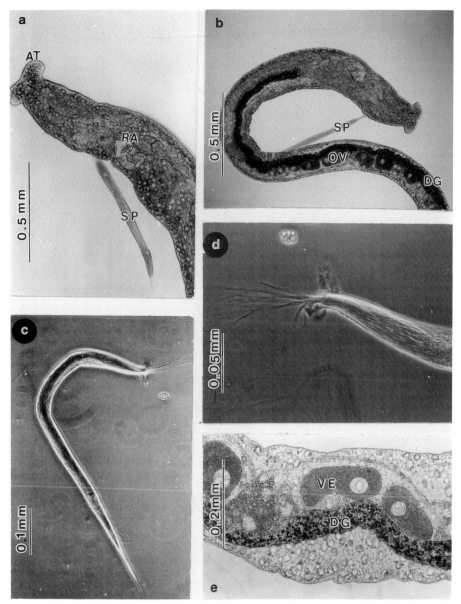

FIGURE 16.2. *Ganitis* sp. *a.* Crawling animal with spermatophore (SP) attached to the left side (dorsal view). Note the single pair of anterior tentacles (AT) and single row of teeth of radula (RA). *b.* Female with attached spermatophore (SP). The ovary (OV) contains a series of vitellogenic eggs (DG, digestive gland). *c.* Spermatophore filled with sperm with the attachment end in the upper right of the picture. *d.* Closeup of attachment end of spermatophore with sperm in dense packets within and with some sperm protruding from the tip. *e.* Closeup of four vitellogenic eggs (VE) in the ovary which parallels the thin tubular digestive gland (DG).

These species do not have a penis, and both form spermatophores. In the Washington species, there is a ciliated groove from the genital pore to the right tentacle (Fig. 16.1c), and the gonads lie ventrally, closely associated with the digestive gland (Fig. 16.3). The ovary is loosely packed with eggs in different stages of vitellogenesis. Only one or two eggs are fully developed at a time (Fig. 16.3b). Immature eggs often appear lightly stained in sections (Fig. 16.3c). The testis lies ventral

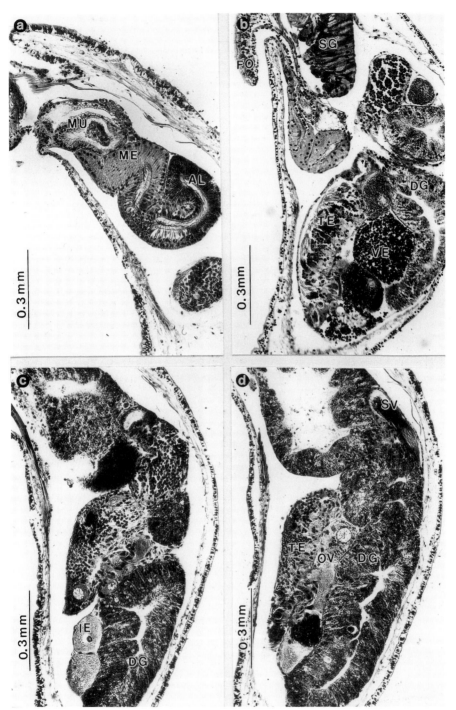

FIGURES 16.3. Light micrographs of longitudinal sagittal sections from a single specimen of *Asperspina* sp. collected at Reid Rock, San Juan Island, Washington. Anterior is at the top of each picture, and dorsal surface is to the right. *a.* Secretory area of hermaphroditic duct (AL, albumen gland; ME, membrane gland; MU, mucous gland). *b.* Hermaphroditic gland with large vitellogenic eggs (VE) and testis (TE) (DG, digestive gland; SG, salivary gland; FO, foot). *c.* Hermaphroditic gland with immature eggs (IE) below digestive gland (DG). *d.* Area of digestive gland (DG) showing testis (TE), ovary (OV), and seminal vesicle (SV).

and to the left of the ovary in the posterior visceral mass (Fig. 16.3d). Mature sperm pass into a seminal vesicle (Fig. 16.3d), where they are packaged and then presumably pass through the male part of the hermaphroditic duct. In sections, the female part of the hermaphroditic duct has distinct areas that correspond to the albumen, membrane, and mucus areas of other opisthobranchs (Fig. 16.3a). A similar reproductive system has been described for *Asperspina riseri* (Morse, 1976). Furthermore, in *A. riseri*, I observed two developing egg capsules containing three embryos. At the end of two months, a crawl-away juvenile emerged from one capsule and was positively identified as *A. riseri* (Morse, 1976).

Members of the Hedylopsidae are monoecious. Two species, *Pseudunela coronuta* (Challis, 1970) and *Hedylopsis spiculifera* Kowalesky, 1901, do not produce spermatophores. However, a penis is present for sperm transfer. *Pseudunela coronuta* has a tubular ovotestis compressed beneath the digestive gland with male and female regions situated in different areas of the same follicles (Challis, 1970). A ciliated hermaphroditic duct leads through the lumens of the albumen and mucous glands, then bifurcates into an oviduct and a vas deferens. The oviduct opens into a ciliated cloaca; a blind-ended sac for allosperm, the bursa copulatrix, and the intestine all connect to the cloaca. The vas deferens courses anteriorly to the base of the right rhinophore as a closed ciliated intraepidermal duct. The ciliated duct runs back from the right rhinophore, where it connects to a large prostate gland and a penial gland before leading to the base of the penis, a complex muscular organ, equipped with a spine. In my observations of this species taken from the type locality in Marau Sound, East Guadalcanal, in the Solomon Islands, the penis structure was associated with the left anterior side of the head-foot complex. According to Wawra (1992), *Hedylopsis spiculifera* is a sequential hermaphrodite with an open seminal groove and without a bursa copulatrix for allosperm. He concludes that the penial spine is used as an hypodermic needle to inject sperm through the body wall during copulation.

Reproduction in the Interstitial Aplacophorans

Aplacophorans in the order Neomeniomorpha are hermaphroditic. Two neomenioid aplacophorans, the type of the genus, *Meiomenia swedmarki* Morse, 1979, and *Meiomenia arenicola*, Salvini-Plawen, 1985, have been studied in detail. The gonads are paired tubular organs that are dorso-medially located above the gut (Hadfield, 1979), and the female portion occupies the greatest part of the gonad. In living specimens (Figs. 16.4a,b) and in sections (Fig. 16.4c), several eggs in varying stages of vitellogenesis and, often, one fully formed egg with large yolk granules (Figs. 16.4c and 16.5a,b), can be seen. In sections of a specimen of *M. arenicola*, a single egg was adjacent to a ciliated opening (Fig. 16.5a) presumably leading to the outside. Because of the interdigitated appearance of the ciliated epithelial cells, the duct appears to have the potential to expand and accommodate the large egg. The male part of the gonad lies anterior to the female part. In both species of *Meiomenia*, well-developed copulatory spicules are associated with coelomoducts near the entrance to the mantle cavity (Figs. 16.4b,d). In *M. swedmarki*, the copulatory spicules are spoon shaped and have distinct grapelike prostatic glands associated with them (Morse, 1979). In *M. arenicola* copulatory spicules taper evenly to a point, and lobate glandular tissue, assumed to be prostatic, is situated at the base of the spicules (Fig. 16.4d). Sheets of muscles and some unidentified tubules are also present.

Both Jon Norenburg and I have observed many living specimens of *Meiomenia arenicola* and have concluded that the copulatory spicules are not always present, even in specimens with various stages of vitellogenic eggs. Whether the spicules are lost during copulation or formed only when sperm is transferred is not known. Because copulatory spicules often are lost in the process

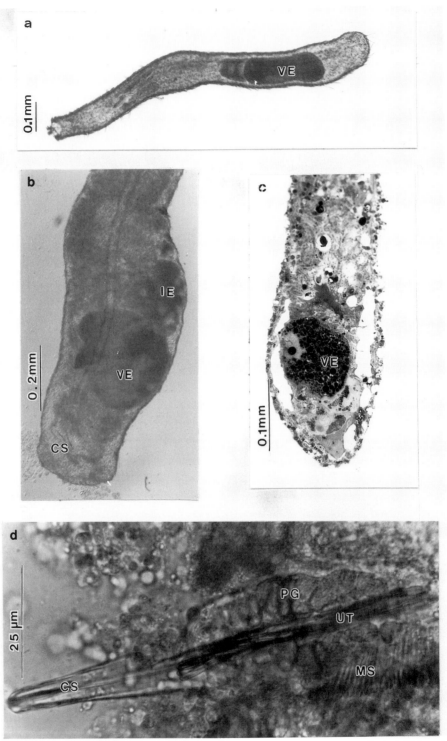

FIGURE 16.4. *a.* Crawling *Meiomenia arenicola* showing large vitellogenic egg (VE) in posterior portion of body. *b. Meiomenia swedmarki* from San Juan Island with large vitellogenic egg (VE) and group of immature eggs (IE). Note copulatory spicules (CS) at posterior end. *c.* One-micron longitudinal frontal Epon section of *Meiomenia* arenicola showing vitellogenic egg (VE). *d.* Squash preparation of copulatory spicule (CS) from *Meiomenia arenicola* showing muscles (MS), prostatic glands (PG), and bundle of unidentified tubules (UT).

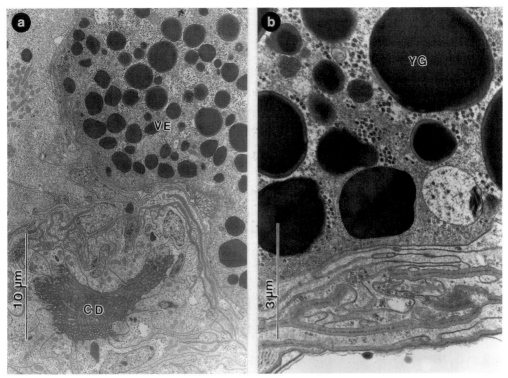

FIGURE 16.5. *a* Electron micrograph of *Meiomenia arenicola* with a mature vitellogenic egg (VE) in the upper right and the ciliated duct (CD) below left in the micrograph. *b*. Electron micrograph of the yolk granules (YG) in the egg of *Meiomenia arenicola*.

of fixation (due to their calcareous nature), or may vary in occurrence, their presence or absence is not a good diagnostic character.

General Discussion and Conclusions

Acochlidiaceans and the aplacophorans that live between sand grains in coarse sand environments show similar adaptations with respect to their reproductive systems. As compared to macrofaunal molluscs, the small size of the interstitial organisms necessitates a reduction; this reduction is seen in cell numbers and organ systems. Low numbers of vitellogenic female sex cells in mature animals are characteristic of both taxa, and, in acochlidiaceans, a tendency to evolve spermatophores rather than male accessory glands and copulatory apparatus is evident. Low numbers of large eggs suggest longer reproductive periods and low reproductive output. This pattern has been confirmed in *Asperspina riseri* (Morse, 1976). The mode of development in the interstitial members of the Aplacophora, however, remains a mystery.

Sperm transfer has been observed in the Acochlidiacea, but little is known about how interstitial aplacophorans exchange sperm. Copulation is assumed to be due to the copulatory spicules. In the acochlidiaceans, there has been a reduction of the male copulatory structures and a definite trend toward the formation of spermatophores in most of the interstitial species. The presence of accessory glands and a penis in several marine interstitial species seems to indicate the ancestral form; the subsequent loss is a successful adaptation for radiation of the acochlidiacean species in the interstitial habitat. It is interesting that the insular species in the family

Acochlidiidae, from Pacific Island habitats where perhaps predation is reduced, are the macrofaunal acochlidiaceans, which have retained the penis and accessory reproductive structures.

Acknowledgments

The author is particularly grateful to Dr. Wilfried Westheide for providing the photos of living *Ganitis* (Figure 16.2) and to Bill Fowle for his help with electron microscopy. This chapter was presented as part of a celebration to honor the life and scientific accomplishments of Dr. C. Gardner Reed, who for many years was a colleague and friend. Christopher set new heights for excellence and new thoughts of the importance of combining research and teaching in the undergraduate academic world. His life was too short. I am grateful to Chris's close friends, Drs. George Shinn, Steve Stricker and Herb Wilson, for organizing this memorable and fitting occasion. This chapter is dedicated to Chris's daughter, Erica Irene, with the hope that through her mother Paula's help, she will acquire her dad's love of science and the natural world.

Literature Cited

Arnaud, P. M., C. Poizat, and L. v. Salvini-Plawen. 1986. Marine interstitial Gastropoda (including one freshwater interstitial species). Stygofauna Mundi 1:153–161.

Challis, D. A. 1968. A new genus and species of the order Acochlidiacea Mollusca: Opisthobranchia) from Melanesia. Trans. Roy. Soc. N. Z., Zool. 10: 191–197.

———. 1970. *Hedylopsis coronuta*; and *Microhedyle verrucosa*, two new Acochlidiacea (Mollusca: Opisthobranchia) from the Solomon Islands Protectorate. Trans. Roy. Soc. N. Z. Biol. Sci. 12: 29–40.

Doe, D. A. 1974. A new species of the order Acochlidiacea (Opisthobranchia: Microhedylidae) from New England. Trans. Am. Micros. Soc. 93: 241–247.

Hadfield, M. G. 1979. Aplacophora. *In* A. C. Giese and J. Pearse, eds. Reproduction of Marine Invertebrates, Vol. 5. Academic Press, New York, pp. 1–25.

Hertling, H. 1930. Uber eine Hedylide von Helgoland und Bemerkungen zur Systematik der Hedyliden. Wiss. Meeres. Helgo. 18: 1–10.

Hyman, L. H. 1967. The Invertebrates. Vol. 6. Mollusca. McGraw-Hill, New York.

Kowalevsky, A. 1901. Les Hedylides, etude anatomique. Mem. Acad. Imp. Sci. St. Petersbourg 12: 1–32.

Marcus, E. 1953. Three Brazilian sand-Opisthobranchia. Bol. Fac. Cienc. Univ. Sao Paulo 18: 165–203.

Marcus, E., and E. Marcus. 1954. Uber Philinoglossacea und Acochlidiacea. Kiel. Meeresfors. 10: 215–223.

Morse, M. P. 1976. *Hedylopsis riseri* sp. n., a new interstitial mollusc from the New England Coast (Opisthobranchia, Acochlidiacea). Zool. Scripta 5: 221–229.

———. 1979. *Meiomenia swedmarki* gen. et sp. n., a new interstitial solenogaster from Washington, USA. Zool. Scripta 8: 249–253.

———. 1987. Distribution and ecological adaptations of interstitial molluscs in Fiji. Am. Malac. Bull. 5: 281–286.

Morse, M. P., and J. L. Norenburg. 1992. Observations on and redescription of *Meiomenia arenicola* Salvini-Plawen, 1985 (Mollusca: Aplacophora), an interstitial solenogaster from Fort Pierce, Florida. Proc-Biol. Soc. Wash. 105: 674–682.

Rankin, J. J. 1979. A freshwater shell-less mollusc from the Caribbean: Structure, biotics, and contribution to a new understanding of the Acochlidioidea. Life Sci. Contr. Roy. Ontario Mus. 116: 1–123.

Robinson, W. E., and M. P. Morse. 1976. Histochemical investigation of the pedal glands and glandular cells of *Unela nahantensis* Doe, 1974 (Opisthobranchia: Acochlidiacea). Trans. Am. Microsc. Soc. 98: 195–203.

Salvini-Plawen, L. v. 1985. New interstitial solenogastres (Mollusca). Stygologia 1: 101–108.

Scheltema, A. H., and M. Tscherkassky. 1993. Aplacophora. *In* R. W. Harrison. ed. Microscopic Anatomy of Invertebrates, Vol. 5. Mollusca. Wiley-Liss, New York, (in press).

Swedmark, B. 1959. On the biology of sexual reproduction of the interstitial fauna of marine sand. Fifteenth Int. Cong. Zool. 5: 1–3.

———. 1964. The interstitial fauna of marine sand. Biol. Rev. 39: 1–42.

———. 1968a. The biology of interstitial Mollusca. Symp. Zool. Soc. London 22: 135–149.

———. 1968b. Deux espèces nouvelles d'Acochlidiacees (Mollusques: Opisthobranches) de la faune interstitielle marine. Cah. Biol. Mar. 9: 175–186.

Wawra, E. 1992. Sperm transfer in Acochlidiacea. *In* F. Giusti and G. Manganelli, eds. Eleventh Int. Malacol. Cong. University of Siena Press, Siena, Italy, p. 103.

Westheide, W., and E. Wawra. 1974. Organisation, Systematik, und Biologie von *Microhedyle cryptophthalma* nov. spec. aus dem Brandungsstrand des Mittlemeeres. Helgo. Meeres. 26: 27–41.

17 The Ancestral Gastropod Larval Form Is Best Approximated by Hatching-Stage Opisthobranch Larvae: Evidence from Comparative Developmental Studies

Louise R. Page

ABSTRACT I have compared developmental timing for eight structures in planktotrophic larvae of prosobranch and opisthobranch gastropods. This comparison, based on data collected from published literature and from my own observations on opisthobranch morphogenesis, is relevant to the ongoing debate about the ancestral gastropod life history. The most obvious result is that developing opisthobranchs become feeding, free-swimming larvae at a stage of morphogenesis that compares to a prehatching developmental stage in caenogastropods (meso- and neogastropods). When planktotrophic caenogastropod larvae hatch, or shortly thereafter, many features of the definitive (postmetamorphic) body have become superimposed on the basic veliger form, which is epitomized by young opisthobranch larvae. The definitive nephridium, deepened mantle cavity, propodial anlage, larval and adult hearts, cephalic eyes and tentacles, and pedal branch of the larval retractor muscle begin to appear either before or soon after hatching in caenogastropods, but these structures are usually initiated during the second half of larval development or during metamorphosis in opisthobranchs with planktotrophic larvae. Furthermore, there are preliminary indications that the 90° torsion and possibly the hyperstrophic shell of opisthobranch larvae may occur during a transient stage of prosobranch embryogenesis. Among archaeogastropods (none of which have feeding larvae), the basic veliger form has been almost completely supplanted by the very rapid development of juvenile/adult structures. The most straightforward interpretation of these results is that opisthobranch life histories have conserved an ancestral larval form, and morphogenetic programs that generate larvae of extant caenogastropods and archaeogastropods are derived from the early opisthobranch pattern. This interpretation is supported by outgroup comparisons at the ultrastructural level between young opisthobranch veligers and polychaete trochophores.

Introduction

Molluscan development has been studied for many years from many different perspectives, but there is still controversy about the ancestral pattern and subsequent diversification of molluscan life histories. Was the initial product of the ancestral ontogeny a larval stage or a crawl-away juvenile? If a larval stage existed, was it a feeding trochophore or something else? I will argue that further comparative studies of molluscan morphogenesis may yet provide answers to these questions. Indeed, the current controversy about the ancestral life history focuses attention on the

Louise R. Page, Department of Biology, University of Victoria, Victoria, British Columbia, V8W 2Y2.

patchy and often fragmentary nature of existing developmental data, particularly data that form the foundation for current ideas about the evolution of the gastropod body plan.

Jägersten (1972), Strathmann (1978), and Nielsen (1985) have argued that the molluscan veliger, which captures food with opposed bands of pre- and postoral cilia, originated from a feeding trochophore. Conversely, Salvini-Plawen (1969, 1980a, 1985) has discounted the similarities between trochophores and veligers as convergences and has promoted the theory of an ancestral, nonfeeding pericalymma (test-cell) larva for the Mollusca. Bandel (1982) has suggested that the original molluscan ontogeny proceeded directly from a swimming, nonfeeding trochophore (with preoral ciliary band only) to the juvenile stage. He suggests that the planktotrophic veliger has been secondarily superimposed on this direct life cycle by acquisition of a postoral ciliary band, food groove, and other features of both hard and soft veliger anatomy. Salvini-Plawen and Bandel justify their rejection of feeding larvae in ancestral molluscs by citing nonfeeding larval types in presumably primitive molluscan groups, including aplacophorans, archaeogastropods, polyplacophorans, and protobranch bivalves. Finally, Chaffee and Lindberg (1986) have concluded that the small size of Cambrian molluscs (most with shell lengths of less than 10 mm) would have precluded successful reproduction via planktotrophic larvae. Indeed, these authors suggest that the first mollusc crawled from the egg case as a juvenile.

For gastropods, questions about evolution of life history have been interwoven for many years with questions about morphological evolution. Doris Crofts (1937, 1955) hoped that developmental study of *Haliotis tuberculata* would yield information about early gastropod evolution, because adults of extant diotocardians exhibit what may be primitive forms of adult gastropod characters. Planktotrophy has not been convincingly demonstrated for larvae of any extant archaeogastropod. Crofts' observations on morphogenetic events accompanying 180° torsion, and her interpretation of the ancestry of various larval muscles, have had great influence on current ideas about gastropod origins and evolution. However, we cannot assume that the developmental program in *Haliotis* is ancestral merely because *adult* archaeogastropods appear to conserve primitive gastropod characters. As Garstang (1922) pointed out long ago in his incisive rebuttal of Haeckel's biogenetic law, developmental stages can be selected independently of adult stages (Raff and Kaufman [1991] provide a contemporary treatment of this topic). For the same reason, the derived postmetamorphic body form of many opisthobranchs is not sufficient reason by itself to ignore opisthobranch morphogenesis in discussions about gastropod evolution. Nevertheless, these prejudices are implicit in traditional notions about the origin and evolution of the gastropod body plan, and they also ramify through the controversy about the ancestral life history of gastropods.

Hadfield and Switzer-Dunlap (1984) recognized a basic life history pattern among opisthobranchs that includes internal fertilization, small eggs deposited within a benthic egg mass, and a planktotrophic larva (also see Thompson, 1967). They suggested that this pattern "undoubtedly represents the condition in one or more forebears derived from early Prosobranchia." Consensus holds that an ancient caenogastropod gave rise to the opisthobranch and pulmonate lineages (Fretter and Graham 1962; Gosliner 1981; Haszprunar 1985; Robertson 1985). The question that should be asked is: Does any extant caenogastropod retain a developmental pattern that could have generated the basic opisthobranch life history? A review of available literature, which is presented below, suggests that the answer is No. Embryogenesis in the basic opisthobranch life history proceeds directly towards the formation of a simple veliger that hatches as soon as the velar feeding apparatus and gut have differentiated. The initial larval phase is concerned mainly with growth. Alternatively, during caenogastropod ontogenies, numerous structures of the definitive body plan are initiated and may even become functional either before or shortly after a feeding veliger hatches. In archaeogastropods, these heterochronies are even more pronounced, and a feeding veliger is never generated.

In this chapter, I present the hypothesis that newly hatched, planktotrophic opisthobranch veligers may be the best extant approximation of the ancestral gastropod larva, and veligers of prosobranchs are generated from derived morphogenetic programs. To support this claim, I will first document evidence showing that prosobranch veligers incorporate many features of the definitive body, even during the early larval stage. Opisthobranch veligers do not display these heterochronies until the second half of the obligatory larval phase or after metamorphosis. To further buttress the hypothesis, I will itemize the many similarities between young planktotrophic opisthobranch larvae and planktotrophic polychaete trochophores. A number of these similarities require ultrastructural observations.

Light and electron micrographs shown in this chapter were obtained according to methods described in Bickell and Chia (1979), Bickell and Kempf (1983), and Page (1992a, formerly Bickell).

Comparisons of Developmental Programs among Gastropod Larvae

Differences between the morphogenetic programs of planktotrophic opisthobranch and caenogastropod larvae are revealed by comparing the times at which eight characters begin to develop relative to hatching and metamorphosis. In the following comparisons of these developmental characters, I will also include relevant information concerning developing archaeogastropod larvae, although none of these are planktotrophic. These data were collected from published literature and from personal observations on developing opisthobranch larvae.

Protonephridia and Definitive Nephridium

The topic of gastropod larval excretory structures has had a long and convoluted history (see Raven, 1958; Fioroni, 1966; Bonar, 1978a). Recent studies have begun to resolve some of this confusion. Bartolomaeus (1989) has reported that the paired "nephrocysts" in larvae of the nudibranch *Aeolidia papillosa* are large duct cells of protonephridia. I can confirm this identification for larvae of the notaspidean *Pleurobranchaea californica* and for three nudibranch larvae: *Doridella steinbergae* (Fig. 17.1), *Rostanga pulchra*, and *Melibe leonina*. The paired protonephridia of these species are present from before hatching until metamorphosis. Protonephridia have been recognized for many years in pulmonate embryos (Raven, 1958).

The rudiment of the adult nephridium (kidney) in hatching nudibranch veligers (*D. steinbergae*, *R. pulchra*, and *M. leonina*) consists of several undifferentiated cells adjacent to the mantle gland. The cells multiply slowly, but they do not begin morphological differentiation until the last half of the larval phase.

Fretter and Graham (1962) have stated that protonephridia are absent in all free-swimming caenogastropod larvae. However, Ruthensteiner and Schaefer (1991) have provided histological evidence for protonephridia in prehatching embryos of *Nassarius reticulatus*. The protonephridial ducts are intimately associated with the large cells of the so-called larval kidneys, which bulge from the neck region of many caenogastropod embryos. Although it has been thought for many years that the larval kidneys are involved in excretion (see Raven, 1958; Fretter and Graham, 1962), Rivest (1992) has recently demonstrated that the protruding superficial cell of each larval kidney complex in *Searlesia dira* absorbs intracapsular albumen proteins. Furthermore, ultra-

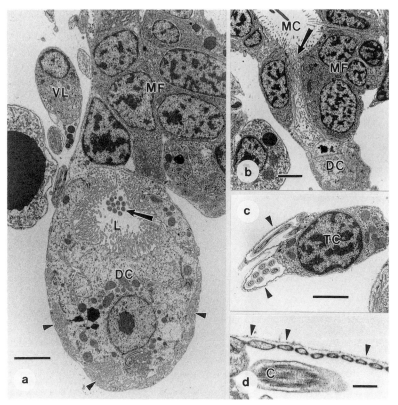

FIGURE 17.1. Transmission electron micrographs (TEM's) of sections through a protonephridium in a larval nudibranch (*Doridella steinbergae*) at 9 days post-hatching. *a.* Large duct cell (DC) adjacent to right mantle fold (MF) and visceral loop connective (VL). Note the central lumen (L), which is lined by microvilli and contains the distal part of the ciliary tuft (arrow). The outer cell membrane is elaborated into numerous fingerlike extensions (arrowheads) (scale bar = 2 μm). *b.* Nephridiopore (arrow) opening into the right mantle cavity (MC) (DC = duct cell. MF = mantle fold; scale bar = 2 μm). *c.* Terminal cell (TC) with a collar of elongate cytoplasmic processes (arrowheads) surrounding the ciliary tuft (scale bar = 2 μm). *d.* Detail of cytoplasmic processes of a terminal cell showing fibrillar material (arrowheads) covering the processes (scale bar = 0.2 μm).

structural observations suggest that other cells of the larval kidney complex in *S. dira* are vestigial protonephridial components (Rivest, 1992).

Unlike the situation in opisthobranch larvae, the protonephridia of *N. reticulatus* and the possibly vestigial protonephridial elements of *S. dira* degenerate prior to hatching (veligers of *S. dira* metamorphose within the egg capsule). Instead, the definitive adult nephridium begins to differentiate before larval hatching in many caenogastropods (Pelseener, 1911; D'Asaro, 1966, 1969; Werner, 1955; Bedford, 1966; Fioroni, 1966; Thiriot-Quiévreux, 1969), or shortly after hatching as in *Strombus gigas* (D'Asaro, 1965) and *Lamellaria perspicua* (Fioroni and Meister, 1976). This nephridium achieves large size and is apparently functional in posthatch planktotrophic veligers of caenogastropods. Histological studies on archaeogastropod larvae have not identified protonephridia or possible candidates.

Larval and Adult Hearts

The larval heart in gastropod larvae is a muscular tube located beneath the floor of the mantle cavity, behind the velum. It begins pulsating shortly before hatching in many caenogastropod larvae (Werner, 1955; Fretter and Graham, 1962; D'Asaro, 1965, 1966, 1969; Bedford, 1966; Fioroni, 1966). According to D'Asaro, the time of differentiation of the adult heart is variable, occurring simultaneously with the larval heart in *Distorsio clathrata* (i.e., before hatching), or within a week of hatching in *Strombus gigas*. The adult heart also begins beating in the early larval stage of veligers belonging to the heteropod genus *Atlanta* (Thiriot-Quiévreux, 1969).

Planktotrophic opisthobranch larvae complete at least half of obligatory larval development before the larval heart begins contractions (Kriegstein, 1977a; Kempf and Willows, 1977; Switzer-Dunlap 1978; Chia and Koss 1978; Bickell and Chia 1979; Todd, 1981; Bickell and Kempf 1983; Paige, 1988; Tsubokawa and Okutani 1991). The adult heart of opisthobranchs, unlike that of caenogastropods, usually starts beating during or after metamorphosis, although Franz (1971) and Tsubokawa and Okutani (1991) report contractions of the adult heart in late-stage larvae of *Acteocina canaliculata* and *Pleurobranchaea japonica*, respectively.

Crofts (1937) and Smith (1935) found that morphogenesis in archaeogastropods proceeds directly to the adult heart without a preceding larval heart.

Propodium

The propodium of gastropods develops as an expansion of the antero-ventral portion of the foot, which eventually gives the foot a planar crawling surface. At the time of hatching, a propodium is not present in typical planktotrophic larvae of either opisthobranchs or caenogastropods, but it begins to form during the early to middle larval phase in caenogastropods (Werner, 1955; Scheltema, 1962; D'Asaro, 1965, 1966). Fretter (1967: 357) states that the propodium of caenogastropod larvae is "active in cleaning the velar edge and other surfaces of the body on which unwanted particles may collect, and also helps in swimming."

In planktotrophic opisthobranch larvae, the expansion of the propodium is one of the last events of obligatory larval morphogenesis (Franz, 1971; Kriegstein, 1977a; Switzer-Dunlap, 1978; Bickell and Kempf, 1983; others reviewed in Todd, 1981). Full propodial development allows crawling and signals metamorphic competence in opisthobranch veligers (see Bonar, 1978a).

Nervous System and Sensory Structures

It is difficult to compare patterns of gastropod neurogenesis because times given for the first appearance of various neural components depend on the resolving power of the technique used for analysis (whole mounts, histological sections ranging from 1 to 10 μm thickness, or ultrathin sections). However, eyespots are easily seen, even in live veligers, due to their pigmentation. As a result, there is excellent documentation of prehatching differentiation of eyespots in plank-totrophic caenogastropod larvae (Pelseener, 1911; Werner, 1955; Scheltema, 1962; D'Asaro, 1965, 1966, 1969; Bedford, 1966; Thiriot-Quiévreux, 1969; Robertson, 1983; among many others). However, in typical planktotrophic opisthobranch larvae, eyespots appear around the midpoint of the obligatory larval phase (Kriegstein, 1977a; Switzer-Dunlap, 1978; Bickell and Kempf, 1983; Tsubokawa and Okutani, 1991; others reviewed by Todd, 1981). Planktotrophic veligers of the cephalaspid *Acteocina canaliculata* are an exception because the right eyespot is present at hatching (Franz, 1971). These eyespots become the definitive cephalic eyes of the postmetamorphic stage; they are not comparable to the posttrochal larval ocelli of polyplaco-phoran and some bivalve larvae.

Planktotrophic caenogastropod larvae usually have one or both cephalic tentacles at hatching (see references given for eyespots), whereas premetamorphic appearance of rhinophores has been described for only three opisthobranch larvae (Thiriot-Quiévreux, 1977; Chia and Koss, 1982;

Tsubokawa and Okutani, 1991). The statocysts appear early in the embryogenesis of both opisthobranchs and caenogastropods.

All major central nervous system ganglia except the buccal ganglia are present in hatching veligers of three caenogastropods studied by D'Asaro (1966, 1969). Even in paraffin sections (8 to 10 μm thickness), many of these ganglia and their connectives could be resolved during later embryogenesis. Bedford (1966) recognized cerebral, pedal, pleural, visceral, and buccal ganglia in *Bembicium nanum*. However, in hatching-stage veligers of *Strombus gigas* (D'Asaro, 1965) and *Lamellaria perspicua* (Fioroni and Meister, 1976) only the cerebral and pedal ganglia were distinguished using paraffin sections.

Hatching opisthobranch veligers have a small pair of cerebral ganglia and an even smaller pair of pedal ganglia (Kriegstein, 1977a,b; Page, 1992a,b); the latter are easily overlooked in histological sections (compare results of Bickell and Kempf [1983] to Page [1992b]). Electron microscopical observations by Schacher et al. (1979) revealed anlagen of several visceral loop ganglia in hatching veligers of *Aplysia californica*. However, each anlage consists of only two to five cells at hatching stage, and these were too small to be resolved in the 1- to 2-μm histological sections used by Kriegstein (1977a,b). Ultrastructural observations on the nudibranch *Melibe leonina* showed a complete visceral loop connective in hatching veligers, but neurons of visceral loop ganglia begin to ingress from the ectoderm slightly later (Page 1992a,b). Subsequent development of the central nervous system in both *Aplysia* and *Melibe* is a long-term process that occurs gradually during the entire course of larval development.

The definitive nervous system of archaeogastropods develops very rapidly during the relatively short period preceding velar loss. According to Crofts (1937), the cerebral, pedal, and pleural ganglia of *Haliotis tuberculata* begin to appear within 2 days of fertilization. By the time the velum is lost at 9 to 14 days postfertilization, all ganglia and connectives of the definitive nervous system are recognizable. Smith (1935) describes a similar pattern of nervous system development in *Patella vulgata*.

Larval Retractor Muscle

All pelagic gastropod veligers have a larval retractor muscle (LRM) that, at least initially, is anchored on the posterior wall of the shell. According to Fretter (1967, 1969, 1972), the LRM of planktotrophic caenogastropod veligers includes cephalic tracts, which insert on the velum as well as the distal foregut, and pedal tracts. Developmental descriptions of *Bursa corrugata*, *Distorsio clathrata*, and *Thais haemastoma floridana* (D'Asaro, 1966, 1969), *Crepidula fornicata* (Werner, 1955), and *Atlanta* spp. (Thiriot-Quiévreux, 1969) indicate that both the cephalic and pedal components of the LRM are present in hatching-stage veligers of these caenogastropods. However, in an ultrastructural study of muscle morphogenesis in planktotrophic nudibranch larvae (Page, in prep.), I found that the pedal component of the LRM does not appear until the midpoint of obligatory larval development. This tract projects initially to the pedal epithelium lining the operculum, but it later includes fibers projecting to the ventral sole of the foot. In younger nudibranch veligers and in newly hatched veligers of *Pleurobranchaea californica*, a ventral branch of the LRM *appears* to extend into the base of the small foot, but in older larvae (particularly after propodial development) the insertion site of this tract becomes recognizable as epithelium lining the ventral lip of the mouth. The early ventral tract in hatching larvae is not a pedal tract.

In larvae of the archaeogastropods *Haliotis tuberculata* and *Calliostoma zizyphinum*, Crofts

(1937, 1955) described a second shell muscle that arises on the ventral border of the shell and inserts on pedal epithelium overlying the operculum. This muscle develops after the LRM. A similar muscle appears in mid-stage nudibranch veligers (called the opercular muscle by Bonar [1976] and the accessory pedal retractor by Bickell and Chia [1979]), but a counterpart in caenogastropod larvae is unknown at present (Fretter, 1969, 1972).

Torsion

The widely accepted scenario for the evolution of the gastropod body plan views 180° torsion as the ancestral condition, mainly because 180° torsion co-occurs with primitive states of other adult characters (bilateral ctenidia, auricles, and nephridia). According to Crofts' (1937, 1955) familiar interpretation, the evolutionary event of torsion is recapitulated by a two-stage morphogenetic process during archaeogastropod development. Archaeogastropods undergo a rapid torsional twist of 90° during the larval phase, but the final 90° of torsion is accomplished much more slowly during the first months after metamorphosis.

Opisthobranchs are said to be "detorted," implying that they arose from an ancient stock of gastropods which was fully torted and that the present condition of partial torsion shown by most extant species is secondary. Yet the anal position in embryos and young veligers of opisthobranchs shows a torsional condition of only 90° or less, regardless of whether development is planktotrophic or lecithotrophic (Thompson, 1958, 1959, 1962; Bonar and Hadfield, 1974; Bickell and Chia, 1979; Soliman, 1991; Tsubokawa and Okutani, 1991; Page, 1992a,b). This is true also of hatching veligers of the cephalaspid *Acteon tornatilis* (Thompson, 1976), although additional torsional displacements are evident in adult organ systems of this species (Fretter and Graham, 1954; Brace, 1977). Kriegstein (1977a) states that newly hatched veligers of the anaspidean *Aplysia californica* show 120° torsion, but the sketch in his Figure 2 shows the anus in the right, lateral position that is typical of other young opisthobranch larvae. If opisthobranchs are secondarily detorted, there is no record during their early ontogeny of the ancestral condition of 180° torsion (Thompson, 1958). Instead, we see an initial condition of 90° torsion with greater or lesser amounts of torsional displacement of the anus occurring *later* in the ontogeny of various species.

D'Asaro (1965, 1966, 1969) has reported that caenogastropod veligers undergo an initial torsional twist of 90°, which is followed by a pause before further torsion occurs. Thus, the preliminary torsional condition in caenogastropods and archaeogastropods is equivalent to that of young, planktotrophic opisthobranch veligers. Salvini-Plawen (1980b) has argued that 90° is the essential synapomorphy of gastropods, because additional torsion occurs after metamorphosis in archaeogastropods. Comparative developmental data support the notion that torsional displacements of greater than 90° are derived conditions among adults, because 90° torsion is the initial common denominator during the development of all gastropod groups. An important implication of this revised view of ontogenetic torsion is that the torsional condition exhibited by young opisthobranch larvae chronicles the true ancestral state for gastropods.

Mantle Cavity

Hatching planktotrophic larvae of caenogastropods have a pronounced mantle cavity (Werner, 1955; D'Asaro, 1965, 1966, 1969; Bedford, 1966; Fretter, 1967, 1969; Thiriot-Quiévreux, 1969; Fioroni and Meister, 1976). On the other hand, hatching, planktotrophic larvae of opisthobranchs have an extremely shallow mantle cavity, with only a few cells bridging between the cephalopodium and the periostracum-secreting cells attached to the shell aperture (Fig. 17.2a). A

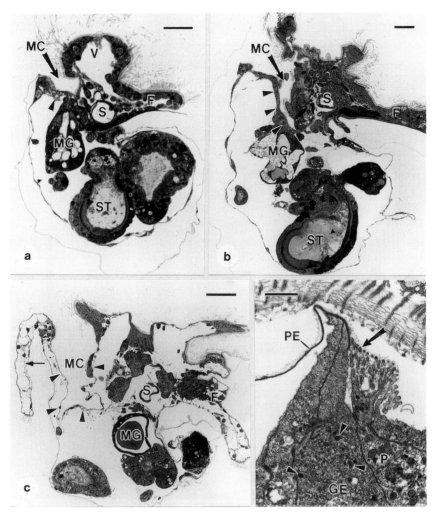

FIGURE 17.2. Photomicrographs of longitudinal sections through the right side of opisthobranch larvae showing the right mantle cavity. (Abbreviations: F = foot; MC = mantle cavity; MG = mantle gland; S = statocyst; ST = stomach; V = velum). *a.* Newly hatched larva of *Melibe leonina* (Nudibranchia) showing an extremely shallow mantle cavity lined by epithelium of the right mantle fold (arrowhead) (scale bar = 20 μm). *b. M. leonina* larva just prior to the mantle retraction stage. Proliferation of the right mantle fold cells (arrowheads) has deepened the mantle cavity (scale bar = 20 μm). *c. Pleurobranchaea californica* (Notaspidea) veliger at 30 days post-hatching showing large size of the right mantle cavity lined by the mantle fold epithelium (arrowheads) (scale bar = 40 μm). FIGURE 17.3 *(bottom right).* TEM of apices of cells involved in periostracum secretion in the larva of *Melibe leonina* at 1-week posthatching. Note the numerous vesicles, some containing electron-dense material (arrowheads) within the growing edge cell (GE). Microvilli (arrow) from proximal cell (P) invest the newly elaborated periostracum (PE). (Terminology after Eyster [1983]; scale bar = 1 μm).

similar situation occurs in hatching planktotrophic larvae of the marine pulmonate *Amphibola crenata*. Little et al. (1985) have stated that young veligers of *A. crenata* lack a mantle cavity altogether. Eventually, however, the right mantle cavity of opisthobranch larvae and of *A. crenata* gradually deepens to a varying, species-specific extent during the larval phase (Figs. 17.2b,c).

Larval Shells

Three distinct shell zones are present in caenogastropods having pelagic larvae (literature reviewed by Jablonski and Lutz, 1980). Protoconch I, also called the embryonic shell, is elaborated by the shell gland during the encapsulated period of embryogenesis. Protoconch II is secreted during the free-swimming larval phase and is often distinctly ornamented and beaked. The teleoconch is the postmetamorphic shell.

In Bandel's (1982) survey of prosobranch morphogenesis, he emphasizes that the peripheral periostracum-secreting cells of the evaginated shell gland detach from the shell aperture after protoconch I has been completed. D'Asaro (1969) describes the same event for *Distorsio clathrata*. Resumption of shell secretion marks the transition to the postmetamorphic teleoconch in archaeogastropods or the protoconch II in posthatching caenogastropod larvae. How this transition occurs requires study. It is not known if the periostracum of the various shell zones is secreted by different populations of mantle cells or by a sequence of differentiation states in a single population.

It has been known since the work of Thompson (1958, 1962) that the periostracum-secreting cells of the mantle detach from the aperture of the larval shell during nudibranch development. In species with planktotrophic development, this event typically occurs after one-half to three-fourths of the obligatory larval phase, when the larval shell has reached its definitive size (Bickell and Kempf, 1983; others reviewed by Todd, 1981). Mantle detachment is obvious in nudibranch larvae because the mantle immediately retracts to the posterior extremity of the shell cavity, where it hypertrophies to form presumptive dorsal epidermis in many species (for an exception see Bonar and Hadfield, 1974; Bonar, 1976).

Arrest of larval shell growth and detachment of the shell from the mantle have been reported also in a number of planktotrophic aplysiid veligers (Switzer-Dunlap, 1978; Paige, 1988). Smith (1967) has stated that shell growth in lecithotrophic veligers of the cephalaspid *Retusa obtusa* is arrested after the visceral mass has been encompassed by the shell gland. In these opisthobranchs, shell growth resumes after metamorphosis.

The detachment of the shell from the mantle during the larval phase of opisthobranchs may be homologous to the detachment of the shell gland from protoconch I that occurs during prosobranch embryogenesis. This suggests that the shell that is elaborated during both the embryonic and the larval phase of opisthobranchs may be homologous to the embryonic protoconch I of caenogastropods and archaeogastropods. The ultrastructure of the periostracum-secreting cell complex in young, planktotrophic larvae of the nudibranchs *Melibe leonina* (Fig. 17.3) and *Doridella steinbergae* is the same as that of the shell gland in embryos of the nudibranch *Aeolidia papillosa*, as described by Eyster (1983). This ultrastructure is very different from that of the periostracum-secreting complex in postmetamorphic *Lymnaea stagnalis*, as described by Saleuddin and Petit (1983).

Obviously, validation of this hypothesis requires many more data than are currently available. I present it here because it may have important implications for the topic of larval shell coiling. Postmetamorphic shells of most extant gastropods, including opisthobranchs, are coiled orthostrophically (see Fretter and Graham [1962] for a description of shell-coiling patterns in gastropods). However, larval shells of opisthobranchs and caenogastropods show a striking dichotomy that has so far defied explanation: opisthobranch larvae usually have hyperstrophically coiled shells (Figs. 17.4a,b), whereas orthostrophic coiling is typical of the larval shell (protoconch II) of most caenogastropods (Fig. 17.4e). Nevertheless, drawings by Pelseener (1911) of embryonic stages of the caenogastropods *Rissoa parva* and *Nassa reticulata* clearly show the protoconch I with an "umbo" on the left side of the cephalopedal mass (Figs. 17.4c,d), a condition typical of the early hyperstrophic shell form in young opisthobranch larvae. Traditionalists might view this

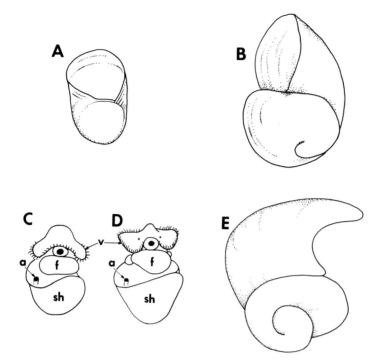

FIGURE 17.4 Embryonic and larval shells of gastropods. All specimens have a dextral soft anatomy, with anus rotated to the right side of the cephalopodium. *a.* Larval shell of a hatching planktotrophic veliger of the nudibranch *Rostanga pulchra* in frontal view; the bulbous apex of shell coil lies toward the left side (adapted from Hurst, 1967). *b.* Hyperstrophic shell of a late stage larva of *Rostanga pulchra* in left lateral view; the apex of shell coil is on left side (adapted from Chia and Koss, 1978). *c, d.* Embryonic shells of *Rissoa parva* and *Nassa reticulata*, respectively, in frontal view; the apex of the shell coil lies toward the left side; compare with *a.* (a = anus; f = foot, sh = shell, v = velum [adapted from Pelseneer, 1911]). *e.* Orthostrophic shell of planktotrophic veliger of the caenogastropod *Strombus gigas* at 8 days posthatching in right lateral view; the apex of shell coil is on right side (adapted from D'Asaro, 1965).

condition as incomplete torsional rotation of the shell, but the possibility of incipient hyperstrophy has been neither considered nor investigated critically. Furthermore, hyperstrophy in posthatching larval shells of architectonids and pyramidellids has been well documented (see Robertson, 1985), although detailed studies of morphogenesis in these prosobranch families are not available. Hadfield and M. Strathmann (1990) have identified incipient hyperstrophy in early larval shells of several trochid archaeogastropods.

Ultrastructural Comparison of Polychaete Trochophores and Young Opisthobranch Veligers

Data given in the previous section indicate that planktotrophic opisthobranch veligers become free-living larvae at an earlier stage of organogenesis than do hatching caenogastropod larvae. What is the polarity of change for these two developmental patterns? The fact that all of the identified differences relate to times at which definitive structures begin to differentiate suggests that the early pattern of morphogenesis in opisthobranchs is ancestral. Another method of determining direction of evolutionary change is to make outgroup comparisons. In this section I have compiled a list of similarities between planktotrophic polychaete trochophores and newly hatched, planktotrophic opisthobranch veligers. Many of these similarities have been overlooked previously because they cannot be resolved without electron microscopical methods.

Pre- and Postoral Ciliary Bands (Prototroch and Metatroch)

The similar form and functioning of the double ciliary band feeding device in spiralian larvae have been thoroughly described elsewhere (Strathmann et al., 1972; Strathmann and Leise, 1979) and will not be repeated here.

Telotroch and Neurotroch

Various molluscan larvae have ciliated regions that may correspond to the telotroch and neurotroch (also called gastrotroch) of polychaete trochophores. Jägersten (1972) has suggested that the prominent ciliary tract extending down the midventral surface of the foot in opisthobranch larvae is homologous with the neurotroch. This tract carries rejected particles away from the mouth (Thompson 1959). Raven (1958) has likened various types of ciliation in the anal region of molluscan trochophores and veligers to the telotroch of polychaete trochophores. Included in this category is a ciliary tuft adjacent to the anus in veligers of docoglossan archaeogastropods (Pelseener, 1911; Smith, 1935) It has not been generally recognized that the prominent anal cells of opisthobranch embryos and young larvae are also ciliated (Fig. 17.5). These cells disappear soon after hatching.

Protonephridia

Protonephridia have been identified in many types of metazoan larvae, including those of many bivalves, polychaetes, and other spiralians. Until recently, the apparent absence of protonephridia in chiton and gastropod larvae, except for embryonic stages of pulmonates (Raven, 1958; Fretter and Graham, 1962), was perplexing. Salvini-Plawen (1969) cited this as evidence that annelid and molluscan larvae are derived convergently. However, electron microscopical studies have now demonstrated protonephridia in larvae of nudibranchs and a chiton (Figs 17.1; Bartolomaeus, 1989).

Ciliary Gutter

A distinctive valvelike structure at the junction of the foregut and midgut has been described for polychaete trochophores (Holborow, 1971; Lacalli, 1984; Heimler, 1988) and for Müller's larva, which is considered to be a feeding larva (Ruppert, 1978). This so-called ciliary gutter (Heimler, 1988) consists of a ring of four highly vacuolated cells and long cilia that project into the lumen of the midgut. A structure similar to the ciliary gutter of polychaete trochophores and Müller's larva is present at the junction of the esophagus and stomach in planktotrophic nudibranch veligers (Fig. 17.6). Swelling of the ciliary membranes is consistently seen for the ciliary gutters of all these larval types (Fig. 17.7), although this may be a fixation artifact.

Primary Larval Nervous System

The nervous system of planktotrophic trochophores belonging to the polychaete *Spirobranchus polyceras* has been reconstructed from serial ultrathin sections by Lacalli (1984). He describes a primary larval nervous system consisting of an apical organ and intraepithelial axon tracts associated with the trochal bands and foregut. The apical organ is a cluster of cell bodies, some of which are multiciliated, that give rise to an apical plexus of neurites containing dense-cored vesicles. Neurons associated with the trochal bands and foregut are described as intraepithelial because their cell bodies and axons occur on the epithelial side of basal laminae and are often surrounded by basal processes of epithelial cells. A later developed adult nervous system consists of cerebral

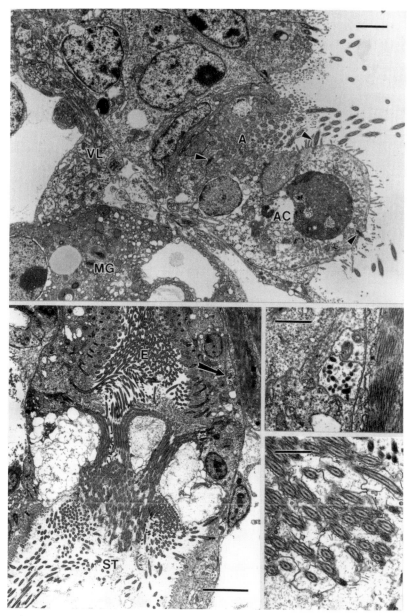

FIGURE 17.5 *(top)*. TEM of a cross-section through the anal region of a newly hatched larva of *Melibe leonina* (Nudibranchia) showing anal cell (AC) with cilia (arrowheads) (A = anus; MG = mantle gland; VL = visceral loop connective; scale bar = 2 μm). FIGURE 17.6 *(bottom left)*. TEM of a longitudinal section through the ciliary gutter located between the esophagus (E) and stomach (ST) in a mid-stage larva of *Doridella steinbergae* (Nudibranchia). Note vacuolated cells and bundle of cilia projecting into stomach. Arrow indicates intraepithelial axons that are shown in higher magnification in Figure 17.8 (scale bar = 5 μm). FIGURE 17.7 *(bottom right)*. TEM of cilia forming the ciliary gutter in a *Doridella steinbergae* larva (scale bar = 1 μm). FIGURE 17.8 *(middle right)*. TEM of intraepithelial axon tracts within foregut epithelium (scale bar = 1 μm).

ganglia, the cerebral commissure, and a pair of ventral nerve cords.

Ultrastructural observations on nudibranch veligers have identified an apical band of cells that possess a bundle of cilia projecting from a deep inpocketing in each cell (Bonar, 1978b; Chia and Koss, 1984). These ciliated cells, together with neighboring nonciliated cells, give rise to a plexus of large axons containing dense-cored vesicles (Page and Kempf, unpub. data). Some of the axons of this plexus show serotoninlike immunoreactivity (Kempf, 1991). In young nudibranch veligers, the apical plexus overlies the cerebral commissure (as does the apical plexus in *Spirobranchus* trochophores), but the distinction between plexus and cerebral commissure becomes difficult to discern in older veligers.

Like polychaete trochophores, opisthobranch veligers also have intraepithelial axon tracts associated with the trochal bands and foregut (Fig. 17.8). The latter are particularly prominent along the ventral margin of the stomodeum.

Discussion

According to Bandel (1982), the feeding gastropod veliger is a caenogenetic stage added to the ontogeny of ancestral gastropods which originally progressed directly from a nonfeeding trochophore to the juvenile stage. Judging from the fact that Bandel restricted his observations to archaeogastropods, caenogastropods, and directly developing pulmonates, his conclusion is understandable. The only larval structure that has been identified for archaeogastropod veligers is the preoral ciliary band, and planktotrophic caenogastropod larvae incorporate many juvenile characters even from the time of hatching. Salvini-Plawen (1980a; 1985) has objected to a common ancestry for polychaete trochophores and molluscan veligers, partly because he claims the pre- and postoral ciliary bands are the only characters held in common. Neither Bandel (1982) nor Salvini-Plawen (1969, 1980a, 1985) considered young opisthobranch larvae.

When developmental data on planktotrophic opisthobranch larvae are compared to those on prosobranch larvae, an alternative interpretation emerges that appears to be more parsimonious than that proposed by Bandel (1982). Extant gastropod ontogenies show a broad range in the extent to which heterochronies have become superimposed on the genesis of the basic veliger form. Newly hatched, planktotrophic opisthobranch veligers show few if any of these heterochronies (although Jägersten [1972] has argued that the veliger shell and foot are examples of ancestral adult structures appearing in the larval stage). Archaeogastropod ontogenies represent the other extreme, with developmental programs for postmetamorphic characters being switched on very precociously within the context of an almost vestigial program for the basic veliger form. Caenogastropod ontogenies that generate feeding larvae occupy intermediate ground between these two extremes.

In opisthobranchs with planktotrophic larvae, nourishment provided to each egg is used to build a basic feeding veliger only. Among prosobranchs, nourishment allotted to eggs allows for construction of the basic veliger form *plus* preliminary forms of a number of juvenile structures. Fioroni (1966) has reviewed the many ways in which prosobranchs provide large energy stores to their eggs. The variety of these strategies, when compared to the extreme conservatism of early opisthobranch development, is another argument for the derived nature of prosobranch ontogenies, even when a planktotrophic veliger is eventually produced.

Chaffee and Lindberg (1986) use a statistical argument, based on probabilities of direct development as a function of adult body size, to conclude that the small size of Cambrian molluscs would have precluded a feeding larval stage. However, this argument is weakened by the fact that there are planktotrophic larvae in the life histories of some extant opisthobranchs with adult sizes of less than 10 mm (Chia, 1971; Chia and Skeel, 1973; Bickell and Chia, 1979).

My interpretation of the ancestry and adaptive radiation of the gastropod life cycle is basically a restatement of that proposed by Jägersten (1972), Strathmann (1978), and Nielsen (1985): Pelagic, planktotrophic larvae are primitive. However, my argument is unconventional in that I have presented evidence that young, planktotrophic opisthobranch larvae may be the best extant approximation of the ancestral gastropod larva, with extant prosobranch veligers being secondary larval types due to varying amounts of heterochronic modifications.

This theory promotes a renewed look at torsion. As Salvini-Plawen (1980b) has suggested previously, perhaps our contemplation of torsion should be phrased as: What was the adaptive value of 90° torsion, rather than 180° torsion, to the ancestral gastropod and why has it been carried further in some species? This in turn should lead us to question the postulate of an exogastric shell coil for the protogastropod. An exogastric shell coil is required for the traditional view of torsion, because 180° torsion neatly results in an endogastric shell. The traditional view also requires the protogastropod to have a posterior mantle cavity housing the ctenidia, so that after torsion the mantle cavity is anterior. This is inconsistent with the fact that extant mono-placophorans and chitons have lateral mantle cavities with gills extending down most of the body length.

The conventional paradigm regarding torsion has spawned the controversy about the extinct bellerophonts; we must accept either perfect 180° torsion at its earliest appearance, or an animal that carried a heavily calcified shell coil directly over its head (see Berg-Madsen and Peel, 1978; Harper and Rollins, 1982; Runnegar and Pojeta, 1985). Might there be a fundamental flaw in the conventional concept of ancestral torsion? A similar reservation has been voiced by Yochelson (1978). Current theories about the origin and evolution of gastropods are based, to a large extent, on morphogenesis in *Haliotis tuberculata* and other archaeogastropods, as interpreted by Crofts (1937, 1955). I argue that archaeogastropods have a derived ontogeny, in which developmental programs for postmetamorphic structures have become extensively superimposed on a residual remnant of the program for a feeding veliger form.

Finally, evidence for the ancestral nature of the early opisthobranch ontogeny, which generates veligers with a hyperstrophic shell, and the possibility of incipient hyperstrophy in embryonic shells of some prosobranchs, should be of interest to paleontologists attempting to make sense of an abundance of gastropodlike Paleozoic fossils with hyperstrophic shells (see Linsley and Kier, 1984; Runnegar and Pojeta, 1985).

Support or falsification of the ideas presented here requires additional detailed studies of molluscan morphogenesis. I hope that my challenge of traditional theories about gastropod design, ontogeny, and evolution will stimulate greater research effort in this area and promote different ways of looking at existing data.

Acknowledgments

I am very grateful to Prof. G. O. Mackie, who has encouraged and supported my research with a grant from the Natural Sciences and Engineering Research Council of Canada, and to Ms. Roswitha Marx for translations of German manuscripts.

Literature Cited

Bandel, K. 1982. Morphologie und Bildung der frühontogenetischen Gehäuse bei conchiferen Mollusken. Facies 7: 1–198.

Bartolomaeus, T. 1989. Larvale Neirenorgane bei *Lepidochiton inereus* (Polyplacophora) und *Aeolidia papillosa* (Gastropoda). Zoomorph. 108: 297–307.

Bedford, L. 1966. The electron microscopy and cytochemistry of oogenesis and the cytochemistry of embryonic development of the prosobranch gastropod *Bembicium nanum* L. J. Embryol. Exp. Morph. 15: 15–37.

Berg-Madsen, V., and J. Peel. 1978. Middle Cambrian monoplacophorans from Bornholm and Australia, and the systematic position of the bellerophontiform molluscs. Lethaia 11: 113–125.

Bickell, L. R., and F.-S. Chia. 1979. Organogenesis and histogenesis in the planktotrophic veliger of *Doridella steinbergae* (Opisthobranchia Nudibranchia). Mar. Biol. 52: 291–313.

Bickell, L. R., and S. C. Kempf. 1983. Larval and metamorphic morphogenesis in the nudibranch *Melibe leonina* (Mollusca: Opisthobranchia). Biol. Bull. 165: 119–138.

Bonar, D. B. 1976. Molluscan metamorphosis: A study in tissue transformation. Am. Zool. 16: 573–591.

———. 1978a. Morphogenesis at metamorphosis in opisthobranch molluscs. *In* F.-S. Chia and M. E. Rice, eds. Settlement and Metamorphosis in Marine Invertebrate Larvae. Elsevier/North-Holland, New York, pp. 177–196.

———. 1978b. Ultrastructure of a cephalic sensory organ in larvae of the gastropod *Phestilla sibogae* (Aoelidiacea, Nudibranchia). Tissue Cell 10: 153–165.

Bonar, D. B., and M. G. Hadfield. 1974. Metamorphosis of the marine gastropod *Phestilla sibogae* Bergh (Nudibranchia: Aeolidacea). I. Light and electron microscope analysis of larval and metamorphic stages. J. Exp. Mar. Biol. Ecol. 16: 227–255.

Brace, R. 1977. Anatomical changes in nervous and vascular systems during the transition from prosobranch to opisthobranch organization. Trans. Zool. Soc. London 34: 1–25.

Chaffee, C., and D. R. Lindberg. 1986. Larval biology of early Cambrian molluscs: The implications of small body size. Bull. Mar. Sci. 39: 536–549.

Chia, F.-S. 1971. Oviposition, fecundity, and larval development of three sacoglossan opisthobranchs from the Northumberland coast, England. Veliger 13: 319–325.

Chia, F.-S., and R. Koss. 1978. Development and metamorphosis of the planktotrophic larvae of *Rostanga pulchra* (Mollusca: Nudibranchia). Mar. Biol. 46: 109–119.

———. 1982. Fine structure of the larval rhinophores of the nudibranch, *Rostanga pulchra* (Mollusca, Opisthobranchia, Nudibranchia). Cell. Tiss. Res. 225: 235–248.

———. 1984. Fine structure of the cephalic sensory organ in the larva of the nudibranch *Rostanga pulchra* (Mollusca, Opisthobranchia, Nudibranchia). Zoomorph. 104: 131–139.

Chia, F.-S. and M. Skeel. 1973. The effect of food consumption on growth, fecundity, and mortality in a sacoglossan opisthobranch, *Olea hansineensis*. Veliger 16: 153–158.

Crofts, D. R. 1937. The development of *Haliotis tuberculata*, with special reference to organogenesis during torsion. Philos. Trans. Roy. Soc. Lond. B 228: 219–268.

———. 1955. Muscle morphogenesis in primitive gastropods and its relation to torsion. Proc. Zool. Soc. Lond. 125: 711–749.

D'Asaro, C. N. 1965. Organogenesis, development, and metamorphosis in the queen conch, *Strombus gigas*, with notes on breeding habits. Bull. Mar. Sci. Gulf Caribb. 15: 359–416.

———. 1966. The egg capsules, embryogenesis, and early organogenesis of a common oyster predator, *Thais haemastoma floridana* (Gastropoda: Prosobranchia). Bull. Mar. Sci. Gulf Caribb. 16: 884–914.

———. 1969. The comparative embryogenesis and early organogenesis of *Bursa corrugata* Perry and *Distorsio clathrata* Lamarck (Gastropoda: Prosobranchia). Malacologia 9: 349–389.

Eyster, L. S. 1983. Ultrastructure of early embryonic shell formation in the opisthobranch gastropod *Aeolidia papillosa*. Biol. Bull. 165: 394–408.

Fioroni, P. 1966. Zur Morphologie und Embryogenese des Darmtraktes und der transitorischen Organe bei Prosobranchiern (Mollusca, Gastropoda). Rev. Suisse Zool. 73: 621–876.

Fioroni, P., and G. Meister. 1976. Zur embryonalen Entwicklung von *Lamellaria perspicua* L. (Gastropoda, Prosobranchia, Mesogastropoda, Lamellariacea). Cah. Biol. Mar. 17: 323–336.

Franz, D. R. 1971. Development and metamorphosis of the gastropod *Acteocina canaliculata* (Say). Trans. Am. Microsc. Soc. 81: 1–11.

Fretter, V. 1967. The prosobranch veliger. Proc. Malacol. Soc. Lond. 37: 357–366.

———. 1969. Aspects of metamorphosis in prosobranch gastropods. Proc. Malacol. Soc. Lond. 38: 375–385.

————. 1972. Metamorphic changes in the velar musculature, head and shell of some prosobranch veligers. J. Mar. Biol. Assoc. 52: 161–177.

Fretter, V., and A. Graham. 1954. Observations on the opisthobranch mollusc *Acteon tornatilis* (L.). J. Mar. Biol. Assoc. 33: 565–585.

————. 1962. British Prosobranch Gastropods. Ray Society, London.

Garstang, W. 1922. The theory of recapitulation: A critical restatement of the biogenetic law. J. Linn. Soc. Lond. 35: 81–101.

Gosliner, T. M. 1981. Origins and relationships of primitive members of the Opisthobranchia (Mollusca, Gastropoda). Biol. J. Linn. Soc. 16: 197–226.

Hadfield, M. G., and M. F. Strathmann. 1990. Heterostrophic shells and pelagic development in trochoideans: Implications for classification, phylogeny and paleoecology. J. Moll. Stud. 56: 239–256.

Hadfield, M. G., and M. Switzer-Dunlap. 1984. Opisthobranchs. *In* A. S. Tompa, N. H. Verdonk and J. A. M. Van Den Biggelaar, eds. The Mollusca, Vol. 7. Reproduction, Academic Press, New York, pp. 209–350.

Harper, J. A., and H. B. Rollins. 1982. Recognition of Monoplacophora and Gastropoda in the fossil record: A functional morphological look at the bellerophont controversy. Proc. Third North American Paleontol Conv. 1: 227–232.

Haszprunar, G. 1985. The Heterobranchia—a new concept of the phylogeny of the higher Gastropoda. Z. Zool. Syst. Evol. 23: 15–37.

Heimler, W. 1988. Larvae. *In* W. Westheide and C. O. Hermans, eds. Ultrastructure of Polychaeta. Microfauna Marina 4: 353–371.

Holborow, P. L. 1971. The fine structure of the trochophore of *Harmothoe imbricata*. In D. J. Crisp, ed. Fourth European Marine Biology Symposium. Cambridge University Press, New York, pp. 237–257.

Hurst, A. 1967. The egg masses and veligers of thirty northeast Pacific opisthobranchs. Veliger 9: 255–288.

Jablonski, D., and R. A. Lutz. 1980. Molluscan larval shell morphology. Ecological and paleontological applications. *In* D. C. Rhoads and R. A. Lutz, eds. Skeletal Growth of Aquatic Organisms. Plenum Press, New York, pp. 323–377.

Jägersten, G. 1972. Evolution of the Metazoan Life Cycle. Academic Press, London.

Kempf, S. C. 1991. The ontogeny of neuronal systems expressing SCP-like and serotonin-like antigens in *Berghia verrucicornis*. Soc. Neurosci. Abst. 17: 1356.

Kempf, S. C., and A. O. D. Willows. 1977. Laboratory culture of the nudibranch *Tritonia diomedea* Bergh (Tritoniidae, Opisthobranchia) and some aspects of its behavioral development. J. Exp. Mar. Biol. Ecol. 30: 261–276.

Kriegstein, A. R. 1977a. Stages in post-hatching development of *Aplysia californica*. J. Exp. Zool. 199: 275–288.

————. 1977b. Development of the nervous system of *Aplysia californica*. Proc. Natl. Acad. Sci. 74: 375–378.

Lacalli, T. C. 1984. Structure and organization of the nervous system of the trochophore larva of *Spirobranchus*. Philos. Trans. Roy. Soc. Lond. B 306: 79–135.

Linsley, R. M., and W. M. Kier. 1984. The Paragastropoda: A proposal for a new class of Paleozoic Mollusca. Malacologia 25: 241–254.

Little, C., P. Stirling, M. P. Pilkington, and J. Pilkington. 1985. Larval development and metamorphosis in the marine pulmonate *Amphibola crenata* (Mollusca: Pulmonata). J. Zool. Lond. 205: 489–510.

Nielsen, C. 1985. Animal phylogeny in the light of the trochaea theory. Biol. J. Linn. Soc. 25: 243–299.

Page, L. R. 1992a. New interpretation of a nudibranch central nervous system based on ultrastructural analysis of neurodevelopment in *Melibe leonina*. I. Cerebral and visceral loop ganglia. Biol. Bull. 182: 348–365.

————. 1992b. New interpretation of a nudibranch central nervous system based on ultrastructural analysis of neurodevelopment in *Melibe leonina*. II. Pedal, pleural, and labial ganglia. Biol. Bull. 182: 366–381.

Paige, J. A. 1988. Biology, metamorphosis, and postlarval development of *Bursatella leachii plei* Rang (Gastropoda: Opisthobranchia). Bull. Mar. Sci. 42: 65–75.

Pelseener, P. 1911. Recherches sur l'embryologie des Gasteropodes. Mem. Acad. Roy. Belg. 3: 1–167.

Raff, R. A., and T. C. Kaufman. 1991. Embryos, Genes, and Evolution. Indiana University Press, Bloomington, IN.

Raven, C. P. 1958. Morphogenesis: The Analysis of Molluscan Development. Pergamon Press, New York.

Rivest, B. R. 1992. Studies on the structure and function of the larval kidney complex of prosobranch gastropods. Biol. Bull. 182: 305–323

Robertson, R. 1983. Observations on the life history of the wentletrap *Epitonium albidum* in the West Indies. Am. Malacol. Bull. 1: 1–12.

———. 1985. Four characters and the higher systematics of gastropods. Am. Malacol. Bull. Spec. Ed. 1: 1–22.

Runnegar, B., and J. Pojeta, Jr. 1985. Origin and diversification of the Mollusca. *In* E. R. Trueman and M. R. Clarke, eds. The Mollusca, Vol. 10. Evolution. Academic Press, New York, pp. 1–57.

Ruppert, E. E. 1978. A review of metamorphosis in Turbellaria larvae. *In* F.-S. Chia and M. E. Rice eds. Settlement and Metamorphosis in Marine Invertebrate Larvae. Elsevier/North-Holland, New York, pp. 65–81.

Ruthensteiner, B., and K. Schaefer. 1991. On the protonephridia and "larval kidneys" of *Nassarius (Hinia) reticulatus* (Linnaeus) (Caenogastropoda). J. Moll. Stud. 57: 323–329.

Saleuddin, S. M., and H. P. Petit. 1983. The mode of formation and the structure of the periostracum. *In* S. M. Saleuddin and K. M. Wilbur, eds. The Mollusca, Vol. 4, Pt. I. Physiology. Academic Press, New York, pp. 199–234.

Salvini-Plawen, L. von 1969. Solenogastres und Caudofoveata (Mollusca, Aculifera): Organisation und phylogenetische Bedeutung. Malacologia 9: 191–216.

———. 1980a. Was ist eine Trochophora?—Eine Analyse der Larventypen mariner Protostomier. Zool. Jahrb., Abt. Anat. Ontog. Tiere 103: 389–423.

———. 1980b. A reconsideration of systematics in the Mollusca (phylogeny and higher classification). Malacologia 19: 249–278.

———. 1985. Early evolution and the primitive groups. *In* E. R. Trueman and M. R. Clarke, eds. The Mollusca, Vol. 10, Evolution. Academic Press, New York, pp. 59–150.

Schacher, S., E. R. Kandel, and R. Woolley. 1979. Development of neurons in the abdominal ganglion of *Aplysia californica*. Develop. Biol. 71: 163–175.

Scheltema, R. S. 1962. Pelagic larvae of New England intertidal gastropods. I. *Nassarius obsoletus* Say and *Nassarius vibex* Say. Trans. Am. Microsc. Soc. 81: 1–11.

Smith, F. G. W. 1935. The development of *Patella vulgata*. Philos. Trans. Roy. Soc. Lond. B 225: 95–125.

Smith, S. T. 1967. The development of *Retusa obtusa* (Montagu) (Gastropoda, Opisthobranchia). Can. J. Zool. 45: 737–763.

Soliman, G. N. 1991. A comparative review of the spawning, development and metamorphosis of prosobranch and opisthobranch gastropods with special reference to those from the northwestern Red Sea. Malacologia 32: 257–271.

Strathmann, R. R. 1978. The evolution and loss of feeding larval stages of marine invertebrates. Evolution 32: 894–906.

Strathmann, R. R., T. L. Jahn, and R. C. Fonesca. 1972. Suspension-feeding by marine invertebrate larvae: Clearance of particles by ciliated bands of a rotifer, pluteus, and trochophore. Biol. Bull. 142: 505–519.

Strathmann, R. R., and E. Leise. 1979. On feeding mechanisms and clearance rates of molluscan veligers. Biol. Bull. 157: 524–535.

Switzer-Dunlap, M. 1978. Larval biology and metamorphosis of aplysiid gastropods. *In* F.-S. Chia and M. E. Rice, eds. Settlement and Metamorphosis of Marine Invertebrate Larvae. Elsevier/North-Holland, New York, pp. 197–206.

Thiriot-Quiévreux, C. 1969. Organogenèse larvaire du genre *Atlanta* (Mollusque Hétéropode). Vie et Milieu 20: 347–395.

———. 1977. Véligère planctotrophe du doridien *Aegires punctilucens* (D'Orbigny) (Mollusca: Nudibranchia: Notodorididae): Description et métamorphose. J. Exp. Mar. Biol. Ecol. 26: 177–190.

Thompson, T. E. 1958. The natural history, embryology, larval biology, and post-larval development of *Adalaria proxima* (Alder and Hancock) (Gastropoda, Opisthobranchia). Philos. Trans. Roy. Soc. Lond. B 242: 1–58.

———. 1959. Feeding in nudibranch larvae. J. Mar. Biol. Assoc. 38: 239–248.

———. 1962. Studies on the ontogeny of *Tritonia hombergi* Cuvier (Gastropoda, Opisthobranchia). Philos. Trans. Roy. Soc. Lond. B 245: 171–281.

———. 1967. Direct development in a nudibranch, *Cadlina laevis*, with a discussion of developmental processes in Opisthobranchia. J. Mar. Biol. Assoc. 47: 1–22.

———. 1976. Biology of Opisthobranch Molluscs, Vol. 1. Ray Society, London.

Todd, C. D. 1981. The ecology of nudibranch molluscs. Oceanogr. Mar. Biol. Ann. Rev. 19: 141–234.

Tsubokawa, R., and T. Okutani. 1991. Early life history of *Pleurobranchaea japonica* Thiele, 1925 (Opisthobranchia: Notaspidea). Veliger 34: 1–13.

Werner, B. 1955. Über die Anatome, die Entwicklung und Biologie des Veligers und der Veliconcha von *Crepidula fornicata* L. (Gastropoda, Prosobranchia). Helgo. Meeres. 5: 169–217.

Yochelson, E. L. 1978. An alternative approach to the interpretation of the phylogeny of ancient mollusks. Malacologia 17: 165–191.

18 Morphology and Ultrastructure of the Larva of the Bryozoan *Tanganella muelleri* (Ctenostomata: Victorellidae)

Russel L. Zimmer and Christopher G. Reed (posthumously)

ABSTRACT The anatomy of the larva of the ctenostome bryozoan *Tanganella muelleri* (Superfamily Victorelloidea) is described on the basis of serial 1-μm sections and transmission electron micrographs. Our study is complementary to and fully compatible with a previous light microscope study by Braem (1951) on the structure and metamorphosis of the same larva. The larva of *T. muelleri* is a composite of primitive and advanced features. Primitive features include the facts that the apical disc contains undifferentiated epithelial cells, the pallial epithelium is shallowly invaginated, the mesoderm is poorly differentiated, and a rudimentary gut is present. Advanced features include the facts that the pallial epithelium provides all the epidermis of the juvenile, the internal sac is of simple cytology and serves only for transitory attachment of the metamorphosing individual, and the ciliated oral epithelium is reduced to a single ring of cells. The presence of a large field of undifferentiated epidermal cells at the oral pole and the putative role of these epidermal cells in forming mesoderm in the juvenile are unique aspects of this larva.

Comparison of this larva with others known for the superfamily reveals extensive differences, at least superficially. Our findings are then evaluated relative to recently proposed phylogenies of ctenostome bryozoans, especially with regard to the possibility that victorelloids are ancestral to the "stoloniferan" superfamilies Vesicularioidea and Valkerioidea.

Introduction

Ctenostome bryozoans have historically been divided into the stoloniferans, an advanced group in which feeding zooids are budded from highly simplified tubular zooids (stolons), and carnosans, which lack stolons and proliferate new feeding zooids directly from existing ones (hence producing "fleshy" colonies). Although Carnosa and Stolonifera have long been recognized as formal taxa, Jebram (1973) concluded that they can be no more than grades, arguing that the stolonate forms are polyphyletic and the carnosans are therefore paraphyletic. d'Hondt (1986) concurred with Jebram's conclusions, but ordered the superfamilies into a significantly different phylogeny. In each of their hierarchies, the carnosan superfamily Victorelloidea has a critical, but different, role in interpreting the phylogenetic radiation of the stoloniferans. Recently, Jebram (1992) proposed that victorelloids may also be ancestral to two or three separate lineages of cheilostome bryozoans. The pivotal position of this small superfamily (ten species in five genera and two families) in these recent phylogenetic discussions—and in earlier ones—makes an analysis of their early life histories essential.

Russel L. Zimmer, Department of Biological Sciences, University of Southern California, Los Angeles, CA 90089-0371.

Despite the potential importance of the group, information on nonadult stages is limited. Toriumi (1944) provided a drawing of the larva of *Victorella pavida*, but the only information on the internal anatomy and metamorphosis of larvae of the family Victorellidae is contained in a single report on *Bulbella abscondita* and *Tanganella muelleri* (Braem, 1951). *Sundanella sibogae*, the sole member of the family Sundanellidae, has been studied in two brief reports: Braem (1939, as *V. sibogae*) reconstructed a sagittal section for a late embryo, and Maturo (pers. comm.) provided information on the shape and size of the larva. Perhaps unexpectedly, there is extensive intergeneric variation among these several victorelloid larvae.

The current study provides detailed information on the anatomy and ultrastructure of the larva of *Tanganella muelleri* and complements Braem's (1951) study based on paraffin sections of the larval morphology and metamorphosis of the same species (as *Victorella muelleri*). We then compare and contrast this larva with others from the superfamily Victorelloidea and with those from the superfamilies Valkerioidea and Vesicularioidea. Unfortunately, broader comparisons and an evaluation of the relative merits of the phylogenies, which were proposed by d'Hondt (1986) and Jebram (1973, 1992) on the basis of adult criteria, are not possible because information on larvae of the relevant groups is inadequate or nonexistent.

Materials and Methods

Adult colonies identified as *Tanganella muelleri* were collected from floats in Edward's Boat Yard at Waquoit Bay, Massachusetts, in August 1984 by C. G. R. Although physical measurements of the bay were not recorded at the time of the collections, during recent summers (1991–92) salinities near Edward's Boat Yard ranged from 2 to 12‰ with rare increases to 25‰, and the temperature approached 20°C in late summer (J. Kremer, pers. comm.).

The identity of the material presents several uncertainties: First, the superfamily Victorelloidea is one of several ctenostome groups for which diagnostic taxonomic features are inadequate. Identification of victorelloids to species, genus, family, and even superfamily may require laboratory culture of the adults (e.g., Jebram and Everitt, 1982). Second, more specifically, the status of *Tanganella muelleri* is confused. It has been variously considered an ecomorph of *Victorella pavida* (e.g., Soule, 1957), distinct from but a congener of *V. pavida* (e.g., Hayward, 1985), or sufficiently different from other species of *Victorella* that it should be placed in its own genus, *Tanganella* (Braem, 1951; Jebram, 1973; d'Hondt, 1983; see also Brattström, 1954). That the two forms are not *conspecific* is supported by several features of adult morphology (e.g., differences in proportions of the digestive tract and in placement of parietal muscles [e.g., Jebram and Everitt, 1982]) and of reproductive biology (an intertentacular organ is present in *V. pavida*, but lacking in *T. muelleri*; the eggs of *V. pavida* are about half the diameter of those of *T. muelleri* and are reported to be freely spawned whereas those of *T. muelleri* are brooded within invaginations of the vestibular lining after being shed [e.g., Braem, 1951; Jebram and Everitt, 1982]). To resolve whether these differences are sufficient to justify two genera is beyond the scope of the current study. We have elected to recognize the genus *Tanganella* since Jebram, the current worker who has done the most work on living victorellids, has argued for its retention. On the bases of placement of the parietal musculature, absence of an intertentacular organ, and the brooding of its embryos, our material corresponds with the description of *T. muelleri,* but not that of *V. pavida*. It should be acknowledged that *Victorella pavida* is a well-known and locally abundant form on the East Coast (Banta, pers. comm.). This species has been reported from the Woods Hole region (Rogick, 1964), but to our knowledge, *V. muelleri* has not. Third, a final problem is that Jebram and Everitt (1982) described a second species of *Tanganella, T. appendiculata* from Lagoon

Pond, Martha's Vineyard, Massachusetts. The entrances of the type locality and that of Waquoit Bay are separated by less than 15 km. It is not clear whether C. G. R. knew of this species in 1984 when he collected the adults, but *T. appendiculata* is reported to brood as many as six embryos at one time, whereas a maximum of three embryos were observed in the current material. Jebram and Everitt (1982) reported *T. appendiculata* and *T. muelleri* may also be distinguished by differences in zooidal proportions *if* the two forms are cultured under parallel conditions and by the fact that the first polypides of the ancestrulae have six and seven tentacles, respectively. We have not studied the ancestrula from our material nor maintained it in culture.

Larvae were collected by exposing brooding adults to light after a period of dark adaptation (usually overnight). Larvae for light and transmission electron microscopic preservations were fixed in 2.5 percent glutaraldehyde in 0.14 M NaCl and 0.2 M Millonig's phosphate buffer (pH 7.4) and postfixed in 2 percent osmium tetroxide in 1.25 percent NaHCO$_3$ (pH 7.4); both fixations were for 1 h at room temperature. Larvae were dehydrated in an ethanol series, exchanged in propylene oxide, and embedded in epoxy resin. Serial 1-μm sections were cut on glass knives, mounted on glass slides, stained with methylene blue and azure II, and studied and photographed using a Zeiss RA microscope equipped with Olympus UV and Zeiss neofluar planapochromatic objectives. Thin (silver-gray) sections cut using a diamond knife on an LKB MT-2b ultramicrotome were collected on 200-mesh copper grids, stained sequentially with uranyl acetate and lead citrate, and examined with a Phillips 300EM electron microscope operated at 60 kV. Transmission electron micrograph images were recorded on Kodak Electron Image film.

Results

General Features

The larva of *Tanganella muelleri* is subspherical, averaging about 65 μm in height (along the oral-aboral axis), about 85 μm in width (right to left), and about 120 μm from anterior to posterior in preserved material (Figs. 18.1–18.3). The aboral hemisphere of the larva is provided with a large apical disc that is surrounded sequentially by pallial cells that are shallowly invaginated, producing an annular pallial sinus, a single ring of elongate supracoronal cells, and the narrow corona (Fig. 18.1). In the oral hemisphere (Figs. 18.1 and 18.2), a pyriform complex is positioned in the anterior midline; elsewhere a band of undifferentiated infracoronal cells separates the corona from a narrow perioral ring of ciliated oral epithelial cells which encircles a small field of undifferentiated oral epithelial cells. In the midline of this field are two closely positioned openings and a small patch of ciliated cells. The more anterior opening leads to an incomplete gut and the other to a simple internal sac.

With few exceptions, the epidermal cells of the larva have branched microvilli whose tips are embedded in a thin filamentous glycocalyx. Laterally, near their apices, adjacent epidermal cells are joined by zonulae adhaerentes. Most cells of the larva contain abundant nutrient reserves in the form of lipid droplets or yolk granules or both.

The larva was tentatively classified as Type E in the larval classification of Zimmer and Woollacott (1977a) and as either Type III or IV in that of d'Hondt (1977a).

Epidermal Components of the Aboral Hemisphere

The apical disc is nearly circular and measures 45–50 μm in diameter. The disc includes at least three cell types, but these have not been mapped in detail. The eccentric, paucicelled neural plate is contiguous with two short parasagittal nerve cords; these lead to a nerve nodule positioned just internal to the ciliated cleft of the pyriform complex (Fig. 18.1). The remainder of the disc consists

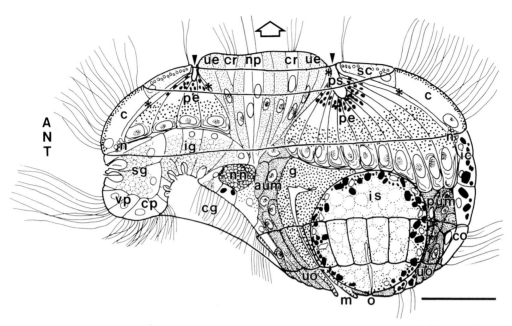

FIGURE 18.1. Composite drawing of a midsagittal section of a *Tanganella muelleri* larva, based on serial 1-μm sections and TEMs. The specimen is viewed from its left side. The large open arrow indicates the direction of swimming and is located at the aboral pole of the larva near the center of the apical disc. The disc is composed of neural plate cells (np), ciliated ray cells (cr), and undifferentiated epidermal cells (ue). Arrowheads identify the openings to the annular pallial sinus (ps) which borders the apical disc. The pallial epithelium lining this sinus consists of an inner ring of suprapallial cell and an outer ring of infrapallial cell (both marked by asterisks), between which is the extensive pallial epithelium proper (pe). Most of the remaining larval surface is covered by successive rings of supracoronal cells (sc), coronal cells (c), infracoronal cells (ic), and ciliated oral epithelial cells (co); the borders of the first three tiers and the approximate boundaries of the individual cells of the ciliated oral epithelium are superimposed upon the section. The bands of infracoronal and ciliated oral cells are interrupted in the anterior midline (ANT) by the pyriform complex, several parts of which are evident (cg, ciliated groove; cp, ciliated plaque; ig, inferior gland cells; sg, superior glandular cells; vp, vibratile plume). The epidermis is completed by a field of undifferentiated oral epithelial cells (uo) which contains the mouth (m) of the gut (g) and the opening (o) of the internal sac (is). Other abbreviations: aum, anterior mass of undifferentiated mesodermal cells; n, equatorial nerve ring; nn, nerve nodule; pum, posterior mass of undifferentiated mesodermal cells. See also Figures 18.2 and 18.3 (scale bar = 25 μm).

largely of numerous undifferentiated epidermal blastemal cells arranged in radial wedges; these alternate with single ciliated ray cells that extend from the neural plate to near the margin of the disc. The blastemal cells are organized as a tall columnar epithelium 20–25 μm in height. The morphology of the basally positioned nuclei (large size, intense basophilia, huge nucleolus), the abundance of free ribosomes, and the relatively paucity of other organelles signal that these cells correspond to the epidermal blastema associated with the apical disc of numerous other gymnolae-mate larvae. A column of undifferentiated mesodermal cells that is in contact with the base of the neural plate is described under internal components; the significance of the undifferentiated mesodermal and epidermal cells is considered in the Discussion.

The pallial epithelium is depressed in a shallow U-shaped furrow (Figs. 18.1, 18.4, and 18.5) rather than being invaginated with the two leaves closely appressed as in most other coronate larvae. This distinctive portion of the epidermis is most extensive in the lateral and posterior quadrants of the larva; the massive aggregation of gland cell of the pyriform complex in the anterior quadrant may restrict the growth of the pallial epithelium here (Fig. 18.1). The aboral and oral margins of the pallial epithelium are composed of single rings of supra- and infrapallial cells that join the pallial epithelium proper to the apical disc and supracoronal cell ring, respectively (Figs. 18.1, 18.3, and 18.4). The pallial sinus reaches a maximum depth of about 10 μm.

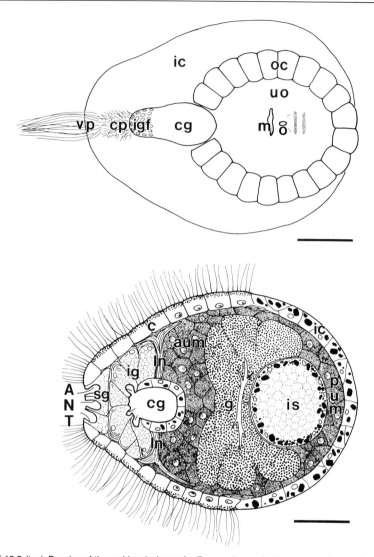

FIGURE 18.2 *(top)*. Drawing of the oral hemisphere of a *Tanganella muelleri* larva, reconstructed from serial sections. Four components of the pyriform complex (vibratile plume [vp], ciliated plaque [cp], inferior glandular field [igf], and ciliated groove [cg]) are centered on the anterior midline. The mouth (m), the opening to the internal sac (o), and several ciliated cells (arrays of dots indicate cilia) are contained in an elliptical field of undifferentiated oral cells (uo). Infracoronal cells (ic) and a "perioral" ring of some twenty large ciliated oral epithelial cells (co) complete the oral hemisphere. Ciliation of the ciliated groove and ciliated oral epithelium are not shown (scale bar = 25 μm). FIGURE 18.3 *(bottom)*. Diagram of transverse section of a *Tanganella muelleri* larva based on 1-μm sections and TEMs. The section passes above the equator (through the oral ends of the coronal cells [c]) in the anterior half and below the equator (through the most aboral infracoronal cells [ic]) posteriorly. Starting at the anterior end (ANT), one sees sequentially superior (sg) and inferior (ig) gland cells, ciliated groove (cg), anterior mass of undifferentiated mesoderm cells (aum), gut (g), internal sac (is), and posterior mass of undifferentiated mesoderm cells (pum). Paired lateral nerves (ln) to the right and left of the ciliated groove bifurcate to give rise to the equatorial nerve ring; additional parts of this nerve ring are seen where coronal and infracoronal cells abut (see also Fig. 18.9). See Figure 18.8 for details of the apical crypts of superior gland cells (scale bar = 25 μm).

Projecting into this U-shaped sinus are enlarged, cuticle-free microvilli of the pallial cells (Figs. 18.3–18.5) and occasional cilia; the latter are probably from either the adjacent ciliated ray cells of the apical disc or from supracoronal cells at the oral margin of the pallial sinus, as we have not seen cilia originate from the pallial tissue. The pallial cells are joined near their apical border by

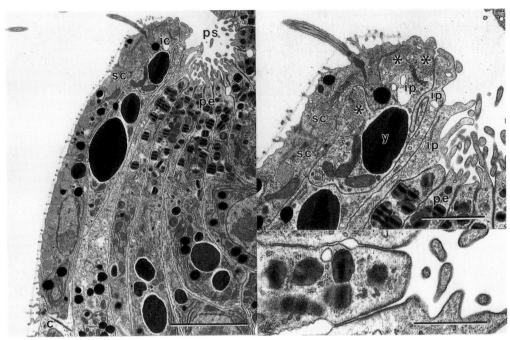

FIGURE 18.4 *(left)*. TEM of radial section through the pallial epithelium and ring of supracoronal cells. The shallow pallial sinus (ps) is lined mostly by cells of the pallial epithelium proper (pe) which are provided with conspicuous microvilli, but no cuticle; note the apical-basal polarization of these cells. The pallial epithelium also includes a single ring of suprapallial cells (not shown) and a single ring of infrapallial cells (ip). The latter joins a single ring of supracoronal cells (sc). Because the section is slightly oblique, portions of three infrapallial and two supracoronal cells are seen. The junction of the lower supracoronal cell with a coronal cell (c) is seen in the lower lefthand corner (scale bar = 5 μm). FIGURE 18.5 *(top right)*. Higher magnification TEM of the same section shown in Figure 18.4. One of the cells of the pallial epithelium proper (pe), the three infrapallial cells (ip), and the two supracoronal cells (sc) are all labeled. The apical regions of the pallial cells proper contain banded secretion vesicles and have thick microvilli which project into the pallial sinus. The infrapallial cells contain large yolk inclusions (y). Bundles of microfilaments (asterisks) which form an infrapallial muscle ring are seen in the outermost infrapallial cell. One of the two cilia of the supracoronal cell that is joined to this infrapallial cell is evident. Note that all cells are joined by zonulae adhaerentes (scale bar = 2 μm). FIGURE 18.6 *(bottom right)*. TEM of secretory vesicles in the apex of a cell of the pallial epithelium proper to show their primary and secondary bandings (scale bar = 1 μm).

one or occasionally two broad zonulae adhaerentes (Fig. 18.5) and are often complexly interfolded laterally.

The pallial epithelium proper, with cells up to 25 μm in height, occupies about one-third of the larval interior, filling much of the aboral hemisphere of the larva and protruding past the larval equator in the posterior half of the larva (Fig. 18.1). The cells of the pallial epithelium proper are prismatic in shape with strongly polarized cytoplasm (Fig. 18.4). Ellipsoidal secretion vesicles about 0.4–0.5 μm in diameter and 0.5–0.8 μm in length are abundant in the apical third or fourth of the cell (Figs. 18.4 and 18.5). The contents of these distinctive, membrane-bound vesicles consist of alternate light and dark bands that number about five or six and are usually aligned with the cell's apex. As seen at high magnification, the light and dark bands are divided into about three and six still finer bands, respectively (Fig. 18.6). The secondary banding is of a uniform periodicity and is parallel (usually) or normal (rarely) to the primary banding. Near their centers, the pallial cells proper have Golgi bodies, numerous small vesicles, mitochondria, and occasional strands of endoplasmic reticulum (Fig. 18.4). A basal zone contains large and small yolk inclusions, abundant but loosely organized sheets of endoplasmic reticulum, and a large nucleus with dense nucleoplasm and an unusually large nucleolus. Throughout the cytoplasm are numerous ribosomes. The wealth of organelles (especially endoplasmic reticulum) and the pattern of their distribution are diagnostic of cells which are precursors of the ancestrular epidermis in other bryozoans.

The supra- and infrapallial cells lack the synthetic machinery and products found in cells of the pallial epithelium proper, but their pale cytoplasm contains larger yolk granules basally (Fig. 18.4). Both infra- and suprapallial cells have aligned bundles of microfilaments in their apical cytoplasm which form composite annular "myofibrils" within each of the rings of cells. The resulting suprapallial ring muscle is poorly developed, but the infrapallial ring muscle is composed of several large bundles of microfilaments (Figs. 18.4 and 18.5).

Supracoronal cells are interposed between the oral margin of the pallial epithelium and the corona (Figs. 18.1 and 18.4). Although in a single ring, the straplike cells are aligned as pickets in a fence to form a broad (15 to 20 μm wide) annulus of the epidermis. Considering the width of these cells (ca. 1.6 μm) and the diameter of the ring they form, we estimate there are about one hundred supracoronal cells. The two cilia of each cell originate near the junction of the supracoronal and infracoronal rings, whereas the nucleus is at the opposite end (Fig. 18.4). The apical cytoplasm of the 2–4-μm-thick cells has numerous vesicles with finely granular contents and a diameter of about 1 μm (Fig. 18.4). The abundance of ribosomes, the electron density of the nucleoplasm, and the size of the nucleolus suggest these cells are relatively undifferentiated and will be retained through metamorphosis, but their postmetamorphic function is unknown.

The corona is slightly wider (20–22 μm) than the supracoronal annulus and also consists of cells in a single band (Fig. 18.1). We have not counted the number of coronal cells directly, but calculate their number as about thirty-six. The large, electron-lucent nuclei are surrounded by abundant mitochondria and storage products (Fig. 18.7). The cells have a more or less rectangular apex with cilia and short capitate microvilli. Each cilium is about 20 μm long and is anchored by two striated processes, a long axial rootlet and a short horizontal one. The latter extends toward the aboral pole of the larva, in a direction opposite that of the effective beat of the cilium.

As their name suggests, intercoronal cells are positioned between adjacent coronal cells. These cells, which are potential receptors of light or other stimuli, are well known from cheilostome larvae, but have been documented for only two ctenostomes. Such cells were not found in the available 1-μm sections of *Tanganella muelleri*, but one was detected in thin sections (Fig. 18.7). This example was located laterally in the larva and appeared to possess a single line of cilia at its apex.

Epidermal Components of the Oral Hemisphere

As in all other gymnolaemates, a pyriform organ occupies the anterior midline of the aboral hemisphere in the larva of *Tanganella muelleri* (Figs. 18.1–18.3). This organ includes a mass of superior gland cells, a vibratile plume, a ciliated plaque, inferior gland cells, and a ciliated furrow in aboral to oral sequence. All of these structures are invariable components of pyriform organs, with the exception of the ciliated plaque, which has probably been overlooked in most cases.

Both the superior and inferior gland cells have deep apical crypts (Figs. 18.1, 18.3, and 18.8). A modified cilium is anchored into the wall of each crypt by a pair of short, modified striated rootlets. Each stereocilium extends a short distance beyond the orifice of the crypt, and its axoneme consists of numerous microtubules (sometimes more than sixty) that occasionally are as doublets, but are usually single. The microtubules lack ATPase arms and, as seen in profile, may be arranged in spiral arrays, but are usually irregularly distributed (Fig. 18.8). Both categories of

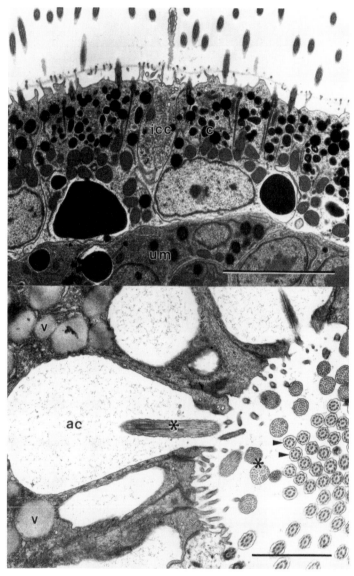

FIGURE 18.7 *(top)*. TEM of transverse section through the corona near its oral margin. Parts of four coronal cells (c) and one intercoronal cell (icc) are seen at the surface. The internal cells are undifferentiated mesodermal cells (um) (scale bar = 2 μm). FIGURE 18.8 *(bottom)*. TEM of longitudinal section through the apices of superior gland cell. Each gland cells has a conspicuous apical crypt (ac) into which the abundant secretory vesicles (v) are apparently emptied. Arising from the wall of each crypt is a single macro- or stereocilium (asterisks). Cross-sections reveal the cores of these macrocilia contain arrays of aligned microtubules which lack dynein arms and only rarely occur as doublets. Cilia with typical axonemes (arrowheads) are from vibratile plume or ciliated bordering cells of the pyriform complex (scale bar = 2 μm).

gland cells are so filled with secretion vesicles that the nucleus, well-developed whorls of endoplasmic reticulum, and most other organelles are compressed in the basal ends of the cells.

The superior gland cells open to the larval surface in a small triangular "field" bounded at its base by the vibratile plume cells and on its sides by the most anterior pair of coronal cells. The vibratile plume, a characteristic feature of the pyriform complex of all gymnolaemate larvae, consists of a more or less coherent bundle of cilia measuring about 40 μm in length. The number of cells contributing to the plume in gymnolaemate larvae is typically four, but we were unable to determine if this is true for *Tanganella muelleri*. Between the plume and the inferior glandular field are a small number of heavily ciliated cells called the ciliated plaque which are thought to be involved in gliding motions of the larva during substrate selection preparatory to metamorphosis. The inferior gland cells, which are grouped in right and left masses (Fig. 18.3), open into a pit at the aboral end of the elongate ciliated cleft or groove.

Except in the anterior midline where the pyriform organ is positioned, the superficial epidermis of the oral hemisphere is formed by three distinctive tissues composed, respectively, of infracoronal, oral ciliated, and oral undifferentiated cells (Figs. 18.1, 18.2, 18.9, and 18.10). Of these, the first two are common to other coronate larvae, but the third is previously undescribed.

Named for its position immediately oral to the corona, the infracoronal tissue of *Tanganella muelleri* forms a 25-µm-wide band several cells in width (Figs. 18.1, 18.2, 18.9, and 18.10). These cells have a number of characteristics of undifferentiated cells (they are unciliated and have a large nucleus and nucleolus, basophilic nucleoplasm, abundant ribosomes, etc.), but also possess large yolk granules and conspicuous bundles of microfilaments. The latter are concentrated in the infracoronal cells adjacent to the corona and are largely parallel the equator, but have a variety of other orientations (Figs. 18.9 and 18.10). Although the infracoronal cells form an equatorial muscle ring, they are otherwise poorly differentiated and as such may play a role in postmetamorphic stages of the life cycle.

Adjacent to the infracoronal tissue is a perioral ciliated ring about 12 µm wide. The ring, which is interrupted in the anterior midline by the ciliated cleft, consists of some twenty multiciliated cells in single file (Figs. 18.1, 18.2, and 18.10). Each of the multiple cilia is provided with a horizontal and an axial striated rootlet; since the former is directed aborally as in coronal cilia, we assume that the effective beat of the cilia of the oral ciliated cells is also toward the oral pole. We equate this perioral ring with the oral ciliated epithelium of other coronate larvae for reasons provided in the Discussion.

The perioral ciliated ring surrounds an elliptical field (about 50 µm along the sagittal midline and 40 µm from side to side) that consists largely of undifferentiated columnar epidermal cells. Centered within the field on the larval midline are, from anterior to posterior, a mouth that leads into an incomplete gut, the opening of the metasomal sac, and two short rows of ciliated cells (Figs. 18.1 and 18.2). The latter produce two narrow (about four cilia wide) transverse bands of cilia, each extending about 8 µm to either side of the midline. The cells producing these cilia bands have abundant yolk reserves in addition to the apical specializations. In contrast, the remaining cells are relatively yolk-poor and have the cytoplasmic and nuclear characteristics of embryonic or blastemal cells (Figs. 18.10 and 18.11). There can be little doubt that this field of undifferentiated cells represents the anlage of some adult component (see the Discussion).

In addition to forming the above superficial components, the oral epithelium is invaginated to form an internal sac. In *Tanganella muelleri* this sac is nearly spherical (Figs. 18.1 and 18.3), with a diameter of about 30 µm. It consists of a single type of gland cell of which all but the basal tenth is filled with secretion vesicles (Fig. 18.12). These membrane-bound vesicles are spherical, about 3 µm in diameter, and contain delicate fibers that are in parallel arrays about 0.6 to 1 µm wide. These bundles are cross-banded at intervals of about 0.5 µm and arranged in swirls through the vesicle. Interestingly, the cross bands of adjacent arrays of fibrils are often in register. Neither the lumen of the sac nor its opening to the surface is apparent even in sections, but the latter is just posterior to the mouth, which leads to an incomplete gut described below. The anterior and lateral faces of the sac are embraced by the gut, and its inner or aboral end is pressed against the basal

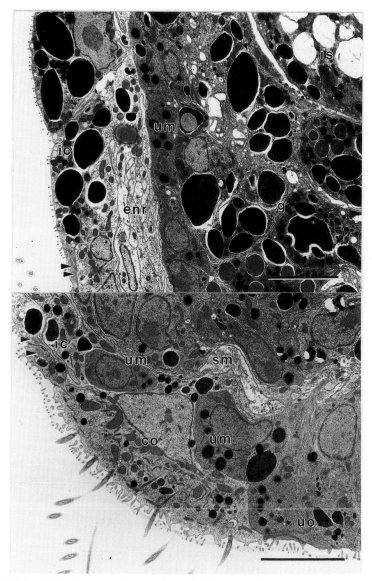

FIGURE 18.9 *(top)*. TEM of a cross-section near the level of the equator, but just oral to the corona. All cells at the surface are infracoronal cells (ic). One of these contains microfilaments which are part of the equatorial muscle ring (arrowhead). The obliquely cut equatorial nerve ring (enr) is partly embedded in the infracoronal cells which are underlain by a single layer of undifferentiated mesodermal cells (um). A portion of the internal sac (is) occupies the upper right corner. The remaining cells, mostly nutrient laden, belong to the rudimentary gut (g) (scale bar = 5 μm). FIGURE 18.10 *(bottom)*. TEM of the posterior, oral portion of a median sagittal section. Counterclockwise at the surface starting in the upper left corner are parts of several infracoronal cells (ic) with bundles of microfilaments (arrowheads), a ciliated oral epithelium cell (co), and undifferentiated oral epithelial cells (uo). Also seen are undifferentiated mesoderm cells (um) and a smooth muscle cell (sm) (scale bar = 5 μm).

surface of the pallial epithelium. Undifferentiated mesodermal cells sheath its posterior face, but are not organized as an epithelium.

Internal Components

Internal components of the larva include an incomplete gut, portions of the nervous system (presumably), certain muscles, and undifferentiated mesodermal cells. Like other coronate lar-

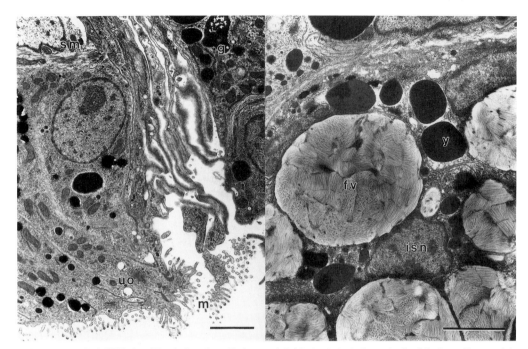

FIGURE 18.11 *(left)*. TEM of a midsagittal section with the mouth (m) just anterior to the oral pole of the larva. The mouth is surrounded by undifferentiated oral epithelium (uo). The cells lining the mouth and gut lumen have apical lamellar folds. Internal are parts of the gut (g) and a smooth muscle cell (sm) (scale bar = 2 μm). FIGURE 18.12 *(right)*. TEM of a cross-section showing a portion of the internal sac and the adjoining gut (g). The internal sac consists of a single layer of prismatic cells with indistinct boundaries and electron-dense nuclei (isn). The cells are largely filled with vesicles containing distinctive, cross-banded fibrils (fv) and ellipsoidal yolk granules (y). At the upper margin of the figure is a portion of the gut (scale bar = 2 μm).

vae, this larva has neither coelomic or blastocoelic compartments nor circulatory and excretory components.

The gut consists of a mouth, esophagus, and stomach, but there is no trace of either an intestine or anal opening. As indicated earlier, the mouth opens just anterior to the internal sac and therefore is positioned near the oral pole of the larva. The thin-walled esophagus leads into a laterally extended "stomach," whose right and left halves are each subdivided into several large lobes (Figs. 18.3 and 18.9). Lamellar folds project from the luminal surface of the gut, especially near the mouth (Fig. 18.11), but cilia and typical microvilli are absent. The cells of the gut are the most yolk-rich of the larva and contain frequent multivesicular bodies and myelin figures (Figs. 18.9 and 18.11). Considering its morphology, the absence of food within its lumen, and the lack of associated muscles, the gut is apparently nonfunctional, as well as incomplete.

In addition to components located at the surface of the larva, the nervous system includes the paraxial nerve cords, nerve nodule, and equatorial nerve ring. The paraxial nerve cords, which are about 4 μm in diameter and 15 μm apart, join the neural plate of the apical disc with the nerve nodule. The latter is a tangled mass of neuronal processes that lies just basal to the ciliated cleft of the glandular-sensory pyriform complex. Extending laterally from this nodule are a pair of fiber tracts that soon bifurcate as anteriorly and posteriorly directed arcs (Fig. 18.3); collectively these four arcs constitute an inconspicuous equatorial nerve ring. The nerve ring lies near the junction of coronal and infracoronal cells and is partially embedded in the latter (Figs. 18.1, 18.3, and 18.9). Although these nerves and the nodule appear to be internal, they may represent only aggregations of axonal and dendritic processes of epidermal cells.

As described earlier, microfilament-based muscles associated with the epidermis include the two annular muscles formed by the supra- and infrapallial cells at the margins of the pallial sinus and one at the larval equator formed by infracoronal elements. Surprisingly, these are some of the best-developed elements of the muscular system. A plexus of smooth muscle fibers, which is located basal to the undifferentiated oral epidermis and is largely anterior to the degenerate gut, is the only conspicuous muscle found in transmission electron micrographs of the larva of *Tanganella muelleri* (Figs. 18.10 and 18.11) and the only muscle recognizable in 1-μm sections. This meshwork may be important in constricting the oral face of the larva and detaching the internal sac after the onset of metamorphosis. Although axial and equatorial ring muscles are usually prominent features of coronate larvae, only the latter was identified in the larva of *T. muelleri* and then only in electron micrographs as delicate isolated fibers centripetal to the equatorial nerve ring.

The mesodermal compartment of *Tanganella muelleri* contains, in addition to muscles, only a single type of cell. The nonmuscle cells are all undifferentiated and have the same cytoplasmic and nuclear characteristics as the undifferentiated oral epithelial cells with which they are contiguous (Figs. 18.1 and 18.10). They are aggregated in two main areas, one anterior to the gut and one posterior to the internal sac (Figs. 18.1 and 18.3), although small groups are seen lateral to the gut and internal sac (Figs. 18.3 and 18.9). The band posterior to the sac extends from the undifferentiated oral epithelium to the basal surface of the invaginated pallial epithelium, whereas the band anterior to the gut extends from the oral undifferentiated epithelium to the apical plate, coursing just posterior to the nerve nodule and paraxial nerves cords. The potential function(s) of these cells are considered in the Discussion.

Discussion

One objective of the current study was to provide data on comparative larval morphology that will be useful in evaluating the relative merits of conflicting interpretations of ctenostome phylogeny. Such analysis of ctenostome larvae is now precluded by the fact that coronate larvae of only four ctenostomes have been studied at the ultrastructural level, and three of these are in one family (Vesiculariidae). The four species are *Alcyonidium gelatinosum* (d'Hondt, 1973, 1974, as *A. polyoum*, family Alcyonidiidae), *Amathia vidovici* (Zimmer and Woollacott, 1993), *Bowerbankia gracilis* (Reed and Cloney, 1982a,b), and *B. imbricata* (d'Hondt 1977b). Earlier light microscope studies, although relatively numerous, covered only a small fraction of ctenostome diversity and were usually limited to examination of intact larvae (see Nielsen, 1971; Zimmer and Woollacott, 1977a). A few early studies utilized paraffin sections, but even these afforded little detail.

Below, we have first made a brief comparison of the larvae of the four species of victorelloids which have been studied. We then provide a detailed integration of our findings on the larva of *Tanganella muelleri* with those of Braem (1951) on the same larva (as *Victorella muelleri*) and relate this to earlier studies. A final comparison is made with the well-studied vesiculariform larva of the superfamily Vesicularioidea and with the only coronate larva known for the superfamily Valkeriidae (that of *Valkeria uva*). The latter comparisons are made since the superfamily Victorelloidea has been identified to be the ancestor of either the vesicularioids or valkerioids (which represent the two major groups of "stoloniferan" ctenostomes) by Jebram (1973) and d'Hondt (1986), respectively.

Comparison of the Larva of *Tanganella muelleri* with Those of Other Victorelloids

Information on larval structure is known for only three other members of the superfamily Victorelloidea. Two of these, *Victorella pavida* and *Bulbella abscondita*, are in the same family

(Victorellidae) as *Tanganella*; the third, *Sundanella sibogae*, is the monotypic representative of the only other family (Sundanellidae). Collectively, these four species represent all but one of the five victorelloid genera.

The simple line drawing of a lateral view of the larva of *Victorella pavida* (Toriumi, 1944) indicates that this larva and that of *Tanganella muelleri* share the same general shape and a large apical disc. However, the corona appears to be considerably wider (apparently having extended orally as well as aborally) than the narrow, equatorially positioned locomotory band of *Tanganella*. The size, nature, and even the existence of the pallial epithelium, supra- and infracoronal cells, ciliated and undifferentiated oral epithelia, mesoderm, internal sac, etc., are not discussed in the Japanese text and are not evident from the drawing, but it is reasonable to suspect that these closely related (and possibly congeneric) species have larvae which share most characteristics.

Bulbella abscondita is the only victorellid other than *Tanganella muelleri* for which there is any information on metamorphosis and for which more than superficial information on larval structure is available (Braem, 1951). Interestingly, its larva differs significantly in morphology from that of *T. muelleri*. The apical disc of *Bulbella* is much smaller than that of *Tanganella*, but shares the same fate (differentiation as the epithelia of the ancestrular lophophore and gut after invagination at metamorphosis). The pallial epithelium of this larva is not invaginated even shallowly, but is the convex, cuticle-covered epidermis of the aboral hemisphere of the dome-shaped larva. The larva has a flattened oral hemisphere so that the narrow corona encircles the lower (oral) edge of the larva, not the equator. Although supra- and infracoronal cells are presumably present, they were not identified by Braem. The oral surface in *Bulbella* appears to be completely ciliated, whereas the corresponding surface in *Tanganella* consists of a narrow ring of ciliated cells and an oval field of nonciliated, undifferentiated cells. As justified below, the ring of cells represents the oral ciliated epithelium, and the oval field, the undifferentiated oral epithelium; there apparently is only ciliated oral epithelium in *Bulbella*. The mesenchyme is aggregated into anterior and posterior masses (between the apical disc and superior gland cells and as a loose accumulation capping the internal sac, respectively), an arrangement similar to that in the larva of *Tanganella*. In both species, the anterior mass is reported to contribute to the mesodermal layers of the polypide. Additional characteristics shared by the two larvae include a rudimentary gut, a simple internal sac, and, most significantly, a metamorphosis in which (1) the pallial epithelium forms the entire cystid epidermis; (2) the internal sac plays only a temporary attachment role and is then histolyzed, and (3) the polypide rudiment's nonmesodermal components originate in the apical disc.

Information on development in *Sundanella sibogae* is very limited. A reconstruction of a late embryo was figured and briefly described by Braem (1939, as *Victorella sibogae*); recently, the larva has been observed and photographed by Maturo (pers. comm.). The elongate shape and large size of this larva place it in sharp contrast to the larvae described for other victorelloids. Maturo reported the larva can exceed 1 mm in length. This is more than eight times the longest dimension of other victorelloid larvae and is to our knowledge the largest dimension of any coronate larva. The fact the larva of *Sundanella* is asymmetrically elongated in its oral-aboral dimension, a feature that is known elsewhere only in vesiculariform larvae, elicits comparisons between these two larval forms since Jebram (1973) concluded vesicularioids originated from victorelloids (see below).

There are several major parallels between the larva of *Sundanella sibogae* and those of *Tanganella muelleri*, *Bulbella abscondita*, and (presumably) *Victorella pavida*: (1) Braem (1939) noted the large size of the apical disc, which suggests it probably contains undifferentiated epidermal cells that contribute the epithelia lining the tentacles and gut of the ancestrula polypide as reported for *T. muelleri* and *B. abscondita*; (2) although the pallial epithelium is of only modest dimension and apparently has been invaginated in such a way that the pallial sinus is obliterated, it is still probable that this tissue forms the entire cystid epidermis, since (3) the small size of the internal sac suggests its role in metamorphosis is limited to temporary attachment. The extent and degree of differentiation of the oral epithelium are not known.

Integration of the Two Studies on the Larva of *Tanganella muelleri*

Although detailed cytological information on the larva of *Tanganella muelleri* is available only from the current work and postmetamorphic roles of tissues are known only from Braem's 1951 paper, the two studies are mutually complementary and together form an integrated whole. No significant inconsonances exist, although a number of minor differences in larval morphology were identified (e.g., Braem did not identify any muscles, ciliated ray cells in the apical plate, infra- and suprapallial cells in the pallial epithelium, or the few ciliated cells located posterior to the opening of the internal sac, but such apparent oversights are understandable considering the techniques available to him).

1. Our study confirms Braem's (1951) observations that the apical disc of *Tanganella muelleri* is large with a small central neural plate surrounded by a broad flange of columnar epithelial cells which we recognize to be undifferentiated. In addition, there are a small number of ciliated ray cells (a typical feature of coronate larvae [e.g., Zimmer and Woollacott, 1989]).

    The disc, after invagination at metamorphosis, contributes the inner layer of a bilaminar vesicle from which the ancestrular polypide is elaborated (Braem, 1951). As in many other coronate larvae (for reviews, see Zimmer and Woollacott, 1977b, or Reed, 1991), this inner layer contributes the epidermis of the tentacles and the gut epithelium, while the outer layer is the source of mesodermal derivatives associated with the tentacles and gut. Considering their highly differentiated nature, the neural plate and ciliated ray cells are probably transitory larval structures as in other coronate larvae (e.g., Woollacott and Zimmer, 1971). Conversely, the abundance and embryonic nature of the undifferentiated cells confirm that they are that part of the disc which contributes the nonmesodermal tissues of the polypide (e.g., Woollacott and Zimmer, 1971; Zimmer and Woollacott, 1977b; Reed, 1991). See items 7 and 10 below concerning the origin of the polypide mesoderm.

2. As reported by Braem (1951), the pallial epithelium is extensive, but only shallowly depressed, leaving a U-shaped pallial sinus. The cytology of the pallial cells supports Braem's observation that they are involved in formation of the cystid epidermis: the abundance and highly polarized distribution of secretory vesicles, mitochondria, and rough endoplasmic reticulum are characteristic of cells which contribute to the cystid epidermis in other species (e.g., Lyke, et al. 1983; Reed, 1991).

3. Although the corona has long been thought to abut the pallial tissue, the two are separated by a narrow ring of mono- or biciliated supracoronal cells in single file in both coronate and cyphonautes larvae (for review, see Zimmer and Woollacott, 1989). The present study reveals that the wide band of larval epidermis separating the pallial and coronal tissues of the larva of *Tanganella muelleri* corresponds well with the supracoronal band of other bryozoan larvae (see below) not only in position, but in consisting of biciliated cells arranged in single file (Braem apparently believed the cells were unciliated). The unusual width of the supracoronal ring in *Tanganella muelleri* results because the apex of each of the approximate one hundred

cells is an unusually narrow trapezoid with a height equal to the width of the ring, about 20 μm.

The functional roles of supracoronal cells (the name is derived from the position of these cells relative to the corona) have been proposed for the coronate larvae of only two ctenostomes and two cheilostomes. In both ctenostomes studied, *Alcyonidium gelatinosum* (d'Hondt, 1974, as *A. polyoum*) and *Bowerbankia gracilis* (Reed and Cloney, 1982b), the supracoronal cells are a transitory larval epidermal tissue. In the ascophoran cheilostome *Watersipora arcuata*, the cells are highly specialized, forming a flange-protected groove that is used by the larva in the transport of mycoplasmalike organisms between successive adult generations (Zimmer and Woollacott, 1983; Boyle et al., 1987); whether these distinctive larval cells have a postmetamorphic function (other than as nutrient material) is unknown. In contrast, the supracoronal cells of the cribrimorph cheilostome *Cribrilina corbicula* are undifferentiated; this led Reed (1991) to speculate that after metamorphosis they may be found as "totipotential cells beneath the ancestrular wall that will determine the budding pattern of the ancestrula." The supracoronal cells of *Tanganella muelleri* are neither highly specialized nor fully embryonic. Of necessity, they form a part of the larval epidermis as in all larvae, but to postulate a postmetamorphic function is premature.

4. Our study confirms Braem's description (1951) that the corona is a relatively narrow band positioned aboral to the larval equator, as in the hypothetical larva proposed by Zimmer and Woollacott (1977a). The number of coronal cells is about thirty-six, which is four more than the assumed primitive number of thirty-two. However, high counts have also been noted in the extremely hypertrophied coronas of at least some vesiculariform larvae (*Bowerbankia imbricata* [d'Hondt, 1977b] and *Amathia vidovici* [Zimmer and Woollacott, 1993]), and in the narrow equatorial corona of *Alcyonidium gelatinosum* (d'Hondt, 1973, as *A. polyoum*).

5. The infracoronal band in *Tanganella muelleri* larvae is unusually wide and consists of cells that are relatively undifferentiated, although provided with abundant microfilaments that collectively form an equatorial muscle ring.

    The coronate larvae of three other ctenostomes (*Alcyonidium gelatinosum,* d'Hondt, 1973, as *A. polyoum*; *Bowerbankia imbricata,* d'Hondt, 1977b; and *Amathia vidovici,* Zimmer and Woollacott, 1993) have a one- or two-cell-wide ring of undifferentiated cells just orad to the corona (hence called infracoronal cells [d'Hondt, 1973]); in addition, four groups of such cells are found in the larva of *B. gracilis* (Reed and Cloney, 1982a). In the first three larvae, infracoronal cells function as the primordium of the ancestrular polypide (specifically, its nonmesodermal components). Braem (1951) suggested no postmetamorphic role for the infracoronal cells of *Tanganella muelleri* and implied that the polypide's nonmesodermal derivatives originate exclusively from apical disc cells. However, the fact that the latter cells and infracoronal ones share ultrastructural characteristics causes pause, since infracoronal cells are known to be a polypide primordium in other ctenostomes and especially since the polypides of the cheilostomes *Watersipora arcuata* and *Cribrilina corbicula* probably have a composite origin from undifferentiated disc and infracoronal cells (Reed, 1991). Obviously it will be important to assess the fate of infracoronal cells in *Tanganella* and its relatives.

6. In other coronate larvae, ciliated cells form all the epidermis between the infracoronal tissue and the opening of the internal sac and are collectively identified as the oral ciliated epithelium. A narrow "perioral" ciliated ring consisting of about twenty epidermal cells in single file borders the infracoronal tissue and surrounds an elliptical field of unciliated cells in the

larva of *Tanganella muelleri*. For reasons documented below, we consider the perioral ring to represent the oral ciliated epithelium of other coronate larvae.

The perioral ciliated ring of *Tanganella muelleri* is topologically equivalent to the ciliated oral epithelium of these other larvae in that its aboral margin borders the infrapallial tissue, but differs in that its oral margin is separated from the sac opening (and the mouth) by an elliptical field of unciliated cells that is apparently unique to *Tanganella* (see 7 below). Precise information on the number and patterning of cells in the oral epithelium has been published for only two ctenostomes, the vesiculariids *Amathia vidovici* (Zimmer and Woollacott, 1993) and *Bowerbankia gracilis* (Reed and Cloney, 1982a). In these two larvae, the number, position, cytology, and arrangement of the oral ciliated cells have parallels with *Tanganella*: on each side of the larva in both species, about ten ciliated cells in single file extend between the infracoronal cells and the elongate opening to the internal sac, an arrangement that is topologically equivalent to a flattened ring.

Searching beyond coronate larvae, could the perioral ring correspond to the ciliated ridges of cyphonautes larvae? Since the ciliated ridges are used in feeding, the presence of a gut in the larva of *Tanganella muelleri* is significant, even if it is nonfunctional. More important, during embryogenesis of *Tanganella*, the oral hemisphere is transitorily invaginated so that the embryo appears to have, at least briefly, the equivalent of the vestibule of a cyphonautes larva (the vestibule, which is involved in feeding, is typically thought to be a unique feature of the planktotrophic cyphonautes). However, the ciliated ridges of cyphonautes bear lateral, frontolateral, and frontal bands of cilia and pass between the mouth and the openings to the internal sac whereas the perioral ring of *Tanganella* is uniformly ciliated and surrounds both of these openings. The topological discordance seems to preclude the two structures sharing the same ontogenetic origin or serving the same functional role.

We conclude that the perioral ring of *Tanganella* is the ciliated oral epidermis of other coronate larvae.

7. An elliptical field of cells centered on the oral pole completes the oral epithelium. Although Braem (1951) implied that the cells of this field lack cilia, a few of the cells posterior to the opening of the internal sac are multiciliated. Our demonstration that the unciliated cells are embryonic in nature indicates they almost certainly serve as an ancestrular primordium. Such a field of undifferentiated epidermal cells surrounding the opening to the internal sac (and the mouth) is known only in the larva of *Tanganella muelleri*.

Under item 10 below, we evaluate the reported role of the anterior half of this undifferentiated oral epithelium as a primordium of ancestrular mesoderm and suggest that its posterior portion has a parallel fate.

8. A larval gut is present, but this is poorly differentiated, unregionated, incomplete, and almost certainly nonfunctional as concluded in both Braem (1951) and the current study. Complete, functional guts are known only for cyphonautes larvae, but, even then, these are transitory larval organs that will be histolyzed at metamorphosis. Although most coronate larvae lack any trace of a digestive system, partial guts (consisting of a pharynx and stomach) are formed during embryogenesis of several species. These usually are histolyzed during metamorphosis, but may degenerate late in embryogenesis so that the larva is anenteric (e.g., *Alcyonidium gelatinosum*, Barrois, 1877; d'Hondt, 1973, as *A. polyoum*), but may be retained through the brief larval stage (e.g., *Tanganella muelleri*). Since cyphonautes are widely accepted to be more primitive than coronate larvae, the presence of a rudimentary gut

in the latter is usually considered to be an ancestral reminiscence (e.g., Jägersten, 1972; Zimmer and Woollacott, 1977a).

9. Braem (1951) documented that the internal sac of *Tanganella muelleri* is small, simply constructed, and serves only for temporary attachment at settlement before its retraction into the larval interior where it is histolyzed. The ultrastructure of the sac is compatible with this transitory role and incompatible with the possibility the sac contributes to the cystid epidermis as in many other species. Since it consists of a single type of gland cells, the internal sac of this larva is "simple" in the terminology of Reed (e.g., Reed, 1991), representing only the neck region of "complex" sacs such as those of *Bugula* spp. (e.g., Woollacott and Zimmer, 1971).

10. The larval interior contains abundant mesoderm cells, but, with the exception of a few muscle cells, these are undifferentiated mesenchymal elements. In other coronate larvae, differentiated as well as undifferentiated mesenchymal cells are present. Where studied, the differentiated cells (which are often polymorphic within a species and of differing types between species) have been assumed to function at various phases of the life cycle (see Woollacott and Zimmer, 1971; d'Hondt, 1974; Reed and Cloney, 1982a,b; Zimmer and Woollacott, 1993). Of immediate interest is the fact that certain differentiated cells provide the primordia of the mesodermal lining of the ancestrular body wall (e.g., the petaliform cells in *Amathia vidovici*, Zimmer and Woollacott, 1993), whereas undifferentiated mesoderm cells serve as the mesoderm of the incipient polypide anlage (e.g., the median band in *Bowerbankia gracilis*, Reed and Cloney, 1982a,b). In *Tanganella muelleri*, the undifferentiated mesoderm is reported to contribute to the polypide mesoderm of the ancestrula, but must also form the cystid mesoderm unless the few muscle cells undergo extensive proliferation and redifferentiation.

Braem (1951) reported that the undifferentiated oral epithelial cells (see 7 above) anterior to the mouth are internalized during metamorphosis and, once inside, join the more anterior aggregation of undifferentiated mesenchymal cells in forming a single layer of cells around the invaginated apical disc. Subsequently, the disc differentiates as the nonmesodermal components of the polypide (tentacular epidermis and gut epithelium), and the surrounding layer forms their coelomothelial linings.

It should be noted that the anterior portion of the larval mesoderm in *Tanganella* has the same position as the median band in the vesiculariform larvae of *Amathia vidovici* and *Bowerbankia gracilis*. For the latter species, Reed and Cloney (1982b) documented that the median band similarly forms the polypide mesoderm (however, no epidermal cells contribute to the mesoderm, and the nonmesodermal parts of the polypide are from the cupiform layer in *B. gracilis*, not undifferentiated cells in the apical disc as in *T. muelleri*).

Braem's observations require that one accept that epidermal cells can become mesodermal elements. Examples of the transformation of one germ layer to another include the production of "ectomesenchyme" during the embryogenesis of many spiralians (e.g., Conklin, 1897), the origin of the heart vesicle of tornaria larvae from the larval epidermis (e.g., Spengel, 1893), and the origin of the digestive tract from ectoderm both during metamorphosis of the larva and during subsequent asexual budding in bryozoans (e.g., Lutaud, 1983).

Most of the remaining undifferentiated mesenchyme cells in the larva of *Tanganella muelleri* form a mass posterior to the internal sac, extending from the pallial epithelium to the undifferentiated oral epithelium. In the absence of other likely sources for the cystid mesoderm, we propose these undifferentiated mesenchyme cells form the coelomic lining of the body wall (cystid), the epidermis of which is derived from the pallial epithelium. We further hypothesize that the posterior portion of the undifferentiated oral epithelium also contributes to the cystid mesoderm after being internalized at metamorphosis. This suggestion provides a plausible role for the rest of the undifferentiated oral epithelium, utilizing a precise parallel with Braem's account that the anterior portion of the undifferentiated oral epithelium contributes to the mesoderm of the polypide.

Further studies are obviously needed to resolve the exact fate of the undifferentiated mesodermal and epidermal tissues in the larva of *Tanganella muelleri*.

11. Braem (1951) identified no muscles in the larva of *Tanganella muelleri*, and we have identified only five. Of these, only two are of "typical" mesodermal derivation, whereas three are composite myoepithelial ring muscles formed within annuli of epidermal tissues. The infrapallial myoepithelial ring muscle is positioned to play a role in the involution of the pallial epithelium during embryogenesis and could also serve as a drawstring at the oral margin of the pallial epithelium during metamorphosis, closing the pallial epithelium under the internal sac after it detaches. The suprapallial myoepithelial ring muscle might serve in completing the extension of the pallial epithelium over the entire imago, by constricting closed near the aboral pole after the apical plate has retracted. The mesodermally derived equatorial ring muscle of typical coronate larvae is represented by only delicate traces in this larva, but a well-developed infracoronal myoepithelial ring muscle has a closely corresponding position. Constriction of the larval equator to effect eversion of the internal sac, a critical role of the equatorial ring muscle, is probably assumed by the infracoronal element.

Although the musculature is highly variable in other coronate larvae, an equatorial muscle ring (see above) and a singular or paired axial muscle are typically present. We did not locate an axial component in *Tanganella muelleri*, but remnants may be present; considering the potential relationship of victorellids with vesicularioids (see below), it may be significant that axial muscles are not found in the larvae of *Bowerbankia gracilis* (Reed, 1980; Reed and Cloney, 1982a) or *Amathia vidovici* (Zimmer and Woollacott, 1993). The only conspicuous mesoderm-derived musculature in *T. muelleri* is a network that parallels the anterior face of the oral hemisphere. This corresponds in position to the anterior constrictor muscle in *Amathia vidovici* (Zimmer and Woollacott, 1993) and the corresponding muscle of *Bowerbankia gracilis* (Reed, 1980), which are assumed to facilitate the involution of the oral surface during the initial phases of metamorphosis.

12. The larva of *Tanganella muelleri* lacks distinctive photoreceptors (e.g., pigment-cup ocelli), and only isolated intercoronal cells were evident. Although conspicuous photoreceptors are typically absent in ctenostome larvae, a pair of potential supracoronal ocelli were reported in the larva of the confamilial *Bulbella abscondita* (Braem, 1951), and abundant, but unpigmented intercoronal cells have been found in the two ctenostomes in which they were specifically sought (Reed and Cloney, 1982a, for *Bowerbankia gracilis* ; Zimmer and Woollacott, 1993, for *Amathia vidovici*).

To summarize the above comparisons, it seems probable that *Bulbella* has the least specialized larva and *Sundanella* the most specialized one in the superfamily Victorelloidea. Our ranking of *Bulbella* is based largely on the fact that its larva has the shape of a lecithotrophic cyphonautes, in which (1) the apical disc is of modest dimensions, (2) the pallial epithelium is not invaginated, and (3) the oral hemisphere is flattened so that the narrow corona is located at the oral margin of the larva rather than at its equator. Further the mesenchyme appears less well organized than in *Tanganella* and specifically is not aggregated as a median band. *Sundanella* is regarded to have the most specialized of the victorelloid larvae in consequence of its unusual asymmetrical elongation in the oral-aboral axis, a feature elsewhere known only in the highly derived vesiculariform larva (see below).

Arguably, the larva of *Tanganella* (and possibly *Victorella*) has more primitive characteristics than that of *Bulbella* although this assumption may be largely a consequence of our more detailed knowledge of the former species. Since the focus of the current chapter is *Tanganella*, it is useful to identify the primitive and advanced characteristics of its larva.

Primitive features of this larva include: (1) The corona is relatively narrow and is largely restricted to its primitive position near the equator (however, the fact that the pallial epithelium is partially invaginated implies that the corona has expanded at least somewhat toward the aboral pole). (2) The pallial epithelium is only shallowly invaginated. (3) A gut, although rudimentary and incomplete, is present. (4) The musculature is poorly developed. (5) Virtually all the mesoderm cells are undifferentiated. (6) There are potential dual rudiments of the precursors of nonmesodermal tissues of the polypide in infracoronal and apical disc tissues. (7) Highly differentiated sense organs such as ocelli or juxtapapillary cells are absent and intercoronal cells are sparse.

However, the larva has several advanced features, including: (1) The corona has more than the presumed primitive number of cells, thirty-two. (2) The cystid epidermis is derived solely from pallial epithelium rather than having a shared origin from the pallial epithelium and the internal sac. (3) There appears to be the equivalent of the median band of mesoderm (we have identified this as an advanced feature, since a median band is found elsewhere only in the specialized vesiculariform larva of vesicularioids, an advanced group of ctenostomes). (4) The internal sac is of simple cytology, provides only transitory attachment at metamorphosis, and then is lysed rather than forming a portion of the cystid epidermis.

Finally, there are two features of questionable assignment: (1) Since undifferentiated oral epithelium has not been found in any other larva and a function has been reported for only the anterior part of it (although we have speculated on the role of the posterior half), it is difficult to determine whether this epithelium is an "invention" of victorelloids (or their immediate relatives) or is a feature of primitive larvae that has been lost or modified in other bryozoan lineages. Unfortunately, no larva of any distant or near ancestor of victorellids has been studied, so the problem remains unresolved, pending future research. (2) Many of the muscles are of myoepithelial rather than mesodermal origin.

Comparisons of Larvae of the Superfamilies Victorelloidea and Valkerioidea (and Its Relatives)

The superfamilies Valkerioidea and Vesicularioidea have been identified as derivatives of the superfamily Victorelloidea by d'Hondt (1986) and Jebram (1973), respectively. (Actually d'Hondt splits the nonvesicularioid stoloniferan ctenostomes into three superfamilies, Valkerioidea, Triticelloidea, and Aeverillioidea). Larvae are known, but only at a superficial level, for a single species in each of four of the eleven genera of nonvesicularioid stoloniferans (*Valkeria uva*, *Triticella koreni*, *Hypophorella expansa*, and *Farella repens*). Recognizing that we are dealing with one of the most highly evolved groups of ctenostomes, it is most unexpected that, of these four examples, only *Valkeria* has a coronate larva: *Farella* and *Hypophorella* have cyphonautes larvae, and *Triticella* has a lecithotrophic cyphonauteslike larva. Although there are several studies of the larva of *Valkeria uva*, none was within the last century, and we know little more about the larva than that it has a short oral-aboral axis (a feature shared with all three victorellid larvae, but not with *Sundanella*) and that Barrois (1877) reported the corona to have extended both orally and aborally (a feature shared only with *Sundanella* of the four known victorelloid larvae). Nothing is known, for example, of the nature of the apical disc, the extent and cytology of the pallial epithelium or internal sac or oral epithelium, the distribution and nature of the mesoderm, or the metamorphosis. In the absence of such detail, it is impossible to make meaningful comparisons between this larva and the coronate larvae of victorelloids (or other ctenostomes), although one could note that the general form of this larva is significantly different from that reported for any other bryozoan.

Comparisons of Larvae of the Superfamilies Victorelloidea and Vesicularioidea

Members of the superfamily Vesicularioidea share a distinctive "vesiculariform" larva characterized by a number of derived features. As far as is known, these derived characteristics are largely unique to the vesicularioids (Reed, 1991; Zimmer and Woollacott, 1993), but a few are shared with members of the superfamily Victorelloidea.

Two composite traits shared by these two superfamilies, but with no other taxon are (1) the pallial epithelium is extensive and apparently contributes the entire ancestrular epidermis, and (2) the internal sac is small, of simple cytology, and plays only a transitory attachment role during metamorphosis before it undergoes degeneration. *Tanganella*, but not *Bulbella*, appears to have the equivalent of the median mesodermal band which is characteristic of vesicularioids; such details of internal structure are not known for the other two victorelloid larvae. High numbers of coronal cells are found in both victorellids and vesicularioids, but this derived trait is also found in other taxa. Two traits common to vesiculariform larvae—the long oral-aboral axis and the relocation of the opening of the internal sac from the oral pole to the anterior face of the larva— appear to be a shared consequence of the fact the coronal cells have undergone progressively greater elongation from the anterior to the posterior face of the larva; to our knowledge, a shift of the internal sac's opening to the anterior face of the larva and, by presumption, an asymmetrical elongation of the coronal cells have occurred elsewhere only in *Sundanella sibogae* (the three other victorelloid larvae have short oral-aboral axes, and at least two of these species have a near axial location of the internal sac opening).

Vesiculariform larvae have several advanced traits not found in the known larvae of victorelloids (or other bryozoans), including extreme hypertrophy of the pallial epithelium, simplification of the apical disc (with loss of any role as a polypide rudiment), and formation of the nonmesodermal components of the polypide from infracoronal or cupiform cells.

On the basis of the distribution of derived characters just documented, one could suggest an ordering of the larvae of *Bulbella*, *Tanganella*, *Victorella*, and *Sundanella* and of the vesiculariform larva that would support Jebram's assertion that victorelloids were the stem group to vesicularioids. Unfortunately, this ordering has little significance since virtually nothing is known of development in the several potential ancestors, sister groups or descendants of these two superfamilies (i.e., the superfamilies Paludicelloidea, Arachnidioidea, and Valkerioidea). As a consequence, with but one exception, we do not know when the derived characteristics originated and therefore whether they are *uniquely* shared by vesicularioids and victorelloids or are also shared by other proposed or potential ancestral or sister groups. One could argue that asymmetric elongation of the corona (and hence the oral-aboral axis) and the consequent anterior displacement of the internal sac originated within the superfamily Vesicularioidea among the immediate ancestors of *Sundanella* and is shared only with vesicularioids. This unique symplesiomorphy would provide support for Jebram's argument that Victorelloidea are ancestral to Vesicularioidea rather than being a sister group as deduced by d'Hondt. However, the occurrence of this symplesiomorphy does not preclude d'Hondt's conclusion that Valkerioidea were (also) derived from Victorelloidea.

Acknowledgments

Christopher Reed collected and embedded the original material in August 1984 at Woods Hole, Massachusetts, and prepared most of the sections and all of the TEMs during his last summer, 1989, at Friday Harbor, Washington. The authors are indebted to Paula Reed for her dedication and darkroom expertise and to Dr. A. O. D. Willows for the use of space and facilities at the University of Washington's Friday Harbor Laboratories. Completion and publication of the report were supported in part by National Science Foundation grant BSR 9008098 to R. L. Z.

Literature Cited

Barrois, J. 1877. Recherches sur l'embryologie des Bryozoaires (Memoire sur l'embryologie des Bryozoaires). Trav. Stn. Zool. Wimereux 1: 1–305.

Boyle, P. J., J. S. Maki, and R. Mitchell. 1987. Mollicute identified in novel association with aquatic invertebrate. Curr. Microbiol. 15: 85–89.

Braem, F. 1939. *Victorella sibogae* Harmer. Zeit. Morphol. Okol. Tiere 36: 267–278.

―――. 1951. Über *Victorella* und einige ihrer nächsten Verwandten, sowie über die Bryozoenfauna des Ryck bei Greifswald. Zoologica 102: 1–59.

Brattström, H. 1954. Undersökningar över Öresund. 36, Notes on *Victorella pavida* Kent. Acta Univ. Lund, N.F. Avd. 50: 1–29.

Conklin, E. G. 1897. The embryology of *Crepidula*. J. Morphol. 13: 1–226

d'Hondt, J. L. 1973. Etude anatomique, histologique, et cytologique de la larve d'*Alcyonidium polyoum* (Hassall, 1841), Bryozoaire Cténostome. Arch. Zool. Exp. Gén. 114: 537–602.

―――. 1974. La métamorphose larvaire et la formation du "cystide" chez *Alcyonidium polyoum* (Hassall, 1841), Bryozoaire Cténostome. Arch. Zool. Exp. Gén. 115: 577–605.

―――. 1977a. Valeur systématique de la structure larvaire et des particularités de la morphogenèse post-larvaire chez les Bryozoaires Gymnolemates. Gegenbaurs Morphol. Jahrb. 123: 463–483.

―――. 1977b. Structure larvaire et histogenèse post-larvaire chez *Bowerbankia imbricata* (Adams, 1798), Bryozoaire Cténostome (Vesicularines). Arch. Zool. Exp. Gén. 118: 211–243.

―――. 1983. Tabular keys for identification of the Recent ctenostomatous Bryozoa. Mem. Instit. Oceanogr. Monaco . 14: 1–134.

―――. 1986. Etat de connaissances sur la position phylogénétique et l'évolution des Bryozoaires. Boll. Zool. 53: 247–269.

Hayward, P. J. 1985. Ctenostome Bryozoans. *In* D. M. Kermack and R. S. K. Barnes, eds. Synopses of the British Fauna, No. 33, E. J. Brill, London.

Jägersten, G. 1972. Evolution of the Metazoan Life Cycle. Academic Press, London.

Jebram, D. 1973. Stolonen-Entwicklung und Systematik bei den Bryozoa Ctenostomata. Z. Zool. Syst. Evolut.-forsch. 11: 1–48.

Jebram, D. H. A. 1992. The polyphyletic origin of the "Cheilostomata" (Bryozoa). Z. Zool. Syst. Evolut.-forsch. 30: 46–52.

Jebram, D., and B. Everitt. 1982. New victorellids (Bryozoan, Ctenostomata) from North America: The use of parallel cultures in bryozoan taxonomy. Biol. Bull. 163: 172–187.

Lutaud, G. 1983. Autozoid morphogenesis in anascan cheilostomes. *In* R. A. Robinson, ed. Treatise on Invertebrate Paleontology, Pt G. Bryozoa., Vol. 1. Geol. Soc. Am., Boulder, Colo., pp. 208–237.

Lyke, E. B., C. G. Reed, and R. M. Woollacott. 1983. Origin of the cystid epidermis during the metamorphosis of three species of gymnolaemate bryozoans. Zoomorph. 102: 99–110.

Nielsen, C. 1971. Entoproct life-cycles and the entoproct/ectoproct relationship. Ophelia 9: 209–341.

Reed, C. G. 1980. The Reproductive Biology, Larval Morphology and Metamorphosis of the Marine Bryozoan *Bowerbankia gracilis* (Vesicularioidea, Ctenostomata). Ph.D. dissertation, Department of Zoology, University of Washington, Seattle, 291 pp.

———. 1991. Bryozoa. *In* Reproduction of Marine Invertebrates: Lophophorates and Echinoderms, Vol. 6. A. C. Giese, J. S. Pearse, and V. B. Pearse, eds. Boxwood Press: Pacific Grove, California, pp. 85–245.

Reed, C. G., and R. A. Cloney. 1982a. The larval morphology of the marine bryozoan *Bowerbankia gracilis* (Ctenostomata: Vesicularioidea). Zoomorph. 101: 23–54.

———. 1982b. The settlement and metamorphosis of the marine bryozoan *Bowerbankia gracilis* (Ctenostomata: Vesicularioidea). Zoomorph. 101: 103–132.

Rogick, M. M. 1964. Phylum Ectoprocta. *In* R. Smith, ed. Keys to Marine Invertebrates, Spaulding Co., Boston, pp. 167–187.

Soule, J. D. 1957. Two species of bryozoan Ctenostomata from the Salton Sea. Bull. So. Calif. Acad. Sci. 56: 13–34.

Spengel, J. W. 1893. Die Enteropneusten des Golfes von Neapel. Fauna Flora Golfes Neapel, Monogr. 18: 1–758.

Toriumi, M. 1944. On the study of freshwater bryozoans (in Japanese). Dobutsugaku zasshi (Zool. Mag.) Tokyo 55: 20–25.

Woollacott, R. M., and R. L. Zimmer. 1971. Attachment and metamorphosis of the cheilo-ctenostome bryozoan *Bugula neritina* (Linné). J. Morph. 134: 351–382.

Zimmer, R. L., and R. M. Woollacott. 1977a. Structure and classification of gymnolaemate larvae. *In* R. M. Woollacott and R. L. Zimmer, eds. Biology of Bryozoans. Academic Press, New York, pp. 57–89.

———. 1977b. Metamorphosis, ancestrulae, and coloniality in bryozoan life cycles. *In* R. M. Woollacott and R. L. Zimmer, eds. Biology of Bryozoans. Academic Press, New York, pp. 91–142.

———. 1983. Mycoplasma-like organisms: Occurrence with the larvae and adults of a marine bryozoan. Science 220: 208–210.

———. 1989. Larval morphology of the ascophoran bryozoan *Watersipora arcuata* (Cheilostomata: Ascophora). J. Morph. 199: 125–150.

———. 1993. Anatomy of the larva of *Amathia vidovici* (Bryozoan: Ctenostomata) and phylogenetic significance of the vesiculariform larva. J. Morph. 215: 1–29.

19 A Hypothesis for the Evolution of the Concentricycloid Water-Vascular System

Daniel A. Janies and Larry R. McEdward

ABSTRACT Concentricycloids were discovered in the deep sea in 1986 and have been described as a new class of echinoderms. Concentricycloids have an unusual water-vascular geometry, in which there are dual circumoral ring canals that are connected by interradial canals. The podia are arranged as a single series on the outer circumoral ring canal. We propose that this unusual water-vascular system evolved by modifications of coelomic morphogenesis. Altered coelomic morphogenesis occurs during the direct development of the starfish, *Pteraster tesselatus*. Coeloms arise, in their juvenile positions, from seven enterocoels that evaginate from separate regions of the archenteron. The water-vascular coelomic system develops from five separate enterocoels (we interpret these first five enterocoels, collectively, as homologous to the five lobes of the typical asteroid hydrocoel). These extend radially, in a transverse orientation, from the central region of the archenteron, and later are joined by lateral evaginations that produce the circumoral ring canal. We postulate that lateral evaginations could be duplicated at an early developmental stage to produce dual circumoral ring canals. The duplicated formation of circumoral rings was likely a key innovation in the evolution of the concentricycloids, which we believe are best characterized as progenetic velatid asteroids.

Introduction

Baker et al. (1986) reported the discovery of an unusual echinoderm on sunken wood collected from the deep sea off the coast of New Zealand. This animal, named *Xyloplax medusiformis*, was classified as an asterozoan (sensu Moore, 1966), but was judged to be sufficiently different from asteroids or ophiuroids to warrant exclusion from these classes. A new class, the Concentricycloidea, was erected to accommodate *Xyloplax*. Subsequently, a second species, *Xyloplax turnerae,* was discovered in the Caribbean (Rowe et al., 1988).

Both species of *Xyloplax* are small ($\approx$ 2–13 mm diam), disc-shaped organisms, which lack radiating arms (Fig. 19.1a). The margin of the body is ringed by small spines, and the aboral surface is covered by scalelike plates. There are three concentric rings of skeletal ossicles on the oral surface: (1) an inner ring that represents a modified mouth-frame skeleton; (2) a middle ring associated with the podia; and (3) a marginal ring that bears spines (Fig. 19.1a). The water-vascular system consists of two circumoral ring canals joined by five, short, interring connecting canals, which are considered to be interradial in location (Fig. 19.1b) (Baker et al., 1986; Rowe et al., 1988). The podia are located on the outer circumoral ring canal and are arranged in a single series around the periphery of the oral surface. A single simple hydropore connects the inner

Daniel A. Janies and Larry R. McEdward, Department of Zoology, University of Florida, Gainesville, FL 32611.

A **B**

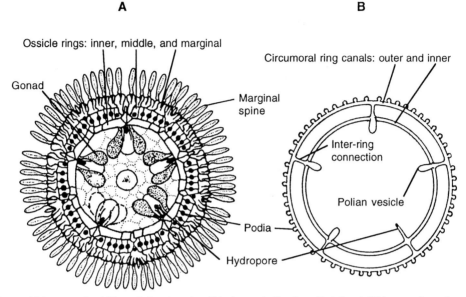

FIGURE 19.1. Adult concentricycloid morphology based on *Xyloplax medusiformis*. *a*. Oral view. *b*. Water-vascular system. (Modified from Baker et al., 1986).

circumoral ring canal to the external sea water. The coelom is divided into two compartments: a main central compartment surrounding the gonads and a small peripheral compartment surrounding the ampullae of the podia. *X. turnerae* has a broad mouth leading to a shallow stomach. *X. medusiformis* has a vestigial gut during development, but lacks a gut as an adult. The entire oral surface of *X. medusiformis* is covered by a velum (Baker et al., 1986; Rowe et al., 1988).

The diagnostic features of the class Concentricycloidea involve the arrangement of the skeletal plates and the water-vascular system. The highly derived water-vascular system was the primary reason given by Rowe et al. (1988) that prevented inclusion of *Xyloplax* within any previously defined class of echinoderms (see also Rowe, 1988). The typical echinoderm (e.g., asteroid) water-vascular system consists of a single circumoral ring canal with an axial connection (i.e., along the oral-aboral axis of the animal) to the external environment. A radial canal extends from the circumoral canal into each of the rays. Lateral canals connect the podia to the radial canals (Fig. 19.2). The double ring canals with interradial connections and a uniserial set of podia located on the outer ring canal of *Xyloplax* represent fundamental differences in the architecture of its water-vascular system.

Smith (1988) questioned the class status of the concentricycloids because the new class was defined by only one or two characters that were judged to be important features of the group. Historically, the assignment of class status in the echinoderms has been problematic and arbitrary (Smith, 1988). Smith (1988) advocated that the concentricycloids should be placed within the asteroids because Baker et al. (1986) failed to set the concentricycloids apart from the asteroids on the basis of overall phenetic or genetic distance. Smith performed a cladistic analysis using selected synapomorphies (shared derived characters) given by Blake (1987) for the asteroid order Velatida and some putative synapomorphies and autapomorphies (unique derived characters) taken from the early, brief published descriptions of *Xyloplax* (Baker et al., 1986) and *Caymanostella* (an asteroid) (Belyaev, 1974; Belyaev and Litvinova, 1977). Smith's cladogram clearly places the concentricycloids as a sister group of the Caymanostellidae, both of which are

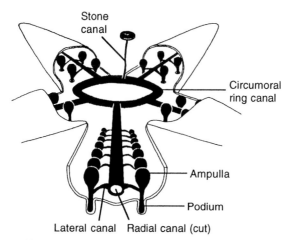

FIGURE 19.2. Morphology of a typical adult asteroid, cut away to reveal water-vascular system.

considered aberrant velatid asteroids (Table 19.1 and Fig. 19.3). In spite of the disagreement over the taxonomic rank of the concentricycloids, there is no question that they are characterized by a novel arrangement of skeletal plates and a novel water-vascular system.

It has been hypothesized that the concentricycloids were derived from an asteroid ancestor, most likely a korethrasterid or caymanostellid, but possibly an asterinid (Rowe et al., 1988; Rowe, 1988; Smith, 1988; Downey, pers. comm.; see Clark and Downey [1992] for descriptions of these asteroid families). There is disagreement over the ordinal-level classification of these three families. Blake (1987), in his reclassification of the Asteroidea, placed the asterinids in the order Valvatida and placed the families Korethrasteridae and Caymanostellidae in the order Velatida (Fig. 19.3). This view has been adopted by many echinoderm systematists (e.g., Smith, 1988; Clark and Downey, 1992). In contrast, Rowe et al. (1988), while acknowledging the similarities between caymanostellids and korethrasterids, consider the caymanostellids to be more closely related to the asterinids and to belong within the order Valvatida.

It is most likely that concentricycloids arose from a caymanostellidlike ancestor (Rowe et al., 1988; Rowe, 1988; Smith, 1988). Synapomorphies (in addition to those listed in Table 19.1 and Fig. 19.3) between caymanostellids and *Xyloplax* spp. include: small size, flattened body form, circular or pentagonal shape, a marginal flange of spines, a simple hydropore instead of a madreporite, a large circular mouth frame, unsuckered podia, and abactinal (aboral surface of animal) plates. Further, there are ecological similarities. Both caymanostellids and concentricycloids are found only in an unusual habitat (i.e., on sunken wood in the deep sea). Whether the caymanostellids are classified as valvatids or velatids influences the assumptions about the general features (especially skeletal structures) from which the concentricycloids were derived. This controversy is a result of the lack of detailed knowledge about the morphology and development of caymanostellids, both species of which are restricted to the deep sea (*Caymanostella admiranda*, 5220 m; *C. spinimarginata*, 1480–6740 m; Clark and Downey, 1992: 345, 348).

Rowe et al. (1988) derived the concentricycloids from a generalized ancestral asteroid ("probably from a common ancestor of certain valvatids," p. 452). Clearly, the evolution of the concentricycloid body plan involved drastic and fundamental modifications of echinoderm design. Rowe et al. (1988) proposed that the radial elements of the water-vascular system and the ambulacral structures (i.e., skeletal, muscular, coelomic, hemal, and nervous structures associated with the radial water-vascular vessel) acquired a circular organization around the periphery of

Table 19.1. Selected characters with primitive and derived conditions in the asteroid superorder Spinulosacea (= orders Velatida and Spinulosida) modified from Smith (1988) and Blake (1987)

Characters and references	Primitive condition in the Spinulosacea	Derived condition in the Spinulosacea
A	Ordinal 35 of Blake (sensu Janies and McEdward, in review)	Origin of large somatocoel from anterior of archenteron	Origin of large somatocoel from posterior of archenteron
B	Ordinal 36 of Blake (sensu McEdward and Janies, in review)	Development via complex feeding or simplified nonfeeding young	Nonfeeding, simplified young only (assumed in Korethrasteridae, Myxasteridae, Caymanostellidae, and *Xyloplax*)
C	Ordinal 38 of Blake	Marginal, abactinal, and actinal ossicles clearly differentiated in adult	These ossicles similar to each other or not differentiated in adult (except Caymanostellidae and *Xyloplax*)
D	Familial 5 of Smith Familial 3 of Blake	Oral ossicles with keellike ridge	Orals well developed but lack keellike ridge and do not form shield over mouth
E	Familial 6 of Smith Familial 6 and Ordinal 1 of Blake	Adambulacural ossicles overlap	Adambulacural ossicles short (except *Xyloplax*, which lacks adambulacurals)
F	Familial 7 of Smith Familial 7 of Blake	Actinal ossicles present	Actinal ossicles present
G	Familial 8 of Smith Familial 8 of Blake	Terminal ossicles in inverted U shape	Terminal ossicles in closed ring
H	Familial 9 of Smith	Marginal ossicles undifferentiated in adult	Inframarginal ossicles with spines (= marginal spines)
I	Familial 10 of Smith	Small peristome; oral plates project into mouth	Very large peristome; oral plates form a circular frame to mouth
J	Familial 11 of Smith Familial 10 of Blake	Madreporite present, axial system opens via a sieve plate	Madreporite lacking; axial system opens via a simple pore
K	Familial 12 of Smith Familial 5 of Blake	Abactinal ossicles with spines; with papillae	Abactinal ossicles platelike; lacking papillae
L	Familial 13 of Smith	Suckered tube feet	Unsuckered tube feet
M	Familial 14 of Smith	Adambulacural ossicles present	Adambulacural ossicles absent
N	Familial 15 of Smith	Ambulacural plating biserial	Ambulacurals separated to form a peripheral ring (sensu Rowe et al., 1988)
O	Familial 12 of Blake	Osculum absent	Enlarged, webbed abactinals form osculum around anal area

[a]The derived conditions of characters are used to build and label (A–O) the family-level cladogram in Figure 19.3. The term *ordinal* corresponds to referenced authors' use of the character to distinguish the order Velatida (character C) or the order Spinuloside (characters A and B) from other asteroid orders. The term *familial* corresponds to referenced authors' use of the character to distinguish among families within the superorder Spinulosacea. The number following these terms (e.g., Ordinal 35) cross-references the use of the character by the referenced author. In discussions of asteroid skeletal morphology, the term *actinal* corresponds to the oral surface whereas the term *abactinal* corresponds to the aboral surface of the adult. Also in this context, the term *oral ossicles* is restricted to mean the mouth angle plates at the head of the ambulacral columns (Blake, 1987).

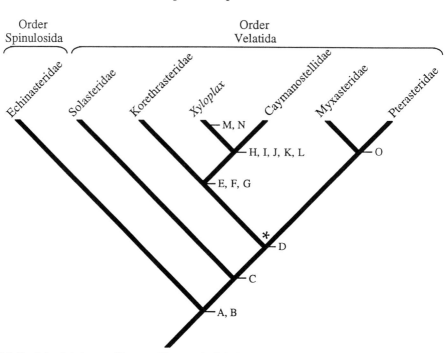

FIGURE 19.3. Family-level cladogram of the asteroid superorder Spinulosacea, modified from Smith (1988) and Blake (1987), showing the origin of the concentricycloid genus *Xyloplax* within the asteroid order Velatida. Table 19.1 lists the selected shared derived characters (synapomorphies), labeled A–O, used to characterize the clades. The node marked with the asterisk represents the hypothesized point of origin for direct development with a reorganized pattern of coelomogenesis important for the evolution of the novel water-vascular morphology seen in *Xyloplax*.

the body by a splitting of the radial canals and a migration toward the periphery (see below). In other words, they derived concentricycloid morphology from the body plan of an adult asteroid with well-developed rays.

Smith (1988) argued that, in addition to sharing a number of unusual morphological features and occupying the same habitat, both concentricycloids and caymanostellids show signs of progenesis. Progenesis is a type of paedomorphosis or juvenilization resulting from an evolutionary change in development such that gonadal maturation is accelerated into the juvenile stage (Gould, 1977). Progenesis can be distinguished from neoteny, the other mechanism of paedomorphosis, because a progenetic descendant is smaller in addition to having a juvenile body plan. For example, *Xyloplax* has a circular body form, marginal spines, and very small size. Smith (1988) drew the comparison between *Xyloplax* and juvenile asteroids and suggested that progenesis could explain the geometry of the water-vascular system, at least the circular arrangement of podia. He claimed that asteroids, at the time of metamorphosis, have the first few podia arranged in a peripheral circle. However, this "peripheral circle" is simply a superficial feature of podial arrangement and is not indicative of a circular water-vascular geometry. At metamorphosis, asteroids generally possess two pairs of podia and a single, unpaired, terminal podium that are arranged radially along each ambulacrum (e.g., Gemmill, 1912, 1914, 1920). These podia are situated close together because of the small size of the juvenile disc and the limited development of the arms. The circular pattern is apparent only because the radially arranged podia have not yet developed into extensive linear series that diverge from the central region of the disc.

Retention of a juvenile condition would maintain the apparent circularity of the podia but would be based on a short, but distinctly radial, water-vascular arrangement. In contrast to Smith's (1988) suggestion, progenesis by itself would not result in a modified organization of the water-vascular system.

We agree that progenesis has played an important role in the evolution of the concentricycloids; however, we believe that the evolution of altered coelomic morphogenesis preceded progenesis. Changes in early coelomic morphogenesis were necessary to produce a novel water-vascular morphology, and progenesis was necessary to shift the juvenile morphology to the definitive body form. One indication that *Xyloplax medusiformis* has modified development is the suggestion that it undergoes direct development while brooded within the gonads (Baker et al., 1986; Rowe et al., 1988). McEdward and Janies (1993) have reviewed the developmental patterns of asteroids and advocated an explicit definition of direct development that is based solely on morphological criteria: development of the juvenile from the embryonic stages, with a complete absence of larval structure or organization (e.g., bilateral symmetry). We also proposed a new term, *mesogen,* to refer to the intermediate stages, between embryo and juvenile, during direct development (McEdward and Janies, 1993). Only one example of direct development has been reported in asteroids, and that is in the velatid starfish, *Pteraster tesselatus* (McEdward, 1992; Janies and McEdward, 1993). Whether concentricycloids, caymanostellids, korethrasterids, or myxasterids undergo direct development is not known. However, features of the direct development of *P. tesselatus* suggest a mechanism whereby a progenetic velatid asteroid could acquire a water-vascular system with two concentric ring canals.

Direct Development of *Pteraster tesselatus*

P. tesselatus has a highly derived pattern of development, in which all larval features have been lost (McEdward, 1992; Janies and McEdward, 1993). Settlement structures such as the brachiolar arms and adhesive disc typical of asteroid larvae are absent. Radial symmetry characterizes all stages of development. The axes of symmetry of the embryo, mesogen, and adult are parallel. The coeloms develop, in their juvenile and adult positions, from seven enterocoels that evaginate from separate regions of the archenteron (Fig. 19.4) (Janies and McEdward, 1993). Although the first five enterocoels develop separately, they are homologous to the lobes of the single hydrocoelic coelom located in the left side of the larvae of other asteroids. The hydrocoel lobes extend radially, in a transverse orientation, from the central region of the archenteron (Fig. 19.4). Each hydrocoel lobe elongates radially and gives rise to a radial canal, a series of lateral canals, ampullae, and the podia of the water-vascular system.

In *P. tesselatus*, the two other enterocoels evaginate from the posterior of the archenteron (Fig. 19.4). The large posterior enterocoel wraps around the gut in a transverse orientation and envelops the hydrocoel lobes. The posterior enterocoel directly produces the oral perivisceral coelom of the adult body plan. Also, the oral and radial perihemal coeloms (which envelop the hemal tissues [see Hyman, 1955]) develop from small diverticula of this large posterior coelom in the same manner as in the juvenile rudiment of an asteroid larva. The posterior origin of the oral perivisceral coelom is a synapomorphy of the superorder Spinulosacea (Table 19.1, Fig. 19.3) (order Spinulosida and order Velatida; Blake, 1987; Janies and McEdward, 1993). The small posterior (i.e., seventh) enterocoel has three fates: aboral perivisceral coelom and axial and oral perihemal coelom. These latter structures develop from the left axocoel (i.e., anterior coelom) and the right somatocoel (i.e., posterior coelom) of typical asteroid larvae.

In most asteroids, development is indirect (McEdward and Janies, 1993). The larval coeloms

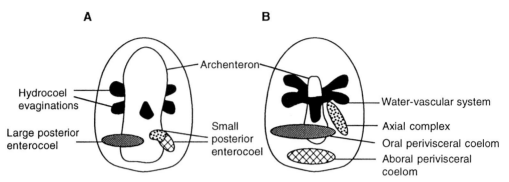

FIGURE 19.4. Diagrammatic representations of enterocoely and coelomic fates in *Pteraster tesselatus*. Anterior of each mesogen is toward the top of the page. *a*. 6 day mesogen *b*. 9 day mesogen.

develop from a pair of lateral enterocoelic evaginations in a bilaterally symmetrical pattern. The enterocoels develop into three sets of coeloms (from anterior to posterior: axocoels, hydrocoels, and somatocoels) (Hörstadius, 1939; Hyman, 1955; Chia and Walker, 1991). The water-vascular coelomic system of the juvenile develops in a sagittal plane, on the left side of the larva from a single coelomic sac, the hydrocoel.

The radical rearrangement of coeloms in *P. tesselatus* likely evolved by means of a simple 90° rotation of the ancestral (larval) spatial pattern for anterior enterocoels relative to the animal-vegetal axis of the archenteron (Janies and McEdward, 1993). This rotation shifted the hydrocoel from its ancestral sagittal orientation on the left side of the body to a transverse orientation encircling the archenteron. Also, in conjunction with the hypothesized rotation and the loss of the preoral coelom, the ancestral axocoel fused with the ancestral right somatocoel to form the small posterior enterocoel. The site of evagination of the large posterior enterocoel, which is common to all spinulosacean asteroids, was unaffected by the relocation of the anterior enterocoels.

A morphogenetic problem was created by the coelomic reorganization in *P. tesselatus*. How do the five independent hydrocoel evaginations from the archenteron (Fig. 19.5a) become connected by a common circumoral ring canal to complete the water-vascular system? In all asteroid larvae, the hydrocoel is a single coelom (Fig. 19.5b,c) which becomes separated from the archenteron well before any water-vascular structures, such as the radial or circumoral canals or podia, are built. Typically, the hydrocoel is crescent shaped (e.g., *Asterias rubens*, *Solaster endeca*; Gemmill, 1912, 1914; Hyman, 1955). In these cases, the circumoral ring canal forms by the closure (i.e., fusion of the ends) of the crescent, which wraps around the gut (Fig. 19.5b). Alternatively, in some asteroids, (e.g., *Henricia* sp.) the hydrocoel is disc-shaped (Masterman, 1902). A circumoral ring is formed by the obliteration of the center of the hydrocoel disc followed by the growth of the esophagus through the space in the center of the disc (Fig. 19.5c) (Masterman, 1902; Hyman, 1955). In either case, the result is the same, a ring of coelom encircling the gut that connects the presumptive radial canals (Fig. 19.2).

In *P. tesselatus*, direct development resulted in an altered spatial relationship of the hydrocoel to the archenteron (Janies and McEdward, 1993). The five enterocoels that develop into the water-vascular system remain connected to the gut, but are independent of each other, during the early development of the radial canals and podia (Figs. 19.4 and 19.5). This unusual type of hydrocoel formation required the evolution of a novel pattern of morphogenesis of the circumoral ring canal. At 8 to 9 days, each radial canal separated from the archenteron by a constriction of the proximal end. Concurrently, the proximal free end of each radial canal produced lateral evagina-

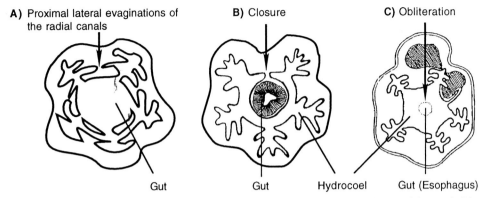

A) Proximal lateral evaginations of the radial canals

B) Closure

C) Obliteration

Gut   Gut   Hydrocoel   Gut (Esophagus)

FIGURE 19.5. A comparison of the known morphogenetic patterns for completion of the circumoral ring canal of the water-vascular system in asteroids. *a.* Drawing of a section through the hydrocoel of *Pteraster tesselatus* showing the morphogenetic pattern in which the circumoral ring canal forms via proximal lateral evaginations of the radial canal. This pattern is known only (thus far) in this species. *b, c.* Sections through the hydrocoel of the juvenile rudiment of typical brachiolarian starfish larvae. The circumoral ring canal forms either via: (*b*) the closure of a crescent-shaped hydrocoel around the gut (i.e., *Solaster, Asterias*) or (*c*) the obliteration of the center of a disc-shaped hydrocoel by the esophagus (i.e., *Henricia*).

tions in the transverse plane of the body (Fig. 19.5a). These evaginations grew around the circumference of the archenteron, met in the interradii, and fused to complete the circumoral ring canal (Janies and McEdward, 1993).

Evolution of The Unique Water-Vascular Morphology of *Xyloplax*

In this section we contrast two scenarios for the evolution of the dual circumoral rings of *Xyloplax*: (1) our hypothesis, based on alterations of coelomic morphogenesis (Fig. 19.6b), and (2) the original interpretation of Rowe et al. (1988) (Fig. 19.6a). These hypotheses lead to substantially different interpretations of concentricycloid morphology.

We suggest that the novel pattern of circumoral ring canal morphogenesis in the development of *P. tesselatus* (Janies and McEdward, 1993) provided a mechanism for the evolution of the unique water-vascular geometry of the concentricycloids. The dual ring canals of the concentricycloids could have evolved through the *duplicated* formation of lateral evaginations of the radial canals. We envision that these lateral evaginations were similar to those that produce the circumoral ring in *P. tesselatus*. In addition, the rays did not develop, and sexual maturation was attained at an early developmental stage: *Xyloplax* is a progenetic velatid asteroid. This interpretation was reached independently by Smith (1988) and Ruppert (pers. comm., based on reexamination of *X. turnerae*).

In our developmental hypothesis, the inner circumoral ring canal of a concentricycloid develops by the fusion of the first (proximal) lateral evaginations of each radial canal (hydrocoel lobe) just as documented in *P. tesselatus* (Janies and McEdward, 1993). The outer circumoral ring develops by the fusion of a second set of (distal) lateral evaginations (Fig. 19.6b). If we assume that the leading edges of the distal lateral evaginations are in fact homologous to the growth zone of the ambulacrum of asteroids, where new podia are produced, then we would predict a distribution of podia matching that described by Rowe et al. (1988); namely, the largest (oldest) podia lie in closest proximity to the interring connections. The elongate "terminal" podium would then represent a novel structure that formed via the fusion of two neighboring lateral evaginations.

Rowe et al. (1988; see also Baker et al., 1986) proposed that the dual circumoral ring morphology of the concentricycloids evolved via modifications in the water-vascular and skeletal

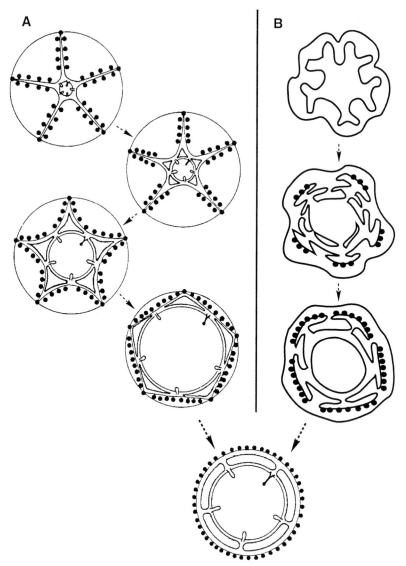

FIGURE 19.6. A comparison of two scenarios for the evolution of the unique water-vascular morphology of the concentricycloids. *a.* In the scenario proposed by Rowe et al. (1988) lacunae develop in the radial canals and the circumoral ring canal of an ancestral adult asteroid water-vascular system. The radial canals (and the associated skeletal elements) split to the point of the Polian vesicles at the interradii. In time, an inner and an outer circumoral water ring form. These circumoral rings widen and move marginally to form the concentricycloid water-vascular geometry. This scenario results in an interradial position of the interring canals. *b.* In the scenario we propose in this chapter, the concentricycloid water-vascular geometry is envisioned to have evolved from a highly derived pentaradial enterocoelic pattern like that seen in the development of the velatid starfish *Pteraster tesselatus*. The inner circumoral ring canal forms via proximal lateral evaginations of the radial canal just as the same structure is formed in *P. tesselatus*. A duplication of this sort of evagination slightly more distal along the radial canal forms the outer circumoral ring and podia. This results in a radial position for the junctions between the dual rings of the concentricycloids.

morphology of an adult asteroid (Fig. 19.6a). In their scenario, lacunae formed at the junction of the circumoral and radial vessels, then expanded laterally and radially to form concentric rings with interring connections in *interradial* positions. The outer ring canal was considered to be made from ancestral radial canals that had split from tip to base. The diameters of the dual circumoral rings widened, and the skeletal elements of the ambulacrum split and moved margin-

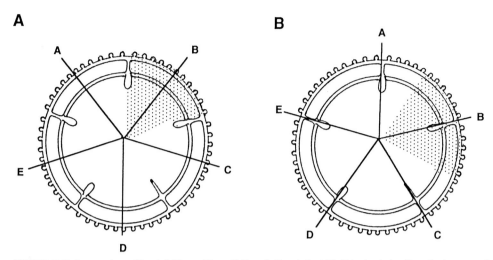

FIGURE 19.7. A comparison of the definitions of the radii (lines A–E and stipple) in *Xyloplax* derived from the two scenarios for the evolution of its water-vascular morphology. *a.* According to the hypothesis of Rowe et al. (1988) each radius lies between the interring connections. *b.* According to our developmental scenario, each radius is defined by the interring connection (because the connection is homologous to the radial canal of an asteroid).

ally to form the middle series of ossicles. The inner ring canal was considered to be homologous to the ancestral, single circumoral canal.

Although our developmental hypothesis arrives at a similar structural arrangement for the units of the concentricycloid water-vascular system, our hypothesis leads to a different interpretation of the homology between adult concentricycloid and adult asteroid body plans. Specifically, we argue that the concentricycloid interring connections are homologous to the radial canals of other asteroids. This leads to a new interpretation of radial and interradial regions of the body in a concentricycloid (Fig. 19.7). We believe that the difference between our interpretation (interring connecting canals are radial in position) and the original interpretation of Baker et al. (1986; see also Rowe et al., 1988; Rowe, 1988) (interring connections are interradial in position) is a direct result of deriving the concentricycloid from a juvenile versus an adult asteroid.

Baker et al. (1986) assumed that in splitting the radial canals, "The pivotal points are the terminal point of each radius; these points remain fixed so that a disc-like body form is derived." Using the terminal points of the radii as reference points is a consequence of attempting to derive the concentricycloid body plan from an adult asteroid body plan: only adults (well-developed postmetamorphic stages) have rays. Keeping the terminal points of the radii fixed requires splits at the bases of the radii and subsequent displacement of the interring connections to interradial locations.

Our scenario differs from the original interpretation because it occurs early in development. We retain the bases of the radii as fixed and derive an outer ring canal by lateral growth from the distal ends of the radial canals before the extension of the rays. In the original suggestion there was radial extension of lacunae relative to a fixed adult ray terminus. An advantage of our scenario is that it keeps the interring connections radial in position and thereby avoids fundamental restructuring of an adult water-vascular organization. Our developmental hypothesis has an additional advantage in that the origin of the dual circumoral ring canal geometry can be explained via morphogenetic mechanisms known to occur in a closely related taxon, the Pterasteridae.

Rowe (1988) argued that *Xyloplax* evolved gradually, in spite of the lack of evidence of intermediate forms. Our hypothesis eliminates the need for the large number of independent small-scale changes in adult morphology that Rowe envisions, because altered development is capable of producing a fundamentally different organization to the body without having to redesign adult structure.

In conclusion, we argue that modifications in coelomic morphogenesis, followed by progenesis, has played an important role in the evolution of both the caymanostellids and the concentricycloids. We have proposed a scenario in which altered coelomic development could produce the radically different water-vascular organization of the concentricycloids. We believe that the information necessary to evaluate hypotheses about the origin of the concentricycloids only will emerge from a comparative study of the morphogenesis of its likely sister groups, the derived velatid asteroids (Fig. 19.3). This clade (korethrasterids, caymanostellids, concentricycloids, pterasterids, and myxasterids) is largely restricted to the deep sea. Consequently, except for *Pteraster tesselatus*, nothing is known about development of the water-vascular system or the morphology of the very young juveniles. It is unlikely that the highly derived morphology of *Caymanostella* or *Xyloplax* evolved by modification of adult skeletal or water-vascular morphology. Rather, radical modifications are more likely to be the result of alterations of morphogenesis, combined with progenesis to produce a novel juvenile morphology as the definitive body form.

Acknowledgments

We thank Maureen Downey and Ed Ruppert for insightful discussions of *Xyloplax* biology.

Literature Cited

Baker, A. N., F. W. E. Rowe, and H. E. S. Clark. 1986. A new class of Echinodermata from New Zealand. Nature 321: 862–864.

Belyaev, G. M. 1974. A new family of abyssal starfishes. Zool. Zhurn.. 53: 1502–1508.

Belyaev, G. M., and N. M. Litvinova, 1977. The second finding of deep-sea starfishes of the family Caymanostellidae. Zool. Zhurn. 56: 1893–6.

Blake, D. B. 1987. A classification and phylogeny of post-Paleozoic sea stars (Asteroidea: Echinodermata). J. Nat. Hist. 21: 481–528.

Chia, F.-S., and C. W. Walker. 1991. Echinodermata: Asteroidea. *In* A. Giese, J. S. Pearse, and V. B. Pearse, eds. Reproduction of Marine Invertebrates. Boxwood, Pacific Grove, Calif., pp. 301–340.

Clark, A. M., and M. E. Downey. 1992. Starfishes of the Atlantic. Chapman and Hall, London.

Gemmill, J. F. 1912. The development of the starfish *Solaster endeca* Forbes. Trans. Zool. Soc. Lond. 20: 1–71.

———. 1914. The development and certain points in the adult structure of the starfish *Asterias rubens*, L. Philos. Trans. Roy. Soc. Lond. B 205: 213–294.

———. 1920. The development of the starfish *Crossaster papposus,* Müller and Troschel. Q. Microsc. Sci. 64: 155–190.

Gould, S. J. 1977. Ontogeny and Phylogeny. The Belknap Press of Harvard University Press, Cambridge, Mass.

Hörstadius. S. 1939. Uber die Entwicklung von *Astropecten aranciacus* L. Pubbl. Staz. Zool. Napoli. 17: 221–312.

Hyman, L. H. 1955. The Invertebrates, Vol. 4. Echinodermata. McGraw-Hill, New York.

Janies, D. A., and L. R. McEdward. 1993. Highly derived coelomic and water-vascular morphogenesis in a starfish with pelagic direct development. Biol. Bull. 184: 56–76.

Masterman, A. T. 1902. The early development of *Cribrella oculata* (Forbes) with remarks on echinoderm development. Trans. Roy. Soc. Edin. 40: 373–418.

McEdward, L. R. 1992. Morphology and development of a unique type of pelagic larva in the starfish *Pteraster tesselatus* (Echinodermata: Asteroidea). Biol. Bull. 182: 177–187.

McEdward, L. R., and D. A. Janies. 1993. Life cycle evolution in asteroids: What is a larva? Biol. Bull. 184: 255–268.

Moore, R. C. (ed.). 1966. Treatise on Invertebrate Palaeontology. Echinodermata. Geological Society of America and University of Kansas Press, Lawrence.

Rowe, F. W. E. 1988. Review of the extant class Concentricycloidea and reinterpretation of the fossil class Cyclocystoidea. *In* R. D. Burke, P. V. Mladenov, P. Lambert, and R. L. Parsley, eds. Echinoderm Biology. A. A. Balkema, Rotterdam, pp. 3–15.

Rowe, F. W. E., A. N. Baker, and H. E. S. Clark. 1988. The morphology, development and taxonomic status of *Xyloplax* Baker, Rowe, and Clark (1986) (Echinodermata: Concentricycloidea), with the description of a new species. Proc. R. Soc. Lond. B 223: 431–439.

Smith, A. B. 1988. To group or not to group: The taxonomic position of *Xyloplax*. *In* R. D. Burke, P. V. Mladenov, P. Lambert, and R. L. Parsley eds. Echinoderm Biology. A. A. Balkema, Rotterdam, pp. 17–23.

Part IV

Larval Dispersal and

Reproductive Ecology

20 Turbulent Transport of Larvae near Wave-Swept Rocky Shores: Does Water Motion Overwhelm Larval Sinking?

M.A.R. Koehl and Thomas M. Powell

ABSTRACT We investigated the potential role that the vertical motion of marine larvae through the water (via sinking or swimming) might play in their dispersal on spatial scales of meters near wave-swept rocky shores. Field releases of particles of different sinking velocities as models of larvae and of dye to track water mixing and transport showed that dilution rates of larvae and dye were similar and that they traveled together across the habitat. Our data suggest that larval vertical motion is overwhelmed by turbulent mixing near wave-beaten shores. Although instantaneous water velocities at wave-swept sites can be quite high (meters per second), the rates of horizontal transport of water and particles across such habitats are much slower (tenths of meters per second).

Introduction
Larval Transport by Moving Water

Benthic marine animals often depend on the water moving around them to disperse their larvae, as described in a number of reviews (e.g., Crisp, 1984; Norcross and Shaw, 1984; Scheltema, 1986; Butman, 1987; Young and Chia, 1987; Levin, 1990; Okubo, in press). Such larval dispersal can have important effects on the population dynamics of benthic species (e.g., Jackson, 1986; Possingham and Roughgarden, 1990) as well as on the gene flow between populations (e.g., Jackson, 1974; Burton and Feldman, 1982; Burton, 1983; Hedgecock, 1979, 1982; Palumbi, 1992). The supply of larvae is also an important factor affecting benthic community structure at some coastal sites (e.g., Bernstein and Jung, 1979; Connell, 1985; Gaines and Roughgarden, 1985; Roughgarden et al., 1987, 1988).

The role of water motion in transporting marine larvae has been investigated at small and large spatial scales. Small-scale water flow (millimeters to centimeters) near the substratum has been shown to affect where larvae settle (e.g., Crisp, 1955; Eckman, 1983, 1987; Hannan, 1984; Butman, 1987), although active substratum selection by settling larvae in flowing water can also play an important role at these small scales (Butman, 1987; Butman et al., 1988; Pawlik et al., 1991). Water flow on larger scales (e.g., internal waves and fronts on spatial scales of tens of meters to kilometers) can be responsible for cross-shelf transport of larvae (Zeldis and Jillet, 1982; Shanks, 1983, 1986; Kingsford and Choat, 1986; Pineda, 1991). Coastal circulation (on spatial scales of one to tens of kilometers) determines the regional distributions of various organisms with planktonic larvae and their retention in bays and estuaries (e.g., Wood and Hargis, 1971;

M.A.R. Koehl, Department of Integrative Biology, University of California, Berkeley, CA 94720.
Thomas M. Powell, Division of Environmental Studies, University of California, Davis, CA 95616.

Rothlisberg, 1982; Cronin and Forward, 1982; Sulkin and van Heukelem, 1982; DeWolf, 1983; Rothlisberg et al., 1983; Levin, 1983; Tegner and Butler, 1985; Emlet, 1986; Johnson et al., 1986; Roughgarden et al., 1987, 1988; Boicourt, 1988). On very large spatial scales, oceanic currents can affect the global distributions of species with long-lived larvae (e.g., Scheltema, 1971, 1975, 1986; Scheltema and Carlton, 1984).

Evidence for the importance of passive dispersal by moving water versus active substratum choice by larvae in determining spatial patterns of invertebrate settlement has been reviewed by Butman (1987) and Pawlik (1992). Butman (1987) suggested that hydrodynamic processes determine larval distributions on large spatial scales (tens of meters to tens of kilometers), whereas active larval behavior can also play a role at small scales (centimeters to meters) near the substratum.

In spite of this wealth of information about hydrodynamic effects on larval dispersal, relatively little is known about the role of water motion in near-shore environments on the spatial scale of meters (0.1–10 m). This scale of water motion provides the transport between the small-scale near-substratum flow and the larger-scale circulation patterns whose effects on larval distributions have been studied already. In spite of the extensive use of rocky-shore communities for basic ecological research, transport on the scale of meters for organisms on wave-swept rocky shores is poorly understood. Furthermore, even physical studies of mechanisms of mass transport in the ocean, such as dispersal of pollutants (e.g., Myers and Harding, 1983) or transport of beach sand (e.g., Bascom, 1980; Basco, 1983), have not focused on the spatial scale of meters near wave-beaten rocks.

Transport of Water and Larvae near Wave-Swept Shores

It is not surprising that water transport on the spatial scale of meters has not been studied for wave-swept rocky shores: water flow at such sites is complicated by the interaction of waves, tidal currents, and complex topography (e.g., Koehl, 1977, 1982, 1984, 1986; Denny, 1988), and waves crashing onto rocks provide a very hostile environment for instrumentation such as current meters or drifters.

One straightforward way of measuring the movement of water and waterborne materials (e.g., dissolved substances, particles, larvae) in challenging environments is by tracking water labeled with dye (e.g., Pritchard and Carpenter, 1960; Okubo, 1971; Riempa, 1985). A patch of water in the ocean (and the materials it carries) can translate and rotate with respect to the substratum (Fig. 20.1a); the patch can also mix with the surrounding water (thereby expanding while becoming more dilute). The rate of horizontal movement of the center of mass of the patch of water is commonly called "advection" by oceanographers. The spread of the patch as it mixes with the adjacent water is called "diffusion," or "turbulent diffusion" (this mixing of the water by turbulent eddies should not be confused with molecular diffusion, due to the random thermal motion of molecules, which is a much slower process).

We have developed a simple dye-tracking technique that can be used to measure such advection and turbulent mixing in the field on the spatial scale of meters (Koehl et al., 1987, 1988, 1993). Although our technique permits us to quantify the transport of dissolved substances carried by the water, we must determine for each habitat whether the transport of dye can provide information about the transport of larvae, which may sink or swim through the water.

It is generally thought that larval swimming is too weak to overcome horizontal currents directly, but that the vertical swimming or sinking of larvae can affect their horizontal transport indirectly by moving them into layers of the water column traveling in different directions or

A.

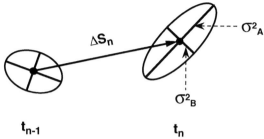

B.

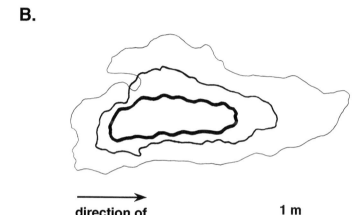

**direction of
wave motion**

1 m

FIGURE 20.1. The fate of patches of waterborne material, such as dye. *a.* Simplified diagram of a patch at two successive times, t_{n-1} and t_n. The centroid of the patch is indicated by the dot, and the distance the centroid traveled is Δs_n. The patch has rotated; hence its spread is indicated by the eigenvalues of the first (σ_A^2) and the second (σ_B^2) principal component axes, as described in the text. *b.* Tracings of the perimeter of an actual dye patch at 2-s intervals (indicated by successively thinner lines) in the surge channel on a calm day (6 August 1989). Expansion of the patch was greatest along the axis of the oscillating flow in the channel, and the patch advected slowly in the direction of wave motion (from left to right in this diagram).

speeds (reviewed in Mileikovsky, 1973; Crisp, 1984; Chia et al., 1984; Young and Chia, 1987; Pawlik, 1992). The speed of sinking and the direction and speed of larval swimming can change as they develop (reviewed by Chia et al., 1984; Young and Chia, 1987). A number of studies have shown that such ontogenetic changes in vertical swimming or sinking by larvae can determine whether they are retained in estuaries (e.g., Graham, 1972; Smith et al., 1978; Boicourt, 1988; Cronin and Forward, 1982; Sulkin and van Heukelem, 1982; Norcross and Shaw, 1984; Laprose and Dodson, 1989; Jacobson et al., 1990). Moreover, the vertical distribution of larvae in the water column can, under calm water conditions, affect the tidal height at which they settle (Grosberg, 1982).

On the other hand, the water motion on wave-swept shores is so rapid and turbulent that we might expect this flow to overwhelm any vertical motion of the larvae. In a well-stirred water column, turbulence is able to counteract the tendency of particles to settle (Csanady, 1983). Evidence that rocky, wave-swept sites are well mixed is provided by the observation that vertical mixing of dye released at the water surface is so rapid that dye concentrations near the bottom (2 m below) can equal those at the surface in a few seconds (Koehl et al., in press). Moreover, analysis of the variation over depth of turbulent kinetic energy in the surf zone suggests strong vertical mixing due to large-scale turbulent vortices (Svendsen, 1987).

The purpose of this study was to determine whether the sinking or downward swimming of larvae in a wave-swept habitat can affect their horizontal dispersal on spatial scales of meters (the scale that determines whether they leave the neighborhood of their parents and enter the larger-scale flows whose effects on larval dispersal have already been studied). Another goal of this work was to assess whether the dispersal of dye is a reasonable tool for estimating the dispersal of particles or larvae in wave-swept habitats.

Materials and Methods

To study the dispersal of waterborne materials in waves, we conducted a series of experiments in which we released dye and particles into the habitat and then tracked their transport and dilution. Because of the technical difficulty of rearing and labeling enough larvae for such field studies, we used particles with different sinking speeds as models of larvae that are negatively buoyant or that swim downward.

Measurement of Particle Sinking Velocities

We chose several types of easily recognizable particles (poppy seeds, glitter [Craft House], and snapdragon seeds) that could be obtained in large quantities and that had sinking velocities similar to those of various marine larvae. We measured particle sinking speeds in sea water (salinity of 33‰) in a cylindrical glass container 16 cm in diameter. Measurements were made in a cold room at 10°C to mimic the temperature of the water at the field site where the particle release experiments were conducted (described below). A jar of each type of particle was well shaken, and then twenty particles were haphazardly picked out of the jar with forceps and placed in sea water. A Pasteur pipette was used to gently deposit a particle below the water surface in the large cylinder. The time for the particle to fall a distance of 5 cm through each of three successive marked intervals in the cylinder was measured to the nearest 0.1 s using a digital stopwatch. All particles had reached terminal velocity before sinking into the first interval, and the mean of the times to fall through the three intervals was used to calculate the sinking speed for each particle. Ten other particles of each type were selected as described above, and the longest dimension of each was measured to the nearest 0.01 mm using an ocular micrometer in a Wild dissecting microscope.

Field Releases of Particles and Fluorescein

All field measurements were made in a large surge channel on Tatoosh Island, Washington, in which the advection and turbulent diffusion were fairly typical of wave-swept rocky shores not subjected to strong unidirectional currents (e.g., Koehl et al., 1993). By confining our measurements to a single channel, ambient flow varied from one experiment to the next while all other parameters could be held constant (i.e., we essentially used that channel as a field version of laboratory wave tank). Water temperature, measured in the channel during each experiment, was 10°C. Experiments were conducted at times during the tidal cycle when water depth in the channel was 2–3 m. Six people ("samplers") were stationed at mapped positions along the south and north shores of the channel.

A mixture of concentrated dye and particles was prepared by placing 55 ml each of dry poppy seeds, glitter, and snapdragon seeds into a can attached to the end of a wooden handle (2 m long) and then adding 220 ml of a solution of fluorescein in sea water (30 g/l of seawater freshly collected from the surge channel); this mixture was stirred until the air bubbles were removed. The can was lowered below the water surface at a defined position in the surge channel (the "zero position," which was one of the six sampling positions mentioned above). The can was then turned over to release its contents and was gently pulled out of the water. A digital stopwatch was started at the instant that the can was turned over, and a signal was called out to the six samplers at 60, 150, 240, 330, and 420 s after dye release. Only one such release was performed per day to assure that dye and particles from earlier experiments had been cleared from the channel. (Indeed, we found that samples of water taken from the channel before each experiment contained no seeds, glitter, or measurable dye.) Experiments were performed on five different days during August and September 1989 to cover a variety of wave and tidal conditions.

Each sampler was equipped with six nets and a wooden handle (2 m long) to which each net could be attached. The sampling nets were small plankton nets (mouth diameter 20 cm, length 50 cm, mesh size 345 µm), each with a 15-ml screw-cap test tube at the downstream end. During each experiment, samples were taken at each of the six positions in the channel at each of the times listed above, and a sample was also taken at the zero position before the mixture of particles and dye was released. A net on the end of the handle was pushed through the water just below the surface for a distance of 2 m; hence the volume of water sampled was about 6.3×10^{-2} m^3. Samples were taken between waves to minimize the effects of ambient flow on the volume of water processed during a sweep of the net; nets were pushed in the opposite direction from the direction of wave movement. After each sample was taken, a cap was immediately put on the test tube, and the net was removed from the handle and rolled up so that no particles could be lost.

At the end of each experiment, all the particles in each net were washed into a paper filter (Mr. Coffee, 8–12 cup size) supported by a wire strainer. A glass grid was laid over each filter, and all the poppy seeds, glitter, and snapdragon seeds on the filter were counted. The test tubes were removed from the nets and stored in the dark until the concentration of fluorescein in the water samples was measured using a Perkin-Elmer Fluorescence Spectrophotometer 204 Å (emission wavelength was 495 nm, excitation wavelength was 513 nm, and both emission and excitation bandpass widths were 10 nm). Any particles in the test tubes were counted and removed before a water sample was put in the spectrophotometer. Samples of water collected from the surge channel before the dye release were used as the blanks for each experiment.

Quantification of Advection and Turbulent Diffusion

We measured advection and turbulent diffusion of the water in the surge channel (on spatial scales of meters to tens of meters) during the particle and fluorescein release experiments described above using the photographic technique reported by Koehl et al. (1993). The expanding blob of fluorescein was photographed at timed intervals from a fixed position on the cliff above the channel. The angle of the camera with the horizontal was measured using a line level and protractor (the camera remained unmoved during the sequence of photographs). The photographs of the dye blobs were projected onto a digitizing tablet (Jandel), the perimeter of each successive blob was traced, and the camera angle was used to correct the coordinates of each point within the blob for parallax (Fig. 20.1b). By making the simplifying assumption that the dye was evenly dispersed within the blob, we calculated the position of the centroid of the blob at each time (t_1, $t_2, \ldots, t_n$). If Δs_n is the distance the centroid has traveled in the nth time interval (i.e., between t_{n-1} and t_n), then the advective speed (U_n) for the nth time interval is calculated as:

$$U_n = (\Delta s_n)/(t_n - t_{n-1})$$

as illustrated in Figure 20.1a.

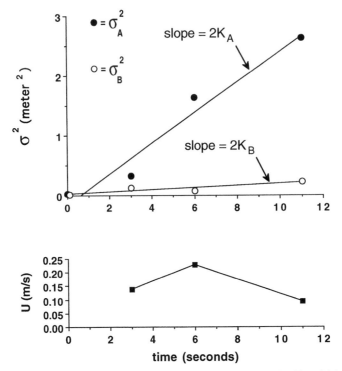

FIGURE 20.2. Example of measurements of mass transport in a surge channel on Tatoosh Island, Washington, using the dye-tracking technique described in the text. The upper graph shows the eigenvalues, σ_A^2 and σ_B^2 (measures of spread of the patch) along the first (dark circles) and second (open circles) principal axes of the dye patch, plotted as functions of time. The lines represent the least-squares, best-fit, straight lines (i.e., linear regressions) calculated through these points. One-half of the slope of such a regression gives the overall diffusivity (K): K_A is the mixing coefficient along the axis of greatest spread of the dye blob, and K_B is the mixing coefficient along the axis perpendicular to the direction of greatest spread. The lower graph shows the speed of advection (U) of the centroid of the patch, plotted as a function of time.

Each dye blob not only spread, but also might have rotated about its centroid (Fig. 20.1a). If we had calculated the rate of spread of the blob with respect to fixed (i.e., non-rotating) axes, then the effects due to expansion could not have been separated from those due to rotation. Therefore, we removed the effect of rotation by using principal component analysis (PCA; e.g., Chatfield and Collins, 1980; Preisendorfer, 1988) to calculate the spread along the two principal axes of the blob. The first principal axis, which we denoted by A, was the direction of greatest spread; the second principal axis, B, was perpendicular to the first. Using the terminology of PCA, the eigenvalues associated with these two directions (i.e., the principal axes) measured the variance (i.e., the spread) along these two axes. We call these two eigenvalues σ_A^2 and σ_B^2; σ_A is a statistical measure of the linear dimension of the blob along its direction of greatest spread, and σ_B is that for the perpendicular direction. We determined overall mixing coefficients (K_A and K_B) for the entire sequence of photographs of a dye blob by calculating linear regressions for plots of σ_A^2 and σ_B^2 versus time (Fig. 20.2); the slopes of these regression lines equal $2K_A$ and $2K_B$, respectively (Okubo, 1980; Koehl et al., 1993).

Results and Discussion
Sinking Rates of Particles

The sinking rates and sizes of the particles we used are given in Table 20.1, where examples of types of larvae that sink or swim at similar speeds are listed.

Table 20.1. Particle sizes and sinking rates, along with larvae with similar speeds

Particle	Width (S.D.) (N = 10) (mm)	Sinking speed (S.D.) (N = 20) (cm · s^{-1})	Larvae with similar speeds[a]	
			Sinking	Swimming[b]
Poppy seed	1.4(0.11)	1.2(0.29)	Crab zoea	Crab zoea (d,h,u)
				Ascidian tadpole
Glitter	1.1(0.07)	1.1(0.13)	Crab zoea	Crab zoea (h)
				Ascidian tadpole
				Sponge amphiblastula (h)
			Bivalve veliger	Bivalve veliger
Snapdragon seed	1.0(0.11)	0.3(0.47)	Coral planula	Coral planula (h,u)
			Bivalve veliger	Bivalve veliger (u)
				Annelid trochophore
			Barnacle cypris	Barnacle nauplius (h,u)
			Crab zoea	Crab zoea (d,h,u)
			Crab megalopa	Crab megalopa
				Lobster phyllosoma (d,u)
				Bryozoan cyphonautes
				Ascidian tadpole

[a]From Chia et al. (1984), who review speed data for particular species.
[b]With direction of swimming indicated (if reported): d = down, h = horizontal, u = up.

Concentrations of Particles and Fluorescein

We released and collected particles and fluorescein near the water surface. If particles had been sinking out of the surface water, we would expect their concentrations in our samples to decrease at a greater rate than that of the dye. In contrast, if the vertical motion of the particles was overwhelmed by the turbulent mixing of the water in the surge channel, we would expect the dye and the particles to be diluted at the same rate by this mixing, and we would expect particles and dye to travel together horizontally.

An example of the changes in concentration (C) of particles and fluorescein at one of the positions in the surge channel is illustrated in Figure 20.3. Both particles and dye showed exponential rates of loss at each position:

$$C_t/C_0 = e^{\lambda t}$$

where C_t is the concentration at time t, C_0 is the highest concentration reached at that position, and λ is the slope of a linear regression of $\ln(C_t/C_0)$ as a function of time (Fig. 20.3), starting at the time when C_0 was measured. We compared the λ's (decay, or loss constants) of the different particles and the dye; we considered that these slopes were significantly different from each other if their 95 percent confidence intervals did not overlap. We only compared loss rates of the different materials within a position on a single day. In seventy of the seventy-five comparisons we made, we found no significant differences between the loss rates of particles with different sinking speeds, or between the particles and the dye. (Note that 95 percent confidence limits imply that, out of seventy-five comparisons selected at random, four should show disagreement simply by chance alone.) Furthermore, in the few cases where the loss rates of certain materials did differ from each other, there was no pattern in the types of materials that showed different λ's, or in the sites or days on which the differences occurred. Therefore, we conclude that there was no

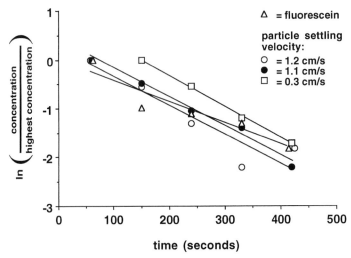

FIGURE 20.3. Plot of the natural log (ln) of the concentration of waterborne material as a function of time at one of the positions (4.5 m from the release point) on 18 August 1989. The concentrations of the particles are given in number per cubic meter of water, whereas the concentration of fluorescein is given in milligrams per cubic meter; all the concentrations are normalized by dividing them by their highest concentration on that day at that position in the channel. The lines represent linear regressions for each type of material (fluorescein: $r^2 = 0.88$, $\lambda = -0.0044$ s^{-1}, 95 percent confidence interval $= 0.0019$; poppy seeds (settling velocity $= 1.2$ cm·s^{-1}): $r^2 = 0.85$, $\lambda = -0.0059$ s^{-1}, 95 percent confidence interval $= 0.0028$; glitter (1.1 cm·s^{-1}): $r^2 = 0.98$, $\lambda = -0.0059$ s^{-1}, 95 percent confidence interval $= 0.0010$; snapdragon seeds (0.3 cm·s^{-1}): $r^2 = 0.92$, $\lambda = -0.0065$ s^{-1}, 95 percent confidence interval $= 0.0004$).

systematic statistical difference in the loss rates of the dye and of the different types of particles.

An example of how particle and fluorescein concentrations varied spatially with time is given in Figure 20.4. Such plots indicate that the dyed patch of water and the particles traveled together across the habitat. These observations, coupled with the similarity of the λ's for dye and particles, suggest that tracking dye blobs should give a reasonable measure of the transport of particles or larvae at turbulent, wave-swept habitats such as our surge channel. These results also suggest that the vertical motion of larvae does not play an important role in their transport on the scale of meters near wave-exposed rocky shores where turbulent mixing is great enough to overwhelm larval motion. This contrasts with the important role that vertical swimming or sinking can play in larval transport in less turbulent waters and with the important role that larval behavior can play at the smaller scale of settlement events, as discussed in the Introduction.

Advection and Turbulent Diffusion

We quantified the transport of water in the wave-swept surge channel by measuring the dispersal of patches of dye. Figure 20.1b illustrates an example of the spread and the horizontal translation of a blob of dye, and Figure 20.2 shows how σ^2 (spread) and U (advective velocity) varied with time. Mixing coefficients (K_A and K_B) and advective velocities for each of our experiments are listed in Table 20.2.

The advective velocities we measured were much lower than the instantaneous water velocities in surge channels on Tatoosh (which often exceed 5 m·s^{-1}; Koehl, 1977, 1984). We have observed this phenomenon at other wave-swept sites as well (Koehl et al., 1987, 1988, 1993). Such slow advection is not surprising if we consider that water oscillates back and forth as a wave shape moves across the habitat (e.g., Bascom 1980; Koehl, 1984, 1986; Denny, 1988), while there is some net pumping of water in the direction of wave motion (i.e., Stokes drift; see

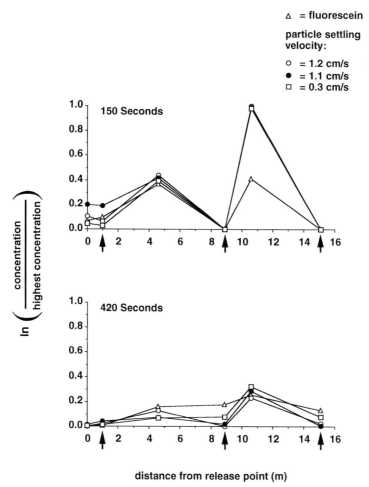

△ = **fluorescein**

particle settling velocity:

○ = **1.2 cm/s**
● = **1.1 cm/s**
□ = **0.3 cm/s**

FIGURE 20.4. Plot of the concentration of particles (number per cubic meter) and fluorescein (milligrams per cubic meter) at different positions in the surge channel at 150 s (upper graph) and 420 s (lower graph) after the mixture of fluorescein and particles was released on 18 August 1989. All the concentrations have been normalized by dividing them by their highest concentration on that day. The dye and particles were released from a position (0 m) on one side of the channel; positions indicated by arrows were along the opposite side of the channel from that release point. Note that material was initially (150 s) transported along one side of the channel, but eventually (420 s) mixed over to the opposite side. (Although the highest concentrations of dye and particles at the 150 s sampling time all co-occurred at the 10.5-m position (indicating that they were traveling together), the relative concentration of dye was lower there than were the relative concentrations of the particles. Since only one sample per position could be taken at each time, we do not know the uncertainty of these concentration measurements and hence cannot evaluate whether this difference was significant. Furthermore, we observed no consistent pattern in our 150 samples as to which type of material showed higher or lower concentrations at particular positions.)

LeMehaute, 1976). Therefore, although wave-swept habitats appear to be characterized by rapid water movement, this impression is quite misleading when considering the net transport of material in such environments.

Our dye patch data in this and in previous studies (Koehl et al., 1987, 1988, 1993) also revealed that mixing in shallow wave-swept habitats was not isotropic. As illustrated in Figure 20.2, the mixing coefficient (K_A) in the direction parallel to the axis of the oscillatory water flow in the channel was much greater than the mixing coefficient (K_B) at right angles to this. One mechanism responsible for this anisotropy is probably shear dispersion (as explained in, e.g., Bowden, 1965; Okubo, 1968, 1980; Koehl et al., 1993).

Table 20.2 Mixing coefficients and advective velocity for experimental conditions

Date (conditions)	Mean loss constant, λ (s⁻¹)	(range of λ's)[a] (s⁻¹)	Mixing coefficient (m² · s⁻¹) K_A	K_B	Advective velocity, U (m · s⁻¹)
8/17/89 (Calm; slack high tide)	−0.010	(−0.002 to −0.022)	0.05	0.02	0.11
8/18/89 (Moderate waves; flooding tide)	−0.007	(−0.003 to −0.013)	0.13	0.01	0.15
9/14/89 (Big waves; ebbing tide)	−0.010	(−0.004 to −0.020)	[b]	[b]	[b]
9/15/89 (Big waves; ebbing tide)	−0.007	(−0.002 to −0.013)	0.15	0.02	0.19
9/16/89 (Calm; flooding tide)	−0.009	(−0.004 to −0.020)	0.05	0.01	0.11

[a]Data for fluorescein and particles at all six sampling sites summarized.
[b]No data due to camera malfunction.

Summary

Flowing water plays an important role in the dispersal of the larvae of benthic animals. The effects of small-scale near-substratum flow and of larger-scale circulation patterns on larval dispersal have been well studied, but water motion on the spatial scale of meters to tens of meters, which provides the transport between these two flow regimes, is less well understood. We have developed a simple dye-tracking technique to quantify water transport at this spatial scale and have used it to characterize the flow at wave-swept rocky shores, where we found advection velocities to be an order of magnitude lower than instantaneous water velocities in waves. By comparing the transport of dye with that of particles of different sinking velocities (which mimic those of various larvae), we ascertained that tracking dye should give a reasonable measure of the transport of larvae at wave-exposed shores. Our data also suggest that the vertical motion of larvae does not play an important role in their transport on spatial scales of meters near wave-swept shores where turbulent mixing is great enough to overwhelm larval motion.

Acknowledgments

This research was supported by National Science Foundation (NSF) grant OCE-8717028 (to M. A. R. K.) and NSF grant OCE-8717678 (to T. M. P.). We are grateful to J. Jed, K. Lynn, K. Sonoda, and S. Zaret for technical assistance, to G. Dairiki for writing the software for dye blob analysis, and to E. Boulding, G. Dairiki, S. Distefano, C. McFadden, F. Mingo, R. Paine, C. Pfister, and T. Wootton for help in the field. We thank R. Alberte for the use of his spectrophotometer and R. Paine for facilitating our work on Tatoosh. Research at Tatoosh is possible only with permission of the Makah Tribal Council and the U.S. Coast Guard.

Literature Cited

Basco, D. R. 1983. Surf-zone currents. Coastal Eng. 7: 331–355.

Bascom, W. 1980. Waves and Beaches. Anchor Press/Doubleday, Garden City, N.J.

Bernstein, B. B., and N. Jung. 1979. Selective pressures and coevolution in a kelp canopy community in southern California. Ecol. Monogr. 49: 335–355.

Boicourt, W. C. 1988. Estuarine larval retention mechanisms on two spatial scales. *In* V. S. Kennedy, ed. Estuarine Comparisons. Academic Press, New York, pp. 445–458.

Bowden, K. F. 1965. Horizontal mixing in the sea due to a shearing current. J. Fluid Mech. 21: 83–95.

Burton, R. S. 1983. Protein polymorphisms and genetic differentiation of marine invertebrate populations. Mar. Biol. Letters 4: 193–206.

Burton, R. S., and M. W. Feldman. 1982. Population genetics of coastal and estuarine invertebrates: Does larval behavior influence population structure? *In* V. S. Kennedy, ed. Estuarine Comparisons. Academic Press, New York, pp. 537–551.

Butman, C. A. 1987. Larval settlement of soft-sediment invertebrates: The spatial scales of pattern explained by active habitat selection and the emerging role of hydrodynamic processes. Oceanogr. Mar. Biol. Ann. Rev. 25: 113–165.

Butman, C. A., J. P. Grassle, and C. M. Webb. 1988. Substrate choices by marine larvae settling in still water and in flume flow. Nature 333: 771–773.

Chatfield, C., and A. J. Collins. 1980. Introduction to Multivariate Analysis. Chapman and Hall, New York.

Connell, J. H. 1985. The consequences of variation in initial settlement vs. post-settlement mortality in rocky intertidal communities. J. Exp. Mar. Biol. Ecol. 93: 11–46.

Chia, F., J. Buckland-Nicks, and C. M. Young. 1984. Locomotion of marine invertebrate larvae: A review. Can. J. Zool. 62: 1205–1222.

Crisp, D. J. 1955. The behavior of barnacle cyprids in relation to water movement over a surface. J. Exp. Biol. 32: 569–590.

———. 1984. Overview of research on marine invertebrate larvae, 1940–1980. *In* D. J. Costlow and R. C. Tipper, eds. Marine Biodeterioration: An Interdisciplinary Study. Naval Institute Press, Washington, D.C., pp. 102–126.

Cronin, T. W., and R. B. Forward. 1982. Tidally timed behavior: Effects on larval distributions in estuaries. *In* V. S. Kennedy, ed. Estuarine Comparisons. Academic Press, New York, pp. 505–521.

Csanady, G. T. 1983. Advection, diffusion and particle settling. *In* E. P Meyers and E. T. Harding, eds. Ocean Disposal of Municipal Wastewater: Impacts on the Coastal Environment. Seagrant College Program, Massachusetts Institute of Technology, Cambridge, Mass., pp. 179–247.

Denny, M. W. 1988. Biology and the Mechanics of the Wave-Swept Environment. Princeton University Press, Princeton, N.J.

DeWolf, P. 1983. Ecological observations on the mechanisms of dispersal of barnacle larvae during planktonic life and settling. Neth. J. Sea Res. 6: 1–129.

Eckman, J. E. 1983. Hydrodynamic processes affecting benthic recruitment. Limnol. Oceanogr. 28: 241–257.

———. 1987. The role of hydrodynamics in recruitment, growth, and survival of *Argopecten irradians* (L.) and *Anomia simplex* (D'Orbigny) within eel grass meadows. J. Exp. Mar. Biol. Ecol. 106: 165–192.

Emlet, R. B. 1986. Larval production, dispersal, and growth in a fjord: A case study on larvae of the sand dollar *Dendraster excentricus*. Mar. Ecol. Prog. Ser. 31: 245–254.

Gaines, S., and J. Roughgarden. 1985. Larval settlement rate: A leading determinant of structure in an ecological community of the marine intertidal zone. Proc. Natl. Acad. Sci. U.S.A. 82: 3707–3711.

Graham, J. J. 1972. Retention of larval herring within the Sheepscot estuary of Maine. Fish. Bull. 70: 299–305.

Grosberg, R. K. 1982. Intertidal zonation of barnacles: The influence of planktonic zonation of larvae on the vertical distribution of adults. Ecology 63: 894–899.

Hannan, C. A. 1984. Planktonic larvae may act like passive particles in the turbulent near-bottom flows. Limnol. Oceanogr. 29: 1108–1116.

Hedgecock, D. 1979. Biochemical genetic variation and evidence of speciation in *Chthamalus* barnacles of the tropical eastern Pacific Ocean. Mar. Biol. 54: 207–214.

———. 1982. Genetic consequences of larval retention. *In* V. S. Kennedy, ed. Estuarine Comparisons. Academic Press, New York, pp. 553–568.

Jacobson, T. R., J. D. Milutinovic, and J. R. Miller. 1990. Observational and model studies of physical processes affecting benthic larval recruitment in Delaware Bay. J. Geophys. Res. 95: 333–345.

Jackson, J. B. C. 1974. Biogeographic consequences of eurytopy and stenotopy among marine bivalves and their evolutionary significance. Amer. Nat. 105: 542–560.

———. 1986. Modes of dispersal of clonal benthic invertebrates: Consequences for species' distributions and genetic structure of local populations. Bull. Mar. Sci. 39: 588–606.

Johnson, D. E., L. W. Botsford, R. D. Methot, and T. Wainwright. 1986. Wind stress and cycles in Dungeness crab (*Cancer magister*) catch off California, Oregon, and Washington. Can. J. Fish. Aquat. Sci. 43: 838–845.

Kingsford, M. J., and J. H. Choat. 1986. Influence of surface slicks on the distribution and onshore movements of small fish. Mar. Biol. 91: 161–171.

Koehl, M. A. R. 1977. Effects of sea anemones on the flow forces they encounter. J. Exp. Biol. 69: 87–105.

———. 1982. The interaction of moving water and sessile organisms. Sci. Am. 247: 124–132.

———. 1984. How do benthic organisms withstand moving water? Am. Zool. 24: 57–70.

———. 1986. Form and function of macroalgae in moving water. *In* T. J. Givnish, ed. On the Economy of Plant Form and Function. Cambridge University Press, Cambridge, pp. 291–314.

Koehl, M. A. R., T. M. Powell, and T. L. Daniel. 1987. Turbulent transport near rocky shores: Implications for larval dispersal. EOS 68: 1750.

———. 1988. Turbulent transport of marine larvae near rocky shores. Am. Zool. 28: 113A.

Koehl, M. A. R., T. M. Powell and G. Dairiki. 1993. Measuring the fate of patches in the water: Larval dispersal. *In* J. Steele, T. M. Powell, and S. A. Levin, eds. Patch Dynamics in Terrestrial, Marine, and Freshwater Ecosystems. Springer-Verlag, Berlin, pp. 50–60.

Laprose, R., and J. J. Dodson. 1989. Ontogeny and importance of tidal vertical migrations in the retention of larval smelt *Osmerus mordax* in a well-mixed estuary. Mar. Biol. Prog. Ser. 55: 101–111.

LeMehaute, B. 1976. An Introduction to Hydrodynamics and Water Waves. Springer-Verlag, New York.

Levin, L. A. 1983. Drift tube studies of bay-ocean water exchange and implications for larval dispersal. Estuaries 6: 363–371.

———. 1990. A review of methods for labeling and tracking marine invertebrate larvae. Ophelia 32: 115–144.

Mileikovsky, S. A. 1973. Speed of active movement of pelagic larvae of marine bottom invertebrates and their ability to regulate their vertical position. Mar. Biol. 23: 11–17.

Myers, E. P., and E. T. Harding. 1983. Ocean Disposal of Municipal Wastewater: Impacts on the Coastal Environment. Seagrant College Program, Massachusetts Institute of Technology, Cambridge, Mass.

Norcross, B. L., and R. F. Shaw. 1984. Oceanic and estuarine transport of fish eggs and larvae: A review. Trans. Am. Fish. Soc. 113: 153–165.

Okubo, A. 1968. Some remarks on the importance of the "shear effect" on horizontal diffusion. J. Oceanogr. Soc. Japan 24: 20–29.

———. 1971. Oceanic diffusion diagrams. Deep-Sea Res. 18: 789–802.

———. 1980. Diffusion and Ecological Problems: Mathematical Models. Springer-Verlag, Berlin.

———. In press. The role of diffusion and related physical processes in dispersal and recruitment of marine populations. *In* P. W. Sammarco and M. Heron, eds. Marine Larval Dispersal and Recruitment : An Interdisciplinary Approach. Springer-Verlag, Berlin.

Palumbi, S. 1992. Marine speciation on a small planet. Trends Ecol. Evol. 7: 114–118.

Pawlik, J. R. 1992. Chemical ecology of the settlement of benthic marine invertebrates. Oceanogr. Mar. Biol. Ann. Rev. 30: 273–335.

Pawlik, J. R., C. A. Butman, and V. R. Starczak. 1991. Hydrodynamic facilitation of gregarious settlement of a reef-building tube worm. Science 251: 422–424.

Pineda, J. 1991. Predictable upwelling and the shoreward transport of planktonic larvae by internal tidal bores. Science 253: 548–551.

Possingham, H., and J. Roughgarden. 1990. Spatial population dynamics of a marine organism with a complex life cycle. Ecology 7: 973–985.

Preisendorfer, R. W. 1988. Principal Component Analysis in Meteorology and Oceanography. Developments in Atmospheric Science, Vol. 17. Elsevier, Amsterdam.

Pritchard, D. W., and J. H. Carpenter. 1960. Measurement of turbulent diffusion in estuarine and inshore waters. Bull. Int. Assoc. Sci. Hydrol. 20: 37–50.

Riempa, H. W. 1985. Current meter records and the problem of the simulation of particle motions in the North Sea near the Dutch coast. Oceanol. Acta 8: 403–412.

Rothlisberg, P. C. 1982. Vertical migration and its effect on dispersal of penaeid shrimp larvae in the Gulf of Carpenteria, Australia. Fish. Bull. 80: 541–554.

Rothlisberg, P. C., J. A. Church, and A. M. G. Forbes. 1983. Modelling the advection of penaeid shrimp larvae in the Gulf of Carpenteria, Australia. J. Mar. Res. 41: 511–538.

Roughgarden, J., S. Gaines, and S. Pacala. 1987. Supply-side ecology: The role of physical transport processes. Proc. Br. Ecol. Soc. Symp. 27: 491–518.

Roughgarden, J., S. Gaines, and H. P. Possingham. 1988. Recruitment dynamics in complex life cycles. Science 241: 1460–1466.

Scheltema, R. S. 1971. The dispersal of the larvae of shoal-water benthic invertebrate species over long distances by oceanic currents. *In* D. J. Crisp, ed. Fourth European Marine Biology Symposium. Cambridge University Press, Cambridge, pp. 7–28.

———. 1975. Relationship of larval dispersal, gene-flow and natural selection to geographic variation of benthic invertebrates in estuaries and along coastal regions. Estuar. Res. 1: 372–391.

———. 1986. On dispersal and planktonic larvae of benthic invertebrates: An eclectic overview and summary of problems. Bull. Mar. Sci. 39: 290–322.

Scheltema, R. S., and J. T. Carlton. 1984. Methods of dispersal among fouling organisms and possible consequences for range extension and geographic variation. *In* J. D. Costlow and R. C. Tipper, eds. Marine Biodeterioration: An Interdisciplinary Study. Naval Institute Press, pp. 127–133.

Shanks, A. L. 1983. Surface slicks associated with tidally forced internal waves may transport pelagic larvae of benthic invertebrates and fishes shoreward. Mar. Ecol. Prog. Ser. 13: 311–315.

———. 1986. Tidal periodicity in the daily settlement of intertidal barnacle larvae and an hypothetical mechanism for the cross-shelf transport of cyprids. Biol Bull. 170: 429–440.

Smith, W. G., J. D. Sibunka, and A. Wells. 1978. Diel movements of larval yellowtail flounder, *Limanda ferruginea*, determined from discrete depth sampling. U.S. Natl. Mar. Fish. Service Fish. Bull. 76: 167–178.

Sulkin, S. D., and W. van Heukelem. 1982. Larval recruitment in the crab *Callinectes sapidus* Rathbun: An amendment to the concept of larval retention in estuaries. *In* V. S. Kennedy, ed. Estuarine Comparisons. Academic Press, New York, pp. 459–475.

Svendsen, I. A. 1987. Analysis of surf zone turbulence. J. Geophys. Res. 92: 5115–5124.

Tegner, M., and R. A. Butler. 1985. Drift-tube study of the dispersal potential of green abalone (*Haliotis fulgens*) larvae in the Southern California Bight: Implications for recovery of depleted populations. Mar. Ecol. Prog. Ser. 26: 73–84.

Wood, L., and W. J. Hargis. 1971. Transport of bivalve larvae in a tidal estuary. *In* D. J. Crisp, ed. Fourth European Marine Biology Symposium. Cambridge University Press, Cambridge, pp. 459–475.

Young, C. M., and F.-S. Chia. 1987. Abundance and distribution of pelagic larvae as influenced by predation, behavior, and hydrographic factors. *In* A. C. Giese, J. S. Pearse, and V. B. Pearse, eds. Reproduction of Marine Invertebrates, Vol. 9. General Aspects: Seeking Unity in Diversity. Blackwell Scientific Publications, Palo Alto, Calif. pp. 385–463.

Zeldis, J. R., and J. B. Jillet. 1982. Aggregation of pelagic *Munida gregaria* (Fabricus) (Decapoda, Anomura) by coastal fronts and internal waves. J. Plankton Res. 4: 839–857.

21 Larval Transport, Food Limitation, Ontogenetic Plasticity, and the Recruitment of Sabellariid Polychaetes

Joseph R. Pawlik and David J. Mense

ABSTRACT Sabellariid polychaetes have long-term planktotrophic larvae whose dispersal is likely to be greatly influenced by hydrodynamic transport processes. Adult distributions of gregarious species, and distributions of nongregarious species that form aggregations under some circumstances, suggest that oceanic and smaller-scale, near-shore circulation patterns strongly influence patterns of larval supply. During offshore transport, larvae may be advected into areas of low phytoplankton abundance. Previous laboratory studies on five species of sabellariids of the genera *Phragmatopoma* and *Sabellaria* indicate that the rate of larval development for these polychaetes is highly dependent on food concentration. In this study, competent (i.e., ready to metamorphose) larvae of *Phragmatopoma lapidosa californica* are demonstrated to undergo a reversible reversion to a precompetent condition in response to starvation. This response occurred rapidly: metamorphosis of larvae sampled from cultures denied food for 4 days dropped from 76 percent to 20 percent. Larvae starved for 6 days regained competence after two days of feeding; those starved for 20 days regained competence after 8 days of feeding. Changes in competence in response to food availability were coincident with changes in larval morphology. Starvation resulted in decreases in body length and energy content, and larval tentacles became shorter and lost cilia associated with putative sensory organs that are potentially involved in substratum selection. Larval morphology returned to normal when feeding was reinitiated, with regeneration paralleling regained competence. Ontogenetic plasticity, coupled with physical defenses against planktonic predation, may help sabellariid larvae survive long periods of larval transport in oligotrophic offshore waters. Plasticity may also prevent metamorphosis when food is scarce and juvenile survivorship would likely be low.

Introduction

Marine invertebrate zooplankton live in an environment that is nutritionally dilute (Conover, 1968) and where the food supply of phytoplankton is temporally and spatially patchy (Steele, 1978; Mackas et al., 1985). For holoplankton, such as copepods, food limitation is more likely to occur offshore than near-shore (Checkley, 1980), with some species displaying greater tolerances for lower food availability than others (Dagg, 1977). For planktotrophic invertebrate larvae, the importance of food supply on larval survival and subsequent recruitment has been the subject of some debate (reviewed in Day and McEdward, 1984; Olson and Olson, 1989). In general (i.e., for

Joseph R. Pawlik, Department of Biological Sciences and Center for Marine Science Research, University of North Carolina at Wilmington, Wilmington, NC 28403-3297.

David J. Mense, Department of Biological Sciences and Center for Marine Research, University of North Carolina at Wilmington, Wilmington, NC 28403-3297; and Department of Marine, Earth and Atmospheric Sciences, North Carolina State University, Raleigh, NC 27695-8208.

the few species that have been investigated), crustacean larvae appear to have little starvation tolerance, while molluscan and echinoderm larvae endure low phytoplankton abundances with a concomitant decrease in growth rate. Little is known of the responses of polychaete larvae to food limitation (see Table 1 of Olson and Olson, 1989).

Marine polychaetes of the family Sabellariidae inhabit tubes made of cemented grains of sand. Found in all seas, there are some fifty species belonging to at least six genera (reviewed in Pawlik and Faulkner, 1988). The group is remarkably cohesive in terms of adult and larval morphology. Moreover, all species that have been studied are dioecious, broadcast spawners with long-term planktotrophic larvae (Eckelbarger, 1978). Approximately twenty species (mostly in the genera *Phragmatopoma*, *Sabellaria*, and *Gunnerea*) settle gregariously to form colonies and reefs in lower intertidal and shallow subtidal environments around the world (see Fig. 1 of Pawlik and Faulkner, 1988); the remainder are mostly or entirely nongregarious, found on shells or rocks at depths ranging from the intertidal to the deep sea.

Sabellariid larvae have long been popular study organisms, particularly for investigations of larval behavior at the time of settlement (e.g., Wilson, 1968; Eckelbarger, 1978; Smith and Chia, 1985; Pawlik, 1986, 1988a,b; Amieva et al., 1987). Research on settlement specificity has explained much about the observed recruitment patterns of the adult worms. Laboratory experiments conducted in still-water have revealed that, while gregarious species preferentially settle on the tube sand of adult conspecifics, nongregarious species settle just as readily on control sand, and both gregarious and nongregarious species delay metamorphosis indefinitely in the absence of substrata (Pawlik, 1986, 1988a,b; Pawlik and Chia, 1991). A near-absolute substratum specificity has been confirmed for the gregarious species *Phragmatopoma lapidosa californica* in choice experiments conducted in laboratory flumes under hydrodynamic conditions similar to those of subtidal environments, with the additional discovery that larval behavior in different flow regimes can alter larval delivery to the substratum (Pawlik et al., 1991; Pawlik and Butman, 1993). Much of the foregoing work, however, has concentrated on the behavior of competent larvae (i.e., mature larvae that are poised to settle) that have come into proximity of the substratum; yet, the precompetent larval life span of sabellariids is estimated at weeks to months (Barry, 1989; see later discussion) and may constitute a significant portion of their lives in the plankton.

In this chapter, we will focus on presettlement factors that may play important roles in the dispersal and ultimately the recruitment of sabellariid larvae. We will summarize available evidence that transport of sabellariid larvae by currents and eddies greatly influences the distribution and gregarious settlement of several species. We will also discuss nutrition-dependent plasticity in the development of sabellariid larvae and the potential importance of this flexibility in extending the larval life span.

Materials and Methods

The ontogenetic plasticity of *Phragmatopoma lapidosa californica* was examined in five laboratory time-course experiments in which three batches of larvae of *P. l. californica* were alternately fed or starved. Subsamples of each batch were assessed for metamorphic competence, body length, tentacle length, and energy content. Results from only two experiments (time courses A and B) are presented because the other experiments were prematurely terminated when cultures became contaminated and large numbers of larvae died. Experiments were conducted at Friday Harbor Laboratories, University of Washington, using larvae from animals collected off Point Loma, San Diego, California (see Pawlik, 1986). Time course A was conducted September to November 1989; time course B ran May to July 1990. Larvae were cultured, measured, and assayed for metamorphic competence as detailed in Pawlik (1986, 1988a). Three batches of larvae, derived from a single spawn of multiple male and female worms, were followed for each time course. Each batch consisted of two 3-l jars each containing about 6000 larvae. Both jars of each batch were sampled for each experiment. As the time course progressed, and the number of larvae dwindled, the two jars were pooled into one. Prior to the onset of a starvation sequence, data were pooled for all batches (see Fig. 21.1 and Table 21.1).

Metamorphic competency was assayed by placing 20–30 larvae from each batch (subsampling from the jars) into each of three dishes containing sand from tubes of adult *P. l. californica* (made from "Ottawa sand") and scoring the number of metamorphosed juveniles and swimming larvae after 24 h (see Pawlik, 1986, 1988a). Body lengths and tentacle lengths were measured on ten narcotized larvae from each batch (subsampling from both jars). Energy content of larvae was periodically determined, using the dichromate oxidation microtechnique of McEdward and Carson (1987) with glucose as the standard. Energy content assays were performed on 100 larvae (time course A) or 50 larvae (time course B) per assay tube, with three to four replicates per batch (subsampling from both jars). Larvae were given several hours to void their digestive tracts prior to oxidation with dichromate, so as not to include the gut contents in energy content determinations. Larvae were washed in three changes of 0.33 M solution of $MgSO_4$ to remove chloride.

Scanning electron micrographs were taken of representative larvae of *Phragmatopoma lapidosa californica* sampled during time course A. Techniques employed in narcotization, fixation, embedding, and microscopy were those of Amieva and Reed (1987).

Results and Discussion

Larval Transport

The distribution of most benthic marine invertebrates can ultimately be traced to the settlement of their larvae. Prior to settlement, larval distributions are influenced by a host of physical and biological factors (Young and Chia, 1987; Pawlik, 1992), with physical transport processes likely dominating over much of the larval life span (Butman, 1987). Oceanic currents can drive long-range larval transport (review in Scheltema, 1986), while in estuarine habitats, larval distributions are influenced by exchange rates, circulation patterns (Boicort, 1982), tides, and stratification due to temperature and salinity differences (Stancyk and Feller, 1986; Young and Chia, 1987). Along continental margins, a variety of physical factors may affect transport, including wind drift (Efford, 1970; Denny, 1987), upwelling (Yoshioka, 1986), internal waves (Shanks, 1986), tidal bores (Pineda, 1991), and meanders and filaments (Bane et al., 1981). Recent models of marine invertebrate recruitment consider larval supply entirely the result of physical transport of passive larvae (e.g., Possingham and Roughgarden, 1990), an assumption that has drawn some criticism in light of the evidence for behavioral responses at the time of settlement (review in Pawlik, 1992; Pawlik and Butman, 1993).

Nevertheless, there is good evidence that sabellariid distributions are strongly influenced by passive transport processes. For *Phragmatopoma*, a genus that is most likely monospecific with subspecies distributed along both Atlantic and Pacific coasts of the New World (Fig. 1 of Pawlik and Faulkner, 1988; Pawlik, 1988b), large reef formations are found adjacent to major oceanic boundary currents: along the coasts of California, Ecuador and Peru, Brazil, and East Florida. *Phragmatopoma* is present, but not abundant, between these locations along each coast. Its distribution, however, appears to have little to do with temperature because it is common in the tropics (Caribbean), as well as in cold temperate areas (southern Chile). *Phragmatopoma lapidosa lapidosa* is not found south of Brazil; this southern range limit coincides with the Subtropical

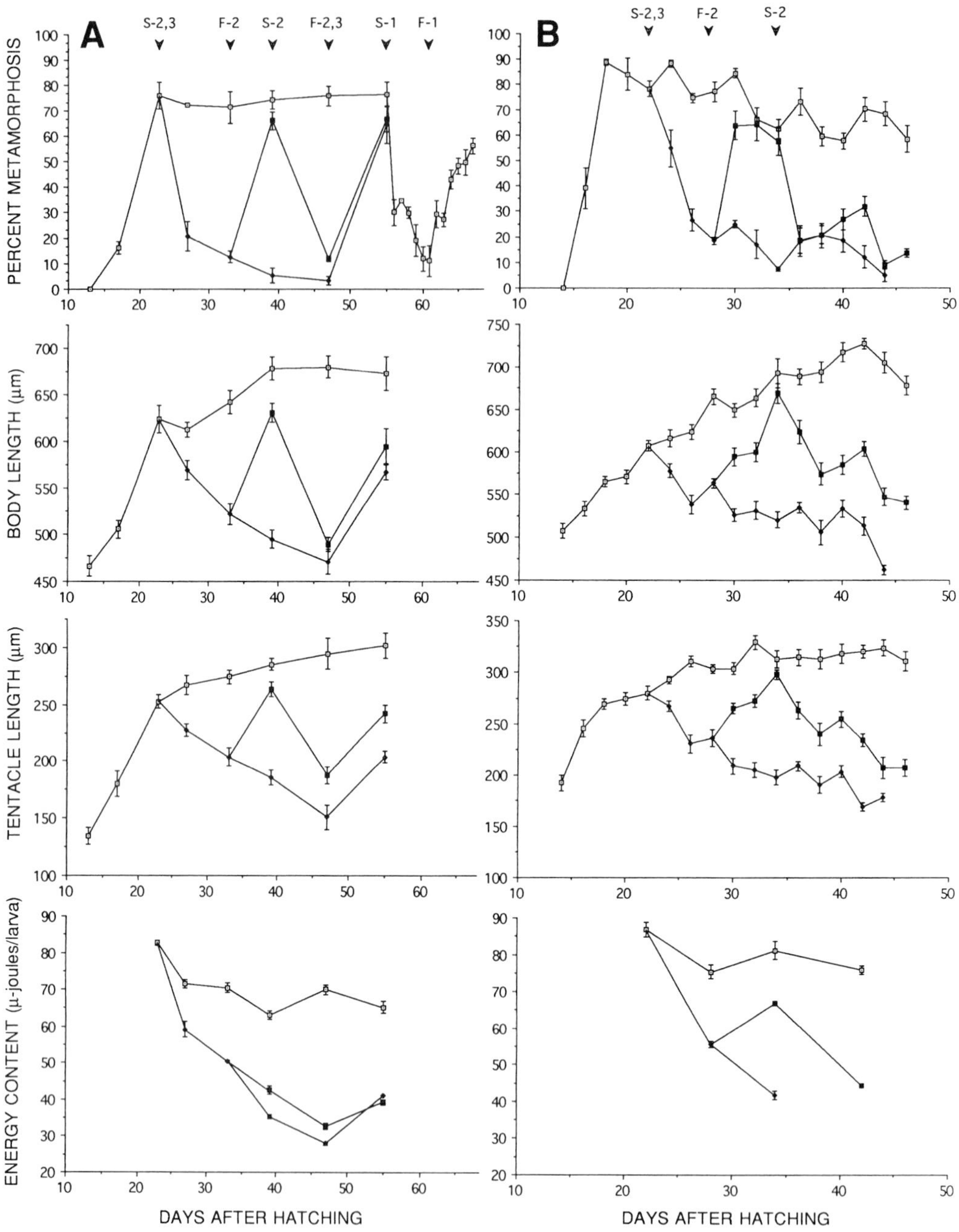

FIGURE 21.1. Ontogenetic plasticity of *Phragmatopoma lapidosa californica* in the laboratory. Two time-course experiments (A and B) in which three batches of larvae (□—□, ■—■, ◆—◆) of *P. l. californica* were alternately fed or starved. Subsamples of each batch were assayed for metamorphic competence, body length, tentacle length, and energy content. Standard errors of the mean are shown for each measurement. Changes in the feeding schedule are indicated above the figures (see Table 21.2), for example, *S-2,3* signifies that starvation began for batches 2 and 3 on that date, *F-2* means that feeding was resumed for batch 2, etc. Data recorded up to the beginning of the first starvation sequence (A: days 1–23; B: days 1–22) are pooled for all batches (all batches treated the same up to this time); from that day until the split between batch 2 and 3 (A: days 23–33; B: days 22–28), data are pooled for batches 2 and 3 (2 and 3 treated the same up to this point); from that day to the end of the time course, data are separate for each batch.

Table 21.1. Time required for development to metamorphic competence for sabellariid polychaetes cultured in the laboratory and for a field population

Species (distribution)	Culture temperature (°C)	Time to maturity	Source
Phragmatopoma lapidosa lapidosa	15	27 days	Pawlik, 1988b
(Florida to Brazil)	20	16 days	Pawlik, 1988b
	22–25 (?)	3–4 weeks	Mauro, 1975
	21–23	14–30 days	Eckelbarger, 1976
Phragmatopoma l. californica	15	23 days	Pawlik, 1988a
(California to Panama)	20	16 days	Pawlik, 1988a
	17–18	34–39 days	Eckelbarger, 1977
	21–23	18–25 days	Eckelbarger, 1977
	?[a]	2–5 months	Barry, 1989
Sabellaria alveolata	15	25 days	Pawlik, 1988a
(North Sea to Mediterranean)	20	14 days	Pawlik, 1988a
	15	6–32 weeks	Wilson, 1968
	?	12 weeks	Cazaux, 1964
Sabellaria floridensis	15	>40 days	Pawlik, 1988b
(Gulf of Mexico, Florida)	20	17 days	Pawlik, 1988b
	21–23	18–27 days	Eckelbarger, 1977
Sabellaria cementarium	15	25 days	Pawlik and Chia, 1991
(North Pacific)	20	27 days	Pawlik and Chia, 1991
	10–14	5–8 weeks	Smith and Chia, 1985

[a]Field population.

Convergence of the Brazil Current and the Falkland Current. The former, southward-flowing current may supply the Brazilian coastline with larvae from colonies further north, and its deflection to the east by the northward-flowing Falkland Current marks the range limit. The same can be said for *P. l. lapidosa* on the northern end of its range; the deflection of the Florida Current eastward to form the Gulf Stream appears to limit its distribution. Small colonies of *P. l. lapidosa* have occasionally been found along the coast of North Carolina (A. McCrary, pers. comm.), perhaps as a result of larval supply from meanders and filaments of Gulf Stream surface water that traveled onto the continental shelf (Bane et al., 1981; Atkinson et al., 1982).

The distribution of *Phragmatopoma* on the Pacific coast of North and South America is less readily explained. The northern limit, off central California, may result from a lack of larval supply from the southward-flowing California Current. Larvae are likely entrained in the currents of the California Bight, as indicated by the large populations of adults off southern California. Strangely, the southern distribution of *P. moerchi/P. virgini* extends to the tip of South America (Pawlik and Faulkner, 1988), despite the Peru Current flowing from the south. It remains unclear how larvae are transported to the southernmost regions of their range. Perhaps localized, nearshore countercurrents are important in maintaining this distribution.

Smaller-scale transport processes may also affect sabellariid distributions. For example, *Sabellaria alveolata*, a gregarious species found in the Northeast Atlantic and Mediterranean, forms large reefs in bays and estuaries (Horne, 1982). Hydrodynamics appear to play a particularly important role in the formation of the large, intertidal reef at the mouth of Mont Saint-Michel Bay in France (Gruet, 1986).

Entrainment may also explain the formation of reefs by otherwise nongregarious sabellariids. *Sabellaria cementarium*, for example, is found from Alaska to northern California. Larvae of this species show no settlement preferences for conspecific tube sand in laboratory experiments (Pawlik and Chia, 1991) and, not surprisingly, are found in single or paired tubes over most of their range. Yet, a large reef of aggregated *S. cementarium* has been reported in an embayment near Coos Bay, Oregon (Posey et al., 1984), and its formation is likely the result of larval entrainment and concentration at the time of settlement. Similarly, *Sabellaria vulgaris*, found along the east coast of North America, forms aggregations only in one area of Delaware Bay between Cape Henlopen and the Mispillion River Jetty (Wells, 1970; Curtis, 1978), again suggesting that larvae are entrained in this shallow, semienclosed region (Galperin and Mellor, 1990).

Food Availability and Ontogenetic Plasticity

Development rates of invertebrate larvae can be influenced by temperature (e.g., Pechenik et al., 1990), but for most planktotrophic species, nutrition appears to play a more important role (Paulay et al., 1985; Olson and Olson, 1989). Of the invertebrate species studied to date, larvae of crustaceans are generally unable to survive after extended periods of low food availability (Lang and Marcy, 1982; Anger, 1987), while molluscan and echinoderm larvae simply slow their rates of growth (Paulay et al., 1985; Boidron-Metairon, 1988; Pechenik et al., 1990).

Larvae of sabellariid polychaetes appear to display the same food-dependent developmental plasticity found in some echinoderms (Paulay et al., 1985; Boidron-Metairon, 1988). Comparisons of the minimum developmental periods reported for five sabellariid species cultured in the laboratory revealed highly variable growth rates for different investigators (Table 21.1). Culture conditions for each of these studies were largely similar, with the exception that different species and concentrations of phytoplankton were used. For *Sabellaria alveolata*, these nutritional differences resulted in development times ranging from 25 days (Pawlik, 1988a) to 6–32 weeks (Wilson, 1968) at 15°C. The latter figure, in fact, is closer to the only estimate available for the development of sabellariid larvae in the field: for *Phragmatopoma lapidosa californica*, Barry (1989) reported a 2–5-month lag between winter storms (which induce adult spawning as worm reefs are damaged) and recruitment of juvenile worms into the intertidal of southern California. It is not evident whether this period was required for larval maturation, or whether larvae matured much earlier, but remained competent in the plankton until they could be advected back to shore. In the laboratory, however, Pawlik (1988a) was able to culture larvae of *P. l. californica* to maturity in 16 days at 20°C and in 23 days at 15°C (Table 21.1).

Plasticity in larval development may go beyond the length of time required to attain metamorphic competence, however. One of us (J. R. P.) has determined that larvae of *Phragmatopoma lapidosa californica* will lose and regain their ability to metamorphose dependent on the availability of food in laboratory cultures (Table 21.2, Fig. 21.1). The response time was remarkably rapid: only 20 percent of larvae sampled from a culture that was denied phytoplankton for 4 days metamorphosed on conspecific tube sand, whereas 76 percent from the same population metamorphosed prior to starvation (Fig. 21.1a, day 23–27). When provided with food again, the ability to metamorphose rapidly returned in the starved population, as levels of metamorphosis similar to those of unstarved larvae reappeared in as few as 2 days (Fig. 21.1b, day 28–30). For one population of larvae, competence was restored three times after two bouts of starvation (Fig. 21.1a).

Table 21.2. Feeding schedule for each batch of larvae (1, 2, and 3) for both time courses A and B

(See Figure 21.1 and Materials and Methods for details.)

	A			B	
Day	Feed	Starve	Day	Feed	Starve
1	1,2,3		1	1,2,3	
23	1	2,3	22	1	2,3
33	1,2	3	28	1,2	3
39	1	2,3	34	1	2,3
47	1,2,3		46	(larvae dying,	
55		1(2,3 exhausted)		experiment ended)	
61	1				

The loss and restoration of metamorphic competence of larvae of *P. l. californica* as a function of food availability were coincident with dramatic changes in larval morphology (Figs. 21.1 and 21.2). The overall length of the larval body decreased over the starvation period, as did the energy content of the larvae. In addition, the larval tentacles decreased in length by about 20 percent during a 10-day period over which time metamorphosis dropped from 76 percent to 12 percent (Fig. 21.1a, day 23–33). As with metamorphic competence, larval morphology was also restored when phytoplankton were again available, and both body and tentacle lengths returned to the sizes of those of unstarved larvae (Fig. 21.1). Strangely, larval energy content did not rebound when starved larvae were fed in time course A, but did rebound in time course B (Fig. 21.1).

Starvation also resulted in ultrastructural changes to the surfaces of the larval body and tentacles of *P. l. californica*, as revealed in the scanning electron micrographs (SEMs) shown in Figure 21.2. The dorsal hump and larval tentacles bear putative chemosensory ciliated structures (sensory tufts) that appear to mediate substratum selection and metamorphosis of this species (Eckelbarger, 1978; Amieva and Reed, 1987; Amieva et al., 1987). The numbers of cilia composing the sensory tufts were greatly reduced after 10 days of starvation (compare Figs. 21.2c and 21.2e with 21.2d and 21.2f), and the distal tips of the larval tentacles appeared to lose their ciliation almost completely. These ciliated structures were also restored when larvae were again provided with food, and SEMs of these larvae were indistinguishable from those of larvae that had never been starved (data not shown).

Larvae of *P. l. californica* were able to survive long periods of starvation, with subsequent loss of metamorphic competence, and then regain the ability to metamorphose when phytoplankton were again available (Fig. 21.1a). After 20 days of starvation, the number of larvae that metamorphosed fell to 3 percent. Larvae were then fed for 8 days, and metamorphosis rebounded to 65 percent. In addition, some larvae were fed and maintained in a competent state for 32 days before being starved, and these responded in much the same way as larvae that were starved immediately after attaining competence (Fig. 21.1a, days 55–67).

Is ontogenetic plasticity of sabellariid larvae important in nature? Given the difficulties in monitoring, sampling, and identifying populations of larvae in the field, the question is unlikely to be answered definitively (see review in Levin, 1990). But the fact that the single field estimate of a sabellariid larval life span (*P. l. californica*, 2–5 months; [Barry, 1989]) is considerably longer than the developmental rate of the same species in the laboratory (23 days at 15°C [Pawlik,

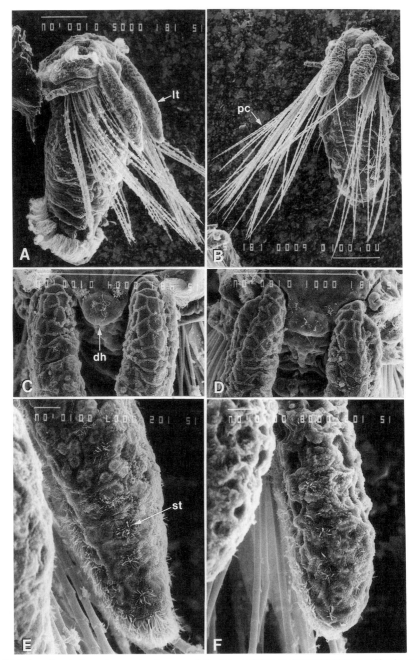

FIGURE 21.2. Scanning electron micrographs of representative larvae of *Phragmatopoma lapidosa californica* sampled from time course A, day 33. The larvae shown in A, C, and E were fed through day 33; those shown in B, D, and F were starved from day 23 through day 33. (scale bar = 100 μm for A–D, 10 μm for E and F.) A and B: full view of larva; note overall change in larval size from A to B (length of provisional chaetae remains constant). C and D: dorsal view, close-up of dorsal hump and proximal end of larval tentacles; note loss of ciliation from C to D. E and F: side view, distal end of larval tentacle; note loss of ciliation, particularly of sensory tufts (dh = dorsal hump, lt = larval tentacle, pc = provisional chaetae, st = sensory tuft). Compare with figures in Eckelbarger (1978); Amieva and Reed (1987); Amieva et al. (1987).

1988a]) suggests that sabellariid growth can be food limited and larvae are probably not developing at an optimum rate in nature. A similar conclusion was reached by Paulay et al. (1985) after rearing larvae of the serpulid polychaete, *Serpula vermicularis*, in natural sea water and sea water augmented with cultured algae. Suboptimal larval growth may permit normal development when food is limiting (as opposed to death, as for many crustacean larvae), but a decrease in growth rate does not ensure survival to recruitment, because the probability of mortality by predation increases with the length of time spent in the plankton (reviewed in Rumrill, 1990).

How might sabellariid larvae survive extended development? Larvae of sabellariid polychaetes are unusual among invertebrate larvae in having been shown to deter planktonic predators with a physical defense mechanism (Wilson, 1929; Pennington and Chia, 1984). When disturbed, a sabellariid larva (greater than 2 to 5 days old) contracts its body longitudinally, thereby erecting two bundles of barbed, provisional chaetae around itself (see Fig. 21.2a and 21.2b, and Fig. 1 in Pennington and Chia [1984]). Larvae of *Sabellaria cementarium* bearing provisional chaetae were eaten significantly less often than nonmotile eggs or larvae that had not yet developed chaetae in laboratory assays with four planktonic predators: a ctenophore, hydromedusa, brachyuran megalopa, and a juvenile fish (Pennington and Chia, 1984). The physical defense afforded by provisional chaetae probably enhances survivorship of sabellariid larvae in the plankton.

Although field evidence for food-dependent ontogenetic plasticity of sabellariid larvae has yet to be obtained, the advantages of the strategy seem apparent. Larvae advected to offshore, oligotrophic regions would slow their development, or reverse it, until transport processes brought them back into more productive near-shore waters where suitable adult habitat might be found. Competent larvae may revert to precompetence in a near-shore area experiencing a periodic oligotrophic condition, such as an upwelling of phytoplankton-poor water (Attwood and Peterson, 1989), thereby forestalling metamorphosis until food is available for early juvenile development. In areas of high phytoplankton abundance, larvae are likely to attain competence quickly and delay metamorphosis until they encounter a suitable substratum. For *Phragmatopoma lapidosa californica*, if not for all gregarious sabellariids, larval survival to recruitment is probably greatly enhanced by the combination of developmental responses to food supply and physical defense and by larval responses to hydrodynamic conditions and the presence of adult conspecifics (Pawlik and Butman, 1993).

Acknowledgments

Support was provided to J. R. P. by a Killam Memorial Postdoctoral Fellowship through the University of Alberta at Edmonton (generously sponsored by Dr. F. S. Chia); by Friday Harbor Laboratories, University of Washington; by a National Science Foundation Presidential Young Investigator Award (OCE-9158065); and by Office of Naval Research grant N00014-92-J-1144). The joint Ph.D. graduate program of North Carolina State University and the University of North Carolina at Wilmington provided financial support to D. J. M. We thank R. Toonen for assistance with larval cultures and for helpful comments on the manuscript.

Literature Cited

Amieva, M. R., and C. G. Reed. 1987. Functional morphology of the larval tentacles of *Phragmatopoma californica* (Polychaeta: Sabellariidae): Composite larval and adult organs of multifunctional significance. Mar. Biol. 95: 243–258.

Amieva, M. R., C. G. Reed, and J. R. Pawlik. 1987. Ultrastructure and behavior of the larva of *Phrag-*

matopoma californica (Polychaeta: Sabellariidae): Identification of sensory organs potentially involved in substrate selection. Mar. Biol. 95: 259–266.

Anger, K. 1987. The D_0 threshold: A critical point in the larval development of decapod crustaceans. J. Exp. Mar. Biol. Ecol. 108: 15–30.

Atkinson, L. P., L. J. Pietrafesa, and E. E. Hofmann. 1982. An evaluation of nutrient sources to Onslow Bay, North Carolina. J. Mar. Res. 40: 679–699.

Attwood, C. G., and W. T. Peterson. 1989. Reduction in fecundity and lipids of the copepod *Calanus australis* (Brodskii) by strongly pulsed upwelling. J. Exp. Mar. Biol. Ecol. 129: 121–131.

Bane, J. M., D. A. Brooks, and K. R. Lorenson. 1981. Synoptic observations of the three-dimensional structure, propagation and evolution of Gulf Stream meanders along the Carolina continental margin. J. Geophys. Res. 86: 6411–6425.

Barry, J. P. 1989. Reproductive response of a marine annelid to winter storms: An analog to fire adaptation in plants? Mar. Ecol. Prog. Ser. 54: 99–107.

Boicourt, W. C. 1982. Estuarine larval retention and mechanisms on two scales. *In* V.S. Kennedy, ed. Estuarine comparisons. Academic Press, New York, pp. 235–246.

Boidron-Metairon, I. F. 1988. Morphological plasticity in laboratory-reared echinoplutei of *Dendraster excentricus* (Eschscholtz) and *Lytechinus variegatus* (Lamarck) in response to food conditions. J. Exp. Mar. Biol. Ecol. 119: 31–41.

Butman, C. A. 1987. Larval settlement of soft-sediment invertebrates: The spatial scales of pattern explained by active habitat selection and the emerging role of hydrodynamical processes. Oceanogr. Mar. Biol. Ann. Rev. 25: 113–165.

Cazaux, C. 1964. Développement larvaire de *Sabellaria alveolata* (Linné). Bull. Inst. Océanogr. Monaco 62(1296): 1–15.

Checkley, D. M. 1980. Food limitation of egg production by a marine, planktonic copepod in the sea off southern California. Limnol. Oceanogr. 25: 991–998.

Conover, R. J. 1968. Zooplankton—life in a nutritionally dilute environment. Am. Zool. 8: 107–118.

Curtis, L. A. 1978. Aspects of the population dynamics of the polychaete *Sabellaria vulgaris* Verrill, in the Delaware Bay. Estuaries 1: 73–84.

Dagg, M. 1977. Some effects of patchy food environments on copepods. Limnol. Oceanogr. 22: 99–107.

Day, R.. and L. McEdward. 1984. Aspects of the physiology and ecology of pelagic larvae of marine benthic invertebrates. *In* K. A. Steidinger and L. M. Walker, eds. Marine Plankton Life Cycle Strategies. CRC Press, Boca Raton, Fla., pp. 93–120.

Denny, M. W. 1987. Life in the maelstrom: The biomechanics of wave-swept rocky shores. Trends Ecol. Evol. 2: 61–66.

Eckelbarger, K. J. 1976. Larval development and population aspects of the reef-building polychaete *Phragmatopoma lapidosa* from the east coast of Florida. Bull. Mar. Sci. 26: 117–132.

———. 1977. Larval development of *Sabellaria floridensis* from Florida and *Phragmatopoma californica* from southern California (Polychaeta: Sabellariidae), with a key to the sabellariid larvae of Florida and a review of development in the family. Bull. Mar. Sci. 27: 241–255.

———. 1978. Metamorphosis and settlement in the Sabellariidae. *In* F. S. Chia and M. E. Rice, eds. Settlement and Metamorphosis of Marine Invertebrate Larvae. Elsevier, New York, pp. 145–164.

Efford, I. E. 1970. Recruitment of sedentary marine populations as exemplified by the sand crab, *Emerita analoga* (Decapoda: Hippidae). Crustaceana 18: 293–308.

Galperin, B., and G. L. Mellor. 1990. A time-dependent, three-dimensional model of the Delaware Bay and river system. Pt. 2. Three-dimensional flow fields and residual circulation. Est. Coast. Shelf Sci. 31: 255–281.

Gruet, Y. 1986. Spatio-temporal changes of sabellarian reefs built by the sedentary polychaete *Sabellaria alveolata* (Linné). P.S.Z.N.I: Marine Ecology 7: 303–319.

Horne, D. J. 1982. The ostracod fauna of an intertidal *Sabellaria* reef at Blue Anchor, Somerset, England. Est. Coast. Shelf Sci. 15: 671–678.

Lang, W. H., and M. Marcy. 1982. Some effects of early starvation on the survival and development of barnacle nauplii, *Balanus improvisus* (Darwin). J. Exp. Mar. Biol. Ecol. 60: 63–70.

Levin, L. A. 1990. A review of methods for labelling and tracking marine invertebrate larvae. Ophelia 32: 115–144.

Mackas, D. L., K. L. Denman, and M. R. Abbott. 1985. Plankton patchiness: Biology in the physical vernacular. Bull. Mar. Sci. 37: 652–674.

Mauro, N. A. 1975. The premetamorphic developmental rate of *Phragmatopoma lapidosa* Kindberg, 1867, compared with that in temperate sabellariids (Polychaeta: Sabellariidae). Bull. Mar. Sci. 25: 387–392.

McEdward, L. R., and S. F. Carson. 1987. Variation in egg organic content and its relationship with egg size in the starfish *Solaster stimpsoni*. Mar. Ecol. Prog. Ser. 37: 159–169.

Olson, R. R., and M. H. Olson. 1989. Food limitation of planktotrophic marine invertebrate larvae: Does it control recruitment success? Ann. Rev. Ecol. Syst. 20: 225–247.

Paulay, G., L. Boring, and R. R. Strathmann. 1985. Food limited growth and development of larvae: Experiments with natural sea water. J. Exp. Mar. Biol. Ecol. 93: 1–10.

Pawlik, J. R. 1986. Chemical induction of larval settlement and metamorphosis in the reef-building tube worm *Phragmatopoma californica* (Polychaeta: Sabellariidae). Mar. Biol. 91: 59–68.

———. 1988a. Larval settlement and metamorphosis of two gregarious sabellariid polychaetes: *Sabellaria alveolata* compared with *Phragmatopoma californica*. J. Mar. Biol. Assoc. 68: 101–124.

———. 1988b. Larval settlement and metamorphosis of sabellariid polychaetes, with special reference to *Phragmatopoma lapidosa*, a reef-building species, and *Sabellaria floridensis*, a non-gregarious species. Bull. Mar. Sci. 43: 41–60.

———. 1992. Chemical ecology of the settlement of benthic marine invertebrates. Oceanogr. Mar. Biol. Ann. Rev. 30: 273–335.

Pawlik, J. R., and C. A. Butman. 1993. Settlement of a marine tube worm as a function of current velocity: Interacting effects of hydrodynamics and behavior. Limnol. Oceanogr. 38: 1730–1740.

Pawlik, J. R., C. A. Butman, and V. R. Starczak. 1991. Hydrodynamic facilitation of gregarious settlement of a reef-building tube worm. Science 251: 421–424.

Pawlik, J. R., and F. S. Chia. 1991. Larval settlement of *Sabellaria cementarium* Moore, and comparisons with other species of sabellariid polychaetes. Can. J. Zool. 69: 765–770.

Pawlik, J. R., and D. J. Faulkner. 1988. The gregarious settlement of sabellariid polychaetes: New perspectives on chemical cues. *In* M. F. Thompson, R. Sarojini, and R. Nagabhushanam, eds. Marine Biodeterioration. Oxford and IBH, New Delhi, pp. 475–487.

Pechenik, J. A., L. S. Eyster, J. Widdows, and B. L. Bayne. 1990. The influence of food concentration and temperature on growth and morphological differentiation of blue mussel *Mytilus edulis* L. larvae. J. Exp. Mar. Biol. Ecol. 135: 47–64.

Pennington, J. T., and F. S. Chia. 1984. Morphological and behavioral defenses of trochophore larvae of *Sabellaria cementarium* (Polychaeta) against four planktonic predators. Biol. Bull. 167: 168–175.

Pineda, J. 1991. Predictable upwelling and the shoreward transport of planktonic larvae by internal tidal bores. Science 253: 548–551.

Posey, M. H., A. M. Pregnall, and R. A. Graham. 1984. A brief description of a subtidal sabellariid (Polychaeta) reef on the southern Oregon coast. Pac. Sci. 38: 28–33.

Possingham, H. P., and J. Roughgarden. 1990. Spatial population dynamics of a marine organism with a complex life cycle. Ecology 71: 973–985.

Rumrill, S. S. 1990. Natural mortality of marine invertebrate larvae. Ophelia 32: 163–198.

Scheltema, R. S. 1986. On dispersal and planktonic larvae of benthic invertebrates: An eclectic overview and summary of problems. Bull. Mar. Sci. 39: 290–322.

Shanks, A. L. 1986. Vertical migration and cross-shelf dispersal of larval *Cancer* spp. and *Randallia ornata* (Crustacea: Brachyura) off the coast of southern California. Mar. Biol. 92: 189–199.

Smith, P. R., and F. S. Chia. 1985. Larval development and metamorphosis of *Sabellaria cementarium* Moore, 1906 (Polychaeta: Sabellariidae). Can. J. Zool. 63: 1037–1049.

Stancyk, S. E., and R. J. Feller. 1986. Transport of non-decapod invertebrate larvae in estuaries: An overview. Bull. Mar. Sci. 39: 257–268.

Steele, J. H. 1978. Spatial Pattern in Plankton Communities. Plenum Press, New York.

Wells, H. W. 1970. *Sabellaria* reef masses in Delaware Bay. Chesapeake Sci. 11: 258–260.

Wilson, D. P. 1929. The larvae of British sabellarians. J. Mar. Biol. Assoc. 16: 221–269.

———. 1968. The settlement behaviour of the larvae of *Sabellaria alveolata* (L.). J. Mar. Biol. Assoc. 48: 387–435.

Yoshioka, P. M. 1986. Chaos and recruitment in the bryozoan, *Membranipora membranacea*. Bull. Mar. Sci. 39: 408–417.

Young, C. M., and F. S. Chia. 1987. Abundance and distribution of pelagic larvae as influenced by predation, behavior, and hydrographic factors. *In* A. C. Giese, J. S. Pearse, and V. B. Pearse, eds. Reproduction of Marine Invertebrates, Vol. 9. Blackwell, Palo Alto, Calif., pp. 385–463.

Daniel R. Brumbaugh, Jordan M. West, Jennifer L. Hintz,

and Frank E. Anderson

ABSTRACT The epiphytic bryozoan *Membranipora membranacea* recruits preferentially to the proximal portion of kelp blades. This portion is both the youngest region of the plant and the leading surface of the blade in water flow. To discriminate between the effects of hydrodynamics and host quality as influences on recruitment, racks containing "water vanes" were used to maintain "normal" and "reversed" blade orientations of *Laminaria groenlandica* with respect to the direction of water flow. Regardless of blade orientation, bryozoans preferentially recruited to the youngest portion of the algal blade; this region has both less damage and less mucus than more distal regions. To test whether tissue age per se or amount of damage and mucus underlies this observed recruitment pattern, we damaged proximal, medial, and distal regions of algal blades, leaving half of each region undamaged as a control. In all regions, recruitment was significantly inhibited by blade damage; therefore, host tissue quality influences recruitment patterns in this species. We also modeled the growth and determined the ultimate sizes of colonies recruiting to different regions of the blade in order to estimate potential fitness differences among these recruits. Because the algal blade grows at the proximal end and erodes at the distal end, colonies recruiting on the proximal region can reach an ultimate size more than two orders of magnitude larger than those recruiting on the medial region and three orders larger than those on the distal region.

Introduction

While many species of sessile marine invertebrates are found on a wide range of surfaces, others are substrate specialists that are restricted to particular types of habitat such as rock, hermit crab shells, or specific types of algae. Such observed habitat specialization has three primary ecological components: physical transport of larvae, larval settlement choice, and the differential survival of recruits on various surfaces. Studies of epiphytic species have identified differential mortality from predation (Bernstein and Jung, 1979; Yoshioka, 1982), competition (Stebbing, 1973), and physical disturbance (Fletcher and Day, 1983; Keough, 1986) as important factors shaping distributions across and within host species. Alternatively, some patterns of epiphytism are shaped primarily by larval substrate specificity (Ryland, 1962; Nishihira, 1967; Hayward and Harvey, 1974a); for instance, Ryland (1962) found that, for two species of bryozoans, distribution patterns

Daniel R. Brumbaugh, Department of Zoology NJ-15, University of Washington, Seattle, WA 98195.
Jordan M. West, Section of Ecology and Systematics, Cornell University, Ithaca, NY 14853.
Jennifer L. Hintz, Department of Biology, University of South Florida, Tampa, FL 33620.
Frank E. Anderson, Institute of Marine Sciences, University of California, Santa Cruz, CA 95064.

in the field could be explained by the larval preferences observed in laboratory choice experiments. Finally, physical factors such as current direction or speed can also affect settlement; Crisp (1955) found that the settlement of barnacle cyprids was sensitive to velocity gradients. This chapter focuses on the latter two factors, larval choice and physical transport, as they relate to the recruitment dynamics of epiphytes on marine algae.

Epiphytic invertebrate species are good subjects for the study of habitat choice because they are more substrate selective than other epibenthic species; hence, researchers have fewer possible evolutionary pressures to consider as explanations for observed patterns of habitat use and specificity. As a result, ecological relationships among epiphytic invertebrates and between the epiphytes and their algal hosts have been well documented by studies ranging from descriptions of species distributions and assemblage dynamics (reviews by Seed and O'Connor, 1981; Seed, 1986), to the demonstration of chemical attraction (Crisp and Williams, 1960; Williams, 1964; Nishihira, 1968) and the characterization of attractants (Kakinuma et al., 1977).

There has been less work, however, on larval choice within an individual host. Just as settlement choice among algae is not a random process, settlement decisions within a given alga are also nonrandom, producing various patterns of epiphytic zonation. Several studies have found that larvae of epiphytic species preferentially settle on the younger portion of kelp blades (Ryland, 1959; Durante and Chia, 1991; Ryland and Stebbing, 1971). For example, working with the kelps *Laminaria digitata* and *L. saccharina,* which have intercalary meristems proximal to their stipes, Ryland and Stebbing (1971) found new colonies of the bryozoan *Scrupocellaria reptans* disproportionately on these young, proximal regions. However, not all epiphytic bryozoans recruit to the youngest algal regions; Ryland (1974) found that most epibionts of *Sargassum natans* avoided recruiting to the youngest, distal tips of this alga, while Hayward and Harvey (1974b) found a tendency of *Alcyonidium hirsutum* and *A. polyoum* to recruit in greatest numbers to the medial portions of *Fucus serratus.*

These studies suggest that settlement of epiphytic larvae may often be guided by age-related algal cues. Larvae may be affected by particular qualities of the algal substratum associated with age such as microtopography, age-specific tissue biochemistry, or the quality of accumulated mucus, detritus, and microorganisms. Researchers have demonstrated that many species of algae vary in phenolic content and antimicrobial activity across their thalli, often in patterns that correlate to distributions of larval recruits (Sieburth and Conover, 1965; Hornsey and Hide, 1976; Al-Ogily and Knight-Jones, 1977). However, because most of these are laboratory experiments and field descriptions (but see Bernstein and Jung, 1979), they generally fail to eliminate the alternative hypothesis that other settlement cues may be important under field conditions. For example, flow character will change as water passes unidirectionally over a surface (Nowell and Jumars, 1984), creating potential hydrodynamic cues or passively depositing invertebrate larvae in ways that correlate with the normal tissue age gradients. Differences in flow conditions over an algal surface might then be expected to produce differences in settlement tendencies along the blade.

Here, we report in situ experimental manipulations designed to separate the effects of water flow and substrate quality on the recruitment of the bryozoan *Membranipora membranacea* to the kelp *Laminaria groenlandica. M. membranacea* is known to preferentially settle on the young, proximal portions of *L. groenlandica* as well as *Nereocystis luetkeana* (Honkalehto, 1983; Thomas, 1988). We manipulated the normal correlation between blade age gradients and hydrodynamics by rigidly affixing algal blades to racks that maintained them at "normal" and "reversed" orientations to current flow. This allowed us, without measuring the flow conditions, to assess the relative importance of flow direction and substrate quality to recruitment dynamics. The next year, an additional treatment was added to determine the effect of blade damage on recruitment. Since amount of tissue damage should increase with age because of the accumulation of wounds from herbivores and physical disturbances, we hypothesized that damage might be a substrate cue used by cyphonautes larvae to assess variations in habitat quality within an algal host. Portions of kelp blades were experimentally damaged to test whether larvae would preferentially recruit to healthy regions over damaged ones.

   M. membranacea grows rapidly, reproduces, and senesces within the normal annual cycle of kelp blade regeneration and loss. Given such an ephemeral substrate, which grows proximally and erodes distally until it is completely lost in the fall, *M. membranacea* should face strong selective pressures for settlement on the younger and longer-living portions of the kelp blade. Using parameters that were empirically derived from the literature, we estimated the benefits of settling on different portions of the algal blade and the implied costs of larval errors.

Materials and Methods
Flow Effect

To assess the effect of flow on bryozoan recruitment, we used an array of *Laminaria groenlandica* blades mounted, on *in situ* racks, in "normal" and "reversed" orientations relative to ambient current direction. In early May 1990, twenty *L. groenlandica* blades varying in length from 50 cm to 90 cm were collected from separate second-year class individuals (Druehl et al., 1987) along the floating docks and breakwater at Friday Harbor Laboratories (FHL), Friday Harbor, Washington (48°32.2′ N, 123°0.8′ W); the blades were maintained in running sea water tables at approximately 12°C. All preexisting *Membranipora membranacea* colonies were gently popped off one side of each blade without any visible damage to the underlying tissue. Tattered algal tissue was trimmed from the distal ends of all blades to prevent shredding during the experiment and to smooth the distal ends to more closely approximate the proximal ends. Each blade was divided into three similarly sized regions, marked by four punched holes near the blade margins (Fig. 22.1a), and traced onto clear plastic for a record of initial area. The regions were defined as "proximal," "medial," and "distal."

   To support the blades in known orientations to the current in the field, we constructed racks from ½″ PVC piping (Fig. 22.2). Each rack consisted of a 60-cm-high, 34-cm-wide current "vane" that kept the submerged rack oriented with the changing current flow. Bolted into each rack were four acrylic support rods (1 × 0.5 × 125 cm), two of which pointed in the direction of the vane and two of which pointed in the opposing upstream direction. The two pairs of rods were separated by a depth of 24 cm. Five such racks were constructed for a total of ten rods at each of two depths. The twenty *L. groenlandica* blades were haphazardly assigned to the support rods ("clean" side up) with either cable ties or monofilament tied through small holes punched in the kelp. In all cases, the proximal end of the blade pointed toward the central axis of the rack. As a result, the blades attached to the rods on the vane side were always oriented with the proximal end upstream ("normal" orientation) while the blades attached to the opposite rods were always oriented with the proximal end downstream ("reverse" orientation) (Fig. 22.2).

   The racks were suspended from the protected (shoreward) side of the floating breakwater at FHL at 4.5-m intervals along the dock. The top blades were approximately 61 cm below the surface, and the bottom blades were approximately 85 cm below the surface. Three racks were deployed on 5 May 1990, and two were deployed on 6 May 1990. During the course of the experiment, periodically sampled current velocities ranged from 0.2 to 12.1 cm/s.

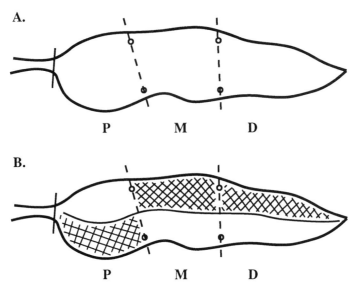

FIGURE 22.1. *a.* Diagram of proximal (P), medial (M), and distal (D) blade regions, demarcated by small marginal holes, used in 1990 experiment; *b.* Similar design for 1991 experiment, with cross-hatching to represent the damage treatment inflicted on half of each blade region.

On 11 May 1990, the racks were recovered, and the number of *M. membranacea* ancestrulae (newly metamorphosed larvae) in each region of the blade was tallied. Final areas of the blade regions were traced onto clear plastic and measured using a computer digitizing program (Sigma Scan, Jandel Scientific) that allowed us to calculate blade growth rates in terms of area added per day (cm^2/day).

Recruitment of ancestrulae was expressed in terms of density rates (recruits/cm^2/day) in order to account for natural variation in blade sizes and slight differences in experimental submersion times. In calculating the densities for the proximal and medial regions (the regions that grew over the course of the experiments), we assumed constant growth rates of the regions and constant settlement rates from the plankton; hence, recruitment densities were estimated using the average (or mid-experimental) size of the regions. For the distal regions, which tended to lose a small amount of surface area over the experimental period, we assumed that recruits were lost in direct proportion to lost kelp tissue, so final sizes were used in calculating recruitment densities.

For these and subsequent analyses, all data were tested for normality (SAS Institute, Inc., 1988) and homoscedasticity (Bartlett's test for homogeneity of variances, Sokal and Rohlf, 1981). A two-way analysis of variation (ANOVA) (SAS Institute, Inc., 1988) blocked by depth, was conducted for each region separately to test for any effect of blade orientation on recruitment. Since data were collected from all three regions of each blade, there may have been blade effects producing correlations in the recruitment rates within blades. As a consequence of this lack of independence, the region variable could not be included as a factor in a single ANOVA. Consequently, differences in recruitment rates between regions were interpreted using a bootstrap procedure that calculated the recruitment means and standard deviations for 1000 simulated data sets constructed from values drawn randomly with replacement from the original data (both orientation treatments were lumped together in this procedure). This method corrected for lack of independence between regions, and estimated the potential amount of variance likely to be seen in repeated experiments under similar conditions.

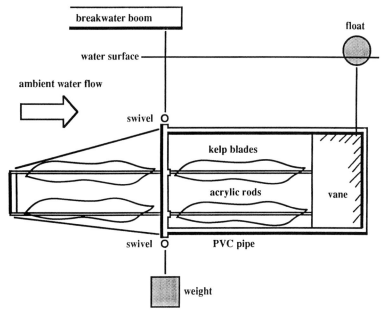

FIGURE 22.2. Diagram of the rack design employed for the 1990 field experiment in which flow direction over blade regions was manipulated. P = proximal end of algal thallus. The same design was employed in 1991, except that we alternated "normal" and "reverse" orientations of the algal blades among the four positions of the rack.

Damage Effect

Using the experimental design described above, we added a "damage" treatment to assess the effect of kelp tissue damage on bryozoan recruitment rates. In April 1991, eighteen second-year class *Laminaria groenlandica* blades were again collected from the floating breakwater at FHL. Preparation and installation of these blades on the racks were similar to that of the previous year, except that "normal" and "reverse" blades regularly alternated positions between rack sides (vane versus nonvane sides) and between depths. Half of each blade region was randomly selected to be scratched with a fine dissecting probe while the other half of the region was left undamaged as a control. The scratches, made as continuous straight lines about 1 cm apart in two orthogonal directions, penetrated through the meristoderm into the cortex of the tissue on the "clean" side of the blade (Fig. 22.1b). The five racks were installed on the afternoon of 22 April 1991 and removed between 29 April and 1 May 1991.

Recruitment rates were calculated as described above and analyzed using a split-level, randomized blocks ANOVA for each region, as well as a multivariate analysis of variation (MANOVA) to analyze the effect of "damage" across all regions. A bootstrap procedure was again used to characterize differences between regions and damage treatments.

Colony Longevity and Size Estimates

To estimate fitness differences between recruits landing in different blade regions, we calculated the longevity of a single point within each region and the maximum potential size achieved by a bryozoan colony growing for that period of time. Since *Membranipora membranacea* is a protandrous hermaphrodite with all zooids capable of reproduction, size and fecundity in *M. membranacea* are highly correlated (Yoshioka, 1982). Therefore, our estimated differences in sizes reflect potential differences in fecundity. Reproduction is possible at small colony sizes (~15 zooids), though it is generally initiated later in the growing season through contact and crowding by conspecifics or damage by predators (Harvell and Grosberg, 1988).

From the initial twenty *Laminaria groenlandica* blade tracings in 1990, we calculated the mean initial blade length and mean midpoint position for each region. We then simulated the movement of these initial midpoints, representing the "average" settling location within each region as the kelp blade tissue grew from the proximal intercalary meristem and eroded from the distal end. Since our experimental data on kelp growth were not extensive enough to estimate the long-term fates of individual points on the blades, we used published growth and erosion rates (defined, respectively, as elongation occurring within 10 cm of the stipe and as tissue loss at the distal end) for *L. groenlandica* from the west coast of Vancouver Island, British Columbia (Druehl et al., 1987). After estimating these rates for consecutive 10-day intervals, we calculated the movement of the mean settling site in each region from 6 May to 15 September. When each settling site reached the distal boundary of the blade, we recorded its longevity, expressed as the number of days since 6 May. From these longevities, we calculated colony sizes using a growth regression from Harvell (1985). The expression (area added [mm^2]) = 1.68 (colony size [mm^2]) + 196.9, estimated from a two-week period of uncrowded growth of Friday Harbor colonies, was transformed into the continuous exponential expression $A_t = A_0 e^{rt}$, where A_0 is the initial area, A_t is the area at time t, and r is the growth rate equal to 0.0755. This estimate for r falls within the range of estimates for *M. membranacea* found by Seed (1976), Thorpe (1979), and Yoshioka (1982). With the formula, we could estimate colony size for any growth period given an initial colony size of 0.5 mm^2.

Results

Blade Growth

In the 1990 experiment, kelp growth was greatest in the proximal region due to growth of the intercalary meristem although the medial region also grew due to continued cell expansion (Fig. 22.3a). The distal region tended to lose tissue to erosion and fragmentation despite our trimming of the most fragile, distal tissue prior to the experimental blade installation. The same pattern was seen in the 1991 experiment (Fig. 22.3b). In addition, the damage treatment had no effect on blade growth in any region (Fig. 22.3b).

Flow and Damage Effects

There was no effect of blade orientation on 1990 recruitment rates (Fig. 22.4a). Recruitment tended to decrease, regardless of orientation and depth, from younger to older regions (Fig. 22.4a; bootstrapping results for lumped orientations: Fig. 22.4b). Recruitment rates were similar in the proximal and medial regions, but decreased significantly (by approximately 50 percent) in the distal region. The bootstrapped means and standard errors (1000 replicates) were essentially identical to the observed values (lumped orientations not shown), differing by less than 1 percent and 3 percent, respectively.

Since the damage treatment had no measurable effect on algal growth, 1991 recruitment rates in each region were calculated using the same growth adjustments regardless of damage treatment. Again, there was no effect of blade orientation (Fig. 22.5a, Table 22.1). As in the previous year, recruitment decreased from younger to older regions, and in this case, all three regions differed significantly from one another (Fig. 22.5a). These recruitment differences among regions were consistent across damage treatments as well (Fig. 22.5b): the pattern of decreasing recruitment across older blade regions is similar in both control and damaged portions of the kelp blades.

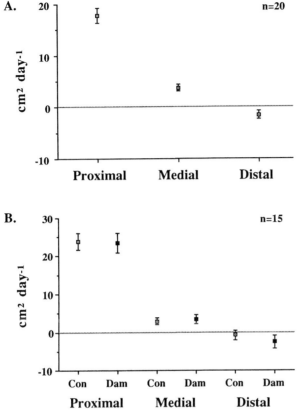

FIGURE 22.3. Mean growth rates of *L. groenlandica* blade regions (±SE): *a.* 1990 experiment; *b.* 1991 experiment with control (Con) and damage (Dam) treatments.

Bootstrapped means and standard errors from 1000 replicates (not shown) were again similar to the observed values, differing by less than 1 percent and 3 percent, respectively, with one exception: the bootstrapped standard error of the distal damaged treatment was 37 percent smaller than the observed value.

The damage treatment significantly decreased the amount of recruitment relative to the control in all regions (Fig. 22.5b, Table 22.1). There were no significant interactions between "damage" and "orientation." The MANOVA results confirmed that the "damage" factor was significant across all regions, while the interaction between "damage" and "orientation" was not. There were not enough degrees of freedom to test the orientation treatment with a multivariate analysis.

Colony Longevity and Size Estimates

Although crude, our estimates of colony longevity and size indicate that the potential differences in size between colonies recruiting to different regions are striking (Table 22.2). Given the assumptions of this model, a cyphonautes settling in the middle of the proximal region, and growing without restriction until reproducing just before death, should achieve a fecundity approximately 170 times that of a larva settling near the middle of the medial region, and 2500 times that of one recruiting to the middle of the distal region.

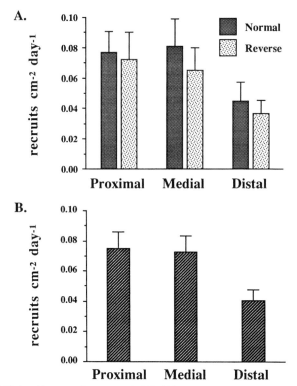

FIGURE 22.4. *a*. Mean recruitment rates (±SE) of *M. membranacea*, measured during the 1990 experiment, to different blade regions and orientations of *L. groenlandica*, N = 10; *b*. Mean bootstrapped recruitment rates (±SE of 1000 replicates) to regions with lumped blade orientations.

Discussion

Flow and Damage Effects

After adjusting our measured recruitment rates for differential growth among the algal blade regions (assuming constant rates of recruitment and blade region growth), we found significant differences in recruitment rates of *Membranipora membranacea* among kelp blade regions in two separate years. During the first year, recruitment differed only between the distal and the two younger regions, with the older region receiving fewer recruits than the proximal and medial regions (Figs. 22.4a and 22.4b). In contrast, during the second year, recruitment rates varied among all regions, with the proximal region receiving the most recruits and the distal the fewest (Figs. 22.5a and 22.5b).

In both years, these patterns held regardless of the manipulated blade orientation. This suggests that ambient hydrodynamic cues or other effects of flow play little role in cyphonautes habitat selection within individual kelp blades. Given the many uncertainties about the specific field conditions under which cyphonautes larvae settle, however, such an inference remains provisional. For example, little is known about natural flow patterns around kelp blades (Koehl and Alberte, 1988) or about larval settling behavior under different flow conditions (Butman et al., 1988). In our experimental design, the attachment of the blades restricted both potential drooping under low flow conditions and flapping under high flow. Such a natural range of kelp movement might offer cyphonautes hydrodynamic cues that were missing in our restrained experimental blades. Unfortunately, such experimental artifacts are common in manipulations of

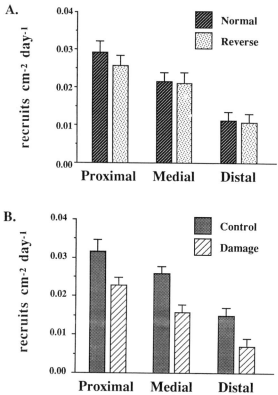

FIGURE 22.5. *a.* Mean recruitment rates (±SE) of *M. membranacea*, measured in the 1991 experiment, to different blade regions and orientations of *L. groenlandica*, normal: *N* = 16, reverse: *N* = 14; *b.* Mean recruitment rates (±SE) to different blade regions and damage treatments, *N* = 15. Mean bootstrapped recruitment rates (±SE of 1000 replicates) to regions and treatments were essentially identical.

Table 22.1. Results of the split-plot, randomized blocks ANOVA for the proximal, medial, and distal regions in the 1991 experiment

Effect	Proximal	Medial	Distal
Orientation	NS[a]	NS	NS
Damage	[b]	[c]	[b]
Orientation times damage	NS	NS	NS

[a]NS − $p > .05$.
[b]$p \leq .05$.
[c]$p \leq .01$.

natural and artificial substrates in fluid environments (Nowell and Jumars, 1984). These limitations notwithstanding, our results indicate that hydrodynamic cues are not necessary for cyphonautes larvae to detect and respond to differences in habitat quality within kelp blades.

In contrast, some intrinsic aspect of the kelp, correlated with tissue age, seems to be critical for the cyphonautes to distinguish habitat quality along the blade length. Possible factors include age-specific tissue biochemistry or microtopography, the character of accumulated mucus, and the successional status of microbial films. It is also possible that different rates of nutrient and gas

Table 22.2. Mean regional midpoints for kelp blades used in 1990, with the estimated logevities of these points and the colony sizes attained during these periods as calculated by the growth model $A_t = A_0\,e^{rt}$, where $A_0 = 0.5$ mm^2 and $r = 0.0755$

	Midpoint (cm)	Longevity (days)	Size (mm^2)	Size (zooids)
Proximal	12.5	128	7,870	37,148
Medial	35.0	60	46	219
Distal	58.0	24	3	15

delivery, due to differences in flow, could indirectly affect regional recruitment by influencing the composition of surface films (Pedersen, 1982, Nowell and Jumars, 1984); different delivery rates would likely favor the production of different species assemblages (Cundell et al., 1977). Our short-term manipulations of flow may not have lasted long enough for different assemblages to develop.

Another factor that often seems to vary with age along the blade is amount of biotic and abiotic damage. We tested the hypothesis that bryozoans would recruit more readily to undamaged kelp tissue than to damaged portions. After experimentally damaging kelp, we found the predicted reduction in recruitment relative to controls in all three regions (Fig. 22.5b, Table 22.1). Though we cannot exclude the possibility that these results were due, at least in part, to early differential survival of settlers on damaged and control portions, we suggest that the pattern is more likely due to the avoidance by cyphonautes of damaged tissue. All stages of bryozoan early colony development, from metamorphosing ancestrulae to juvenile colonies, were seen in both treatments; this suggests that, within the first week after settlement, there were no major effects of kelp damage on colony growth or survivorship. Laboratory studies have also reported observations of choosy, exploratory behavior among cyphonautes of several bryozoan species (Ryland, 1976; Honkalehto, 1983).

The components of algal damage that might inhibit settlement remain unknown. When damaged, many algae exude copious amounts of mucus, and it is possible that larvae avoid heavily covered surfaces. Ryland (1959) suggests that larvae of *Alcyonidium hirsutum* avoided settling on the reproductive tips of *Fucus serratus* because of mucus secretions from the conceptacles. He also cites Bock's (1954, cited in Ryland, 1959) suggestion that the unusually heavy recruitment of epiphytic invertebrates observed on brackish Baltic *Laminaria* blades is due, in part, to a reduced mucus content of the thalli. During 1990, we observed that distal regions of blades tended to collect more sediment than proximal regions, regardless of flow direction (unpubl. data; see also Seed, 1976). If the behavior of passively deposited particles is independent of the surrounding hydrodynamics, then another gradient must underlie the sedimentation pattern. Greater exudation or accumulation of mucus on the distal portions of the *Laminaria groenlandica* blades because of damage or normal aging might result in increased distal retention of sediment and detritus. Cyphonautes may avoid heavily sedimented regions, avoid mucous regions retaining sediment, or settle evenly but suffer a higher mortality on regions with heavy sediment or mucus. In the 1991 damage experiment, however, there were no obvious accumulations of sediment associated with the mucus-producing damaged portions; hence, sedimentation is not necessary for effects of decreased recruitment.

Any effect of mucus in influencing settlement and metamorphosis might also be related to dissolved secondary compounds within the mucus. Secondary metabolites, in the form of phloroglucinol-based phenolics, are ubiquitous in kelp (Steinberg, 1985; Van Alstyne and Paul, 1991). Changes in the secretion of phenolics with damage could influence larval settlement directly, or indirectly through effects on microbial assemblages (Hornsey and Hide, 1976; Al-Ogily and Knight-Jones, 1977; Cundell et al., 1977). P. Yoshioka (pers. comm.) found that mucus from the surface of the kelp *Macrocystis pyrifera*, when smeared onto glass beakers, actually induced greater settlement and metamorphosis of *M. membranacea* cyphonautes. If mucus is indeed implicated in recruitment inhibition, the actual effect may be due to phenolic compounds released into the mucus in response to damage. Although we did not directly observe the time course of mucus release after artificial wounding, we suggest that heavy mucus secretion after cutting lasts approximately 1 day, the length of the algal "curing" (healing) period in studies using pieces of algae in laboratory settlement experiments (Ryland, 1959; Nishihira, 1968; Durante and Chia, 1991). If the damage effect observed in this experiment were due solely to settlement interference from excess mucus, one should expect to see an effect proportional in magnitude to the fraction of time the mucus was present. In our case, this proportion would be 1 day mucus secretion per the ~8-days experimental period, or 12.5 percent; however, the proportional change in recruitment rates between treatments in all regions was greater than 12.5 percent, implying that other mechanisms besides interference by mucus were involved.

Colony Longevity and Size Estimates

The ratios in potential fecundity between bryozoans recruiting to neighboring regions are very great, ranging from approximately 170:1 between mean proximal and medial recruits to 15:1 between mean medial and mean distal recruits (Table 22.2). Although these values were calculated with numerous and simple assumptions, they nevertheless indicate the potential strength of selection for larval habitat choice, especially during low recruitment years when colonies can experience unrestricted growth. Given the assumption of constant exponential growth with no reductions due to competition, predation, or aging, it is remarkable how well the predicted colony sizes fit with those observed in the field. For example, the estimated area of the average proximal recruit, 7870 mm^2, corresponds to a circular colony with a 10-cm diameter. Colonies of this approximate size and location are common in low-density populations in the field, lending support to the assertion that we have not grossly over-extrapolated growth rates.

Our calculated longevities also correspond relatively well with the average life span of *M. membranacea* in the field around Friday Harbor, Washington. For example, we calculated a longevity of 128 days for the mean proximal colony that recruited 6 May (Table 22.2). Thus, this hypothetical colony would die with distal blade erosion around 10 September, a date that conforms with natural bryozoan senescence in September (Harvell and Grosberg, 1988). Had the kelp elongation and erosion rates used to model the changing blade length and colony positions also included the blade expansion observed in our medial regions (Figs. 22.3a and 22.3b), the maximum colony longevities would have increased. Since the proximal colonies might then experience senescence before distal blade senescence, the differences among maximum regional sizes would be somewhat less. However, as the recruitment season progresses, new distal recruits will experience much greater competition from preexisting colonies than recruits to more proximal regions. The possible effects of this early curtailing of growth among distal colonies (Harvell et al., 1990) have not been included in our model, but we would expect them to amplify the estimated fitness differences.

If colony longevity, duration, size, and fecundity are directly related to the original position of recruitment on the kelp blade, then rapid evolution for highly specific habitat choice among these epiphytic bryozoans is likely to be favored. The less intensive intraspecific competition for space on newly created tissue (Ryland and Stebbing, 1971; Stebbing, 1972; Buss, 1979), as well as the decreasing probability of mortality by predators and physical disturbance with increasing size (Hughes, 1984; Jackson, 1985) should also favor such evolution. Consequently, it seems problematic that only 40 percent of the bryozoans recruit to the proximal region despite the general larval ability to judge habitat quality (Figs. 22.4a,b and 22.55a,b).

However, it is possible that some factors might select for bryozoans that do not recruit so close to the proximal end of the blade. For example, if the algal substrate grows faster underneath recruits than they can reattach themselves to the algae, young colonies in the proximal growth zone may be at high risk of being popped off the blade surface (Thomas, 1988). Consequently, the optimal habitat choice would be one that balances this risk with the foregone growth opportunities of more distal recruitment. Though researchers have discussed this possibility for some time, we know of no reliable data regarding this phenomenon.

Other constraints on the evolution of greater specificity might be the absence of adequate cues (Strathmann and Branscomb, 1979) or the need to adapt to a number of different host species (or a wide geographic region) with conflicting cues (Strathmann et al., 1981). One possible indication of the difficulty in simultaneously optimizing specificity for different host species is the tendency for different populations of a species to specialize on specific hosts even when other hosts are available (Ryland, 1962; Knight-Jones et al., 1975; Doyle, 1975; MacKay and Doyle, 1978). Another possibility is that larval settlement cues are often found in coarse-grained patches which may not accurately represent the overall quality of sites. For example, if larvae use physical damage per se as a cue, they may recruit to healthy patches of tissue surrounded by many necrotic patches near the distal ends of kelp blades or avoid mostly healthy proximal regions because of encountering an injured patch of algal tissue. Both laboratory flume and field studies that manipulate densities and grain sizes of "good" and "bad" patches may be able to test this scenario (Butman et al., 1988). The ability of larvae to detect available cues may also be limited; Hayward and Harvey (1974b) also observed that the rugophilic *Alcyonidium* larvae, rather than recruiting to their favored grooves, were often fatally attracted to necrotic pits that were soon shed from the algal surface.

Finally, there are probably times when it is advantageous to be less selective. For example, as planktonic larvae approach the end of their larval competency period and the probability of encountering any appropriate habitat decreases, selection should favor individuals that are able to decrease their level of choosiness accordingly. Some theory has predicted and many laboratory studies have shown such a lessening of settlement cue specificity with age among larvae (Knight-Jones, 1953; Wisely, 1960; Doyle, 1975). The scenario of a planktonic population of cyphonautes of mixed ages with mixed specificities could possibly explain the distribution of recruits away from the presumed proximal optimum.

Acknowledgments

We wish to thank R. Strathmann, S. Rumrill, and L. Tear for their help in the initial phases of this project, and M. Hart, C. D. Harvell, R. Paine, C. Pfister, J. Ruesink, D. Shapiro, and R. Strathmann for their comments on the manuscript. L. Provencher and G. Gilchrist also provided valuable assistance with the statistical analyses. A. O. D. Willows provided use of the facilities and support at Friday Harbor Laboratories. During the course of this work, the authors were supported by a National Science Foundation (NSF) Graduate Fellowship to D. R. B., NSF grant OCE-8817498 to C. D. Harvell (J. M. W.), the FHL Ray Fund (J. M. W. and J. L. H.), and the FHL Marine Science Fund (D. R. B.).

Literature Cited

Al-Ogily, S. M., and E. W. Knight-Jones. 1977. Anti-fouling role of antibiotics produced by marine algae and bryozoans. Nature 265: 728–29.

Bernstein, B. D., and N. Jung. 1979. Selective pressures and coevolution in a kelp canopy community in Southern California. Ecol. Monogr. 49: 335–55.

Bock, K. S. 1954. Einige Zahlen sur Bewuchsdichte von Epizoen auf Laminarien aus der ostlichen Kieler Bucht. Veroff. Inst. Meeresf. Bremerhaven. 3: 42–5.

Buss, L. W. 1979. Habitat selection, directional growth and spatial refuges: why colonial animals have more hiding places. In G. Larwood and B. R. Rosen, eds. Biology and Systematics of Colonial Organisms. Academic Press, London, pp. 459–97.

Butman, C. A., Grassle, J. P., and C. M. Webb. 1988. Substrate choices made by marine larvae settling in still water and in a flume flow. Nature 333: 771–73.

Crisp, D. J. 1955. The behaviour of barnacle cyprids in relation to water movement over a surface. J. Exp. Biol. 32: 569–90.

Crisp, D. J., and G. B. Williams. 1960. Effect of extracts from fucoids in promoting settlement of epiphytic Polyzoa. Nature 188: 1206–07.

Cundell, A. M., Sleeter, T. D., and R. Mitchell. 1977. Microbial populations associated with the surface of the brown alga Ascophyllum nodosum. Microb. Ecol. 4: 81–91.

Doyle, R. W. 1975. Settlement of planktonic larvae: A theory of habitat selection in varying environments. Am. Nat. 109: 113–26.

Druehl, L. D., Cabot, E. L., and K. E. Lloyd. 1987. Seasonal growth of Laminaria groenlandica as a function of plant age. Can. J. Bot. 65: 1599–1604.

Durante, K. M., and F. S. Chia. 1991. Epiphytism on Agarum fimbriatum: Can herbivore preferences explain distributions of epiphytic bryozoans? Mar. Ecol. Prog. Ser. 77: 279–87.

Fletcher, W. J., and R. W. Day. 1983. The distribution of epifauna on Ecklonia radiata (C. Agardh) and the effect of disturbance. J. Exp. Mar. Biol. Ecol. 71: 205–20.

Harvell, C. D. 1985. Partial predation, inducible defense and the population biology of a marine bryozoan. Ph.D. dissertation, University of Washington, Seattle.

Harvell, C. D., Caswell, H., and P. Simpson. 1990. Density effects in a colonial monoculture: experimental studies with a marine bryozoan (Membranipora membranacea L.). Oecologia 82: 227–37.

Harvell, C. D., and R. K. Grosberg. 1988. The timing of sexual maturity in colonial animals: An empirical study. Ecology 69: 1855–64.

Hayward, P. J., and P. H. Harvey. 1974a. Growth and mortality of the bryozoan Alcyonidium hirsutum (Fleming) on Fucus serratus L. J. Mar. Biol. Assoc. 54: 677–84.

Hayward, P. J., and P. H. Harvey. 1974b. The distribution of settled larvae of the bryozoans Alcyonidium hirsutum (Fleming) and Alcyonidium polyoum (Hassall) on Fucus serratus L. J. Mar. Biol. Assoc. 54: 665–76.

Honkalehto, T. 1983. Settlement and orientation of Membranipora (Bryozoa) larvae on Laminaria. Course project for Advanced Invertebrate Zoology, Spring 1983, Friday Harbor Laboratories, Friday Harbor, Washington.

Hornsey, I. S., and D. Hide. 1976. The production of antimicrobial compounds by British marine algae. III. Distribution of antimicrobial activity within the algal thallus. Br. Phycol. J. 11: 175–81.

Hughes, T. P. 1984. Population dynamics based on individual size rather than age: A general model with a coral reef example. Am. Nat. 123: 778–95.

Jackson, J. B. C. 1985. Distribution and ecology of clonal and aclonal benthic invertebrates. *In* J. B. C. Jackson, L. W. Buss, and R. E. Cook, eds. Population Biology and Evolution of Clonal Organisms. Yale University Press, New Haven, Conn., pp. 297–355.

Kakinuma, Y., Nishihira, M., Kato, T., and A. S. Kumanireng. 1977. Bio-assay of the substances obtained from *Sargassum tortile* on the settlement of the planula of *Coryne uchidai*. Bull. Biol. Stn. Asamushi 16: 11–20.

Keough, M. J. 1986. The distribution of a bryozoan on seagrass blades: Settlement, growth, and mortality. Ecology 67: 846–57.

Knight-Jones, E. W. 1953. Decreased discrimination during settling after prolonged planktonic life in larvae of *Spirorbis borealis* (Serpulidae). J. Mar. Biol. Assoc. 32: 33745.

Knight-Jones, E. W., Knight-Jones, P., and S. M. Al-Ogily. 1975. Ecological isolation in the Spirorbidae. *In* L. H. Barnes, ed. Proc. Ninth Europ. Symp. Mar. Biol. Aberdeen University Press, Aberdeen, Scotland, pp. 539–61.

Koehl, M. A. R., and R. S. Alberte. 1988. Flow, flapping, and photosynthesis of *Nereocystis luetkeana*: A functional comparison of undulate and flat blade morphologies. Mar. Biol. 99: 435–44.

MacKay, T. F. C., and R. W. Doyle, 1978. An ecological genetic analysis of the settling behavior of a marine polychaete. Heredity 40: 1–12.

Nishihira, M. 1967. Observations on the selection of algal substrata by hydrozoan larvae, *Sertularella miurensis* in nature. Bull. Biol. Stn. Asamushi 13: 35–48.

———. 1968. Brief experiments on the effects of algal extracts in promoting the settlement of the larvae of *Coryne uchidai* Stechow (Hydrozoa). Bull. Biol. Stn. Asamushi 13: 91–101.

Nowell, A. R. M., and P. A. Jumars. 1984. Flow environments of aquatic benthos. Ann. Rev. Ecol. Syst. 15: 303–28.

Pedersen, K. 1982. Factors regulating microbial biofilm development in a system with slowly flowing sea water. Appl. Environ. Microbiol. 44: 1196–1204.

Ryland, J. S. 1959. Experiments on the selection of algal substrates by polyzoan larvae. J. Exp. Biol. 36: 613–31.

———. 1962. The association between Polyzoa and algal substrata. J. Anim. Ecol. 31: 331–38.

———. 1974. Observations on some epibionts of gulf weed, *Sargassum natans* (L.) Meyen. J. exp. mar. Biol. Ecol. 14: 17–25.

———. 1976. Physiology and ecology of marine bryozoans. Adv. Mar. Biol. 14: 285–443.

Ryland, J. S., and A. R. D. Stebbing. 1971. Settlement and orientated growth in epiphytic and epizoic bryozoans. *In* L. D. J. Crisp, ed. Proc. Fourth European Mar. Biol. Symp. Cambridge University Press, Cambridge, pp. 105–23.

SAS Institute, Inc. 1988. SAS/STAT User's Guide, Release 6. 03 Edition. SAS Institute, Inc., Cary, N.C.

Seed, R. 1976. Observations on the ecology of *Membranipora* (Bryozoa) and a major predator *Doridella steinbergae* (Nudibranchiata) along the fronds of *Laminaria saccharina* at Friday Harbor, Washington. J. Exp. Mar. Biol. Ecol. 24: 1–17.

———. 1986. Ecological patterns in the epifaunal communities of coastal macroalgae. *In* P. G. Moore and R. Seed, eds. The Ecology of Rocky Coasts. Columbia University Press, New York, pp. 22–35.

Seed, R., and R. J. O'Connor. 1981. Community organization in marine algal epifaunas. Ann. Rev. Ecol. Syst. 12: 49–74.

Sieburth, J. M., and J. T. Conover. 1965. *Sargassum* tannin, an antibiotic which retards fouling. Nature 208: 52–53.

Sokal, R. R., and F. J. Rohlf. 1981. Biometry. W. H. Freeman and Co., New York.

Stebbing, A. R. D. 1972. Preferential settlement of a bryozoan and serpulid larvae on the younger parts of *Laminaria* fronds. J. Mar. Biol. Assoc. 52: 765–72.

———. 1973. Competition for space between the epiphytes of *Fucus serratus* L. J. Mar. Biol. Assoc. 53: 247–61.

Steinberg, P. D. 1985. Feeding preferences of *Tegula funebralis* and chemical defenses of marine brown algae. Ecol. Monogr. 55: 333–349.

Strathmann, R. R., and E. S. Branscomb. 1979. Adequacy of cues to favorable sites used by settling larvae of two intertidal barnacles. *In* S. E. Stancyk, ed. Reproductive Ecology of Marine Invertebrates. University of South Carolina Press, Columbia, pp. 77–89.

Strathmann, R.R., Branscomb, E.S., and K. Vedder. 1981. Fatal errors in set as a cost of dispersal and the influence of intertidal flora on set of barnacles. Oecologia 48: 13–18.

Thomas, A. 1988. The effect of *Nereocystis luetkeana* blade growth on the settlement and recruitment of the epiphytic bryozoan *Membranipora membranacea*. Course project for Advanced Invertebrate Zoology, Spring 1988, Friday Harbor Laboratories, Friday Harbor, Washington.

Thorpe, J. P. 1979. A model using deterministic equations to describe some possible parameters affecting growth rate and fecundity in Bryozoa. *In* G. P. Larwood and M. B. Abbott, eds. Advances in Bryozoology. Academic Press, London, pp. 113–20.

Van Alstyne, K. L., and V. J. Paul. 1991. The biogeography of polyphenolic compounds in marine macroalgae: temperate brown algal defenses deter feeding by tropical herbivorous fish. Oecologia 84: 158–63.

Williams, G. B. 1964. The effect of extracts of *Fucus serratus* in promoting the settlement of larvae of *Spirorbis borealis* (Polychaeta). J. Mar. Biol. Assoc. 44: 397–414.

Wisely, B. 1960. Observations on the settling behaviour of larvae of the tube worm *Spirorbis borealis* Daudin (Polychaeta). Aust. J. Mar. Freshwat. Res. 11: 55–72.

Yoshioka, P. M. 1982. Role of planktonic and benthic factors in the population dynamics of the bryozoan *Membranipora membranacea*. Ecology 63: 457–68.

23   Dispersal of Soft-Bottom Benthos: Migration through the Water

Column or through the Sediment?

W. Herbert Wilson, Jr.

ABSTRACT Experiments were performed in a soft-sediment community in False Bay, San Juan Island, Washington, to examine the route of dispersal of the resident organisms. The dominant organisms were (1) the tube-building spionid polychaetes, *Pygospio elegans* and *Pseudopolydora kempi*; (2) a guild of burrowing oligochaetes; and (3) epifaunal ostracodes. Azoic patches of sediment of differing size were prepared, and the recolonization of these patches by adult infauna was monitored. Only the central areas of the patches were sampled. The data indicate that colonization rate into the central area of patches is independent of the total size of the patch. Immigration into azoic patches appears to occur largely by vertical migration rather than by lateral migration; that is, the infauna appear to enter the water column and then descend onto the sediment surface rather than to crawl across or within the sediment surface to reach the unoccupied substratum. These data indicate that dispersal of adult infauna, even tube-dwelling ones, may be far more frequent than previously thought.

Introduction

Invertebrates living in soft-sediment communities are frequently classified into functional groups based on motility (Woodin, 1976; Brenchley, 1981). For instance, many infauna such as spionid and maldanid polychaetes build tubes and are considered sedentary organisms. Others, such as bivalves and thalassinid crustaceans maintain burrows in the sediment and are also considered sedentary. Mobile infauna include infaunal burrowers such as opheliid polychaetes and naticid gastropods and mobile epifaunal species such as ostracodes and many amphipods. The assumption is generally made that sedentary species, both those dwelling in tubes and those dwelling in "permanent" burrows, rarely move after settlement. Infaunal mobile species are presumed to migrate exclusively by lateral dispersal through the sediment while epifaunal species either disperse laterally across the surface of the sediment or may disperse vertically by entering the overlying water column and then dispersing laterally. With present information, infauna are generally considered less mobile than epifauna because it is believed that infauna do not undergo vertical migrations and do not often migrate laterally (but see Smith and Brumsickle, 1989).

This model of adult movement is contraindicated by the observation that many presumed sedentary organisms can be found in the plankton. Such organisms include apodous holothurians (Costello, 1946), infaunal bivalves (Dauer et al., 1982), burrow- and tube-dwelling polychaetes (Dean, 1978a,b; Farke and Berghuis, 1979; Dauer et al., 1980, 1982), and many macrofaunal crustaceans (Dauer et al., 1982). In addition, Wilson (1983) showed that tubicolous spionid polychaetes readily relocate in a density-dependent fashion. When densities were artificially

W. Herbert Wilson, Jr., Department of Biology, Colby College, Waterville, ME 04901.

increased, emigration by adults rapidly returned the densities to preexisting levels. Furthermore, areas in which density was artificially lowered were colonized rapidly, converging on ambient density. It was presumed, but not clearly demonstrated, that migration occurs by lateral migration across the surface of the sediment (Wilson, 1983).

In this chapter, I describe a series of experiments designed to determine the rate of colonization of patches of defaunated sediment. Specifically, the experiments were designed to yield insight into the route of immigration: lateral or vertical. I show that the data are consistent with most immigration occurring by animals that are present in the water column which subsequently resettle on the sediment surface.

Materials and Methods

This research was performed in the summer of 1991, in False Bay, San Juan Island, Washington. (For a detailed map of the area, see Pamatmat [1968].) False Bay is a large embayment, approximately 1 km wide and 1 km broad. Most of the water in the bay empties at low tide.

The particular study site used in this study was located in the upper reaches of False Bay at the +1.0-m tide level. The same area was used in previous research (Wilson, 1981, 1983). The sediment in this portion of the bay is a well-sorted fine sand; virtually all of a sediment sample passes through a 250-μm screen.

The infaunal community at this tidal height is characterized by high density but low diversity. Only six taxa are found regularly. The most abundant taxon is a guild of oligochaetes. No attempt was made to identify these to species; instead, all oligochaetes were counted as a single taxon. The other regularly occurring organisms were (1) the spionid polychaetes, *Pygospio elegans* Claparède and *Pseudopolydora kempi* (Southern); (2) the amphipod crustacean, *Corophium spinicorne* Stimpson; (3) the cumacean crustacean, *Cumella vulgaris* Hart; and (4) a guild of podocopid ostracodes. Both of the spionids maintain vertical tubes which extend to 25 mm below the sediment surface. *Corophium* maintain burrows in the top 50 mm of the sediment. The oligochaetes burrow beneath the sediment surface and do not maintain a "permanent" burrow. The ostracodes scurry across the surface of the sediment. *Cumella* lies buried in the sediment with only its anterior end protruding. Occasionally, however, these cumaceans leave the sediment and swim in the water column, settling down onto the sediment surface at a different location later. Thus, despite the low diversity of organisms in this community, several different functional groups, based on motility, co-occur.

The experimental approach involved creating patches of defaunated sediment in the field and then monitoring recolonization. Sediment was collected from the field in 5-gal buckets and brought to the Friday Harbor Laboratories. All of the supernatant sea water was poured off and replaced with freshwater. The buckets were placed outside the laboratory for at least a week. Prior to commencing an experiment, all of the sediment was sieved to remove any tubes. Visual examination of the sediment indicated that all of the animals had been killed by this procedure.

Matrices of defaunated patches of differing sizes were established as follows. For each matrix, four pairs of wooden dowels were driven into the sediment in a rectangular array of four rows. Each pair of dowels was 3 m apart. Adjacent dowels in either row were 1 m apart. A piece of twine was tautly strung between each pair of dowels. Beginning 75 cm from one dowel, an applicator stick was pushed into the sediment to mark four positions at 50-cm intervals. Each stick marked the center of an experimental area.

Four different cores were used to establish the experimental treatments. The diameters of these cores were: 4.5 cm (designated the A core), 9.5 cm (B core), 16 cm (C core), and 28.5 cm (D core). Within each row, one of the four cores was centered around the applicator stick and pressed into the sediment. Using a trowel and scoopula, the ambient sediment enclosed in the core was removed to a depth of 50 mm. Azoic sediment, prepared in the laboratory, was then used to replace the sediment within the space defined by the core. After the azoic sediment had been introduced, the core was removed, leaving a patch of defaunated sediment with no barriers to lateral immigration through the sediment. The absence of any type of enclosure insured that no hydrodynamic artifacts would be generated (Hulberg and Oliver, 1980). Each matrix was laid out in a Latin square design (Sokal and Rohlf, 1981).

To ensure that the center of each defaunated area could be precisely relocated, I set up two additional pairs of dowels in the study area; these dowels were within 5 m of the 3-week site. I marked four positions between each pair of dowels at 50-cm intervals as described above and then pushed ten applicator sticks (diameter of 2 mm) into the space surrounded by an A core. After 1 week, I returned to these control dowels and determined again the four experimental positions between each pair of dowels and used the A core to sample each position. The number of applicator sticks sampled was noted. I also sampled the area around the A core to search for any applicator sticks that were not recovered when the position was relocated.

Three experimental matrices were used in this study. The first was erected on 30 June, the second on 7 July, and the third on 14 July 1991. All of the matrices were sampled on 21 July, thus providing colonization data after 1, 2, and 3 weeks. Each defaunated area was sampled with a series of concentric cores. For example, a D-core-sized defaunated area was sampled by first removing the center with the A core, then the B core, then the C core; finally all of the sediment remaining was collected as the D core. For A cores, only the A core was used since none of the sediment outside of it had been defaunated. This procedure permits the comparison of equal areas of defaunated areas which differ in the amount of buffer of defaunated sediment surrounding them. If colonization occurs by lateral movements of infauna, one would expect that the A cores from the A-core-sized areas would have more immigrants than the A cores nested within the B, C, and D cores. However, if colonization occurs via the water column, by waterborne adults, then one would expect no difference in the number in each A core, regardless of the amount of defaunated buffer around it. At the time of sampling of the matrices, five cores (0.008 m²) were taken of the ambient community to establish the natural densities within each matrix. All of the samples were returned to the Friday Harbor Laboratories where each was sieved independently through a 500-μm screen. Approximately half of the samples were sieved on 21 July. The remainder were kept in buckets which were placed in sea water tables (water temperature of 10°C) to keep the samples cool. These samples were sieved on the morning of 22 July. The relatively coarse screen was chosen so that recently settled larvae would pass through. The explicit purpose of this experiment was to examine colonization of defaunated sediment by adult or juvenile infauna.

The samples were preserved in 5 percent formalin in sea water. Sorting of the samples was done with a Wild stereomicroscope at 12× magnification. Only heads and whole organisms were counted.

Statistical analyses were performed on both the A cores and the B cores for each of the three experiments. The C cores (only two replicates per treatment for each matrix) and the D cores (only one replicate per treatment for each matrix) were not analyzed because of insufficient replication. Using a one-way analysis of variance (ANOVA) model, the numbers of each of the six common taxa were analyzed for the A cores as a function of the size of the total defaunated area. The B cores were analyzed with a one-way ANOVA in similar fashion. Scheffé post hoc comparisons

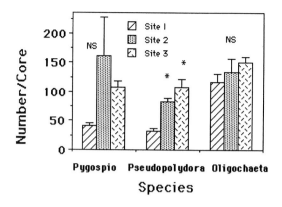

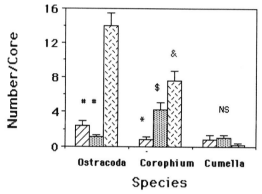

FIGURE 23.1. Comparison of natural densities of the six most common taxa at the three experimental sites. Site 1 is the site of the 1-week experiment, Site 2 the 2-week experiment, and Site 3 the 3-week experiment. Means which share the same symbol are not statistically different ($p > .05$). NS means that there is no statistical difference among any of the three means.

were calculated for each taxon. All of the statistical analyses were done on a Macintosh SE30 using SuperAnova, Version 1.1 (Abacus Concepts).

For both the A and B cores, two-way analyses of variance were performed, testing the effect of position (the size of the total defaunated area) and duration (1, 2, or 3 weeks). In all cases, the interactive term was found to be insignificant. The analyses were then recalculated, removing the interaction term. Comparisons of the average number of immigrants for each taxon were made between the 1-, 2-, and 3-week experiments with Scheffé contrasts.

Results

Figure 23.1 presents a comparison of the ambient densities of the six common taxa at the three experimental sites. Although the three sites were located within 8 m of each other, significant differences in abundance emerged for some taxa. *Pseudopolydora* was more abundant at the Week 2 and Week 3 sites relative to the Week 1 site. Ostracodes were significantly more abundant in the Week 3 site. *Corophium* abundance was significantly least at the Week 1 site, intermediate at the Week 2 site, and greatest at the Week 3 site. *Pygospio*, oligochaetes, and *Cumella* means did not differ among experimental sites.

The results of the control experiment with applicator sticks demonstrated that experimental areas could be precisely located. Of the four groups of ten applicator sticks emplaced in A cores on 14 July, I recovered nine, eight, nine, and ten of the applicator sticks on 21 July. The sticks that

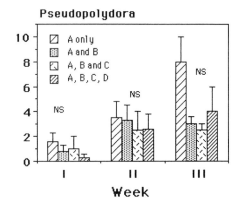

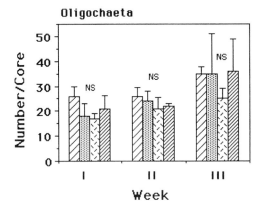

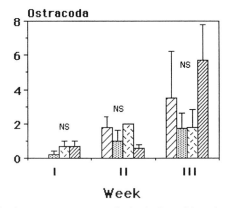

FIGURE 23.2. Results of analysis of the inner (A cores) areas for the four most common taxa in the 1-, 2-, and 3-week experiments. NS means that there is no statistical difference among any of the three means ($p > .05$).

were missed were found to be at the most 4 mm from the outer edge of the A-sized corer. Thus, these data show that I could precisely relocate experimental areas.

Figure 23.2 presents the data for colonization of the central areas, the A cores, for the four most common taxa in the three experiments. *Pygospio* abundance did not differ among the treatments; *Pygospio* density was no greater when the A core was surrounded by the natural community compared to treatments where the A core was encircled by areas of defaunated sediment of three different sizes. These data suggest that colonization of the A cores by *Pygospio* occurred by adults, swept into the water column, which resettled elsewhere. Similar results were found for *Pseudopolydora*, oligochaetes, and ostracodes; the number of these taxa colonizing the A core treatment only were not significantly greater than the remaining treatments with larger areas. Both *Corophium* and *Cumella* showed low abundance in the experimental areas, largely as a function of their relative rarity in the community; the analyses of their experimental abundances will not be presented although they agreed with results for the other taxa.

Figure 23.3 presents the data comparing the abundances of adult colonists among the three different B areas. The A core had been removed from each of these B areas in sampling. There are therefore three treatments, the B ring surrounded by unmanipulated sediment, the B ring surrounded by the C ring, and the B ring surrounded by a C and D ring. For the Week 1 experiment, there were significantly more *Pygospio* in the B only treatment compared to the other two

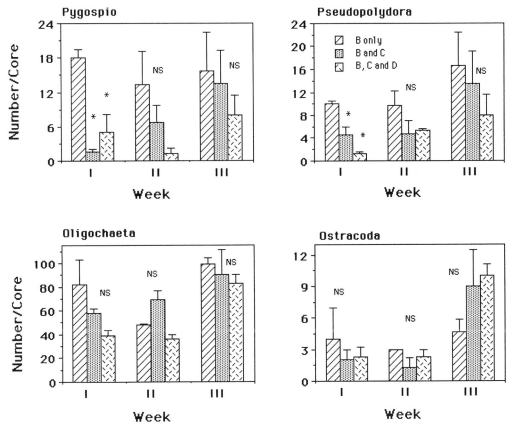

FIGURE 23.3. Results of analysis of the B rings (B core minus the inner A core) for the four major taxa in the 1-, 2-, and 3-week experiments. Means which share the same symbol are not statistically different ($p > .05$) but are different from any mean not sharing that symbol ($p \leq .05$). NS means that there is no statistical difference among any of the three means.

treatments. These data suggest that the presence of buffer zones of defaunated sediment (a C ring or a C and D ring) resulted in decreased adult recruitment; therefore, much of the colonization occurred by lateral migration. However, these patterns were not reflected in the Week 2 and Week 3 experiments where there were no significant differences among treatments. (It is perhaps notable that the nonsignificant trend of the means for the Week 2 and 3 experiments is consistent with immigration occurring laterally rather than vertically.) The same general pattern holds for *Pseudopolydora*. Significantly greater numbers were found in the B only treatment compared to the other two treatments in the Week 1 experiment, which did not differ from each other. However, no statistical differences were found in the Week 2 and Week 3 experiments although, as with *Pygospio*, the trends are in the direction that would be predicted by colonization by crawling laterally from unmanipulated sediment. For the infaunal oligochaetes and the epifaunal ostracodes, no significant differences were found among any of the treatments within each experiment.

Analysis of the pooled data by two-way ANOVA from all three experiments yields the results in Table 23.1. In all cases, the interaction term was nonsignificant, indicating that the two effects were orthogonal. The ANOVAs were recalculated, removing the interaction term from the model, and Scheffé post hoc contrasts were used to compare means among the experiments. The results of these contrasts are given in Figure 23.4. As shown in Fig. 23.4a, there were significantly more

Table 23.1. Results of two-way ANOVAs testing the effect of position (the area of defaunated sediment around the A cores and B cores) and duration (length of the experiment)

	Analysis of A cores		
Species	Position	Duration	Position × duration
Pygospio	*a*	*b*	NS*c*
Pseudopolydora	NS	*b*	NS
Oligochaeta	NS	*b*	NS
Ostracoda	NS	*a*	NS
	Analysis of B cores		
Pygospio	*a*	NS	NS
Pseudopolydora	NS	NS	NS
Oligochaeta	NS	*b*	NS
Ostracoda	NS	*b*	NS

a$p \leq .001$.
b$p \leq .01$.
*c*NS − $p > .05$.

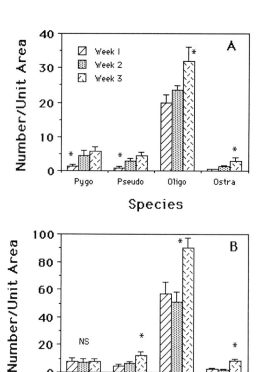

FIGURE 23.4. Comparison of the densities of each of the four major taxa as a function of time. A provides data for the A cores and B gives data for the B rings. Means which share the same symbol are not statistically different ($p > .05$) but are different from any mean not sharing that symbol ($p \leq .05$). NS means that there is no statistical difference among any of the three means ($p > .05$).

Pygospio found in the Week 2 and Week 3 experiments. The same pattern was found for *Pseudopolydora* with more animals found after 2 and 3 weeks compared to the 1-week experiment. For both oligochaetes and ostracodes, there was no difference in density between the 1- and 2-week means, but both were significantly less than the respective means for the 3-week experiment. The data also indicate that the rate of colonization differed among the taxa. For the A cores in the Week 3 experiment, *Pygospio* and *Pseudopolydora* had reached only 26.4 percent and 20.9 percent, respectively, of the natural densities. Oligochaetes reached 40.0 percent of ambient density after 3 weeks while ostracodes reached 43.7 percent of natural density. The same general pattern was obtained for the B areas: *Pygospio* and *Pseudopolydora* abundances within the experimental areas were only 10.2 percent and 16.5 percent, respectively, of ambient density after 3 weeks Higher immigration rates by oligochaetes and ostracodes resulted in densities that were 150 percent and 100 percent of normal densities, respectively, by the end of 3 weeks. Although there were significant differences in the rates of colonization for the different species, there is a strong positive relationship between the number of colonizations and the length of time available for colonization (Fig. 23.4, Table 23.1).

Discussion

Many marine invertebrates display a complex life cycle in which the adult stage is benthic while a larval stage occurs in the plankton. The classic interpretation of such a life cycle (Thorson, 1950) ascribes a dispersal function to the larvae. Such a model works well for sessile, epifaunal organisms where adult dispersal is clearly impossible. For virtually all soft-sediment invertebrates living in a mobile substrate, some movement after settlement is possible. There is a considerable literature, mostly anecdotal in nature, indicating that adults of infaunal invertebrates may be encountered in the plankton (e.g., Costello, 1946; Dean, 1978a,b; Dauer et al., 1982). It is difficult to know from these accounts how frequent such forays into the water column are, although the estimate of a net export of 1.08×10^6 of infaunal invertebrates per day in the Lafayette River, Virginia (Dauer et al., 1982), suggests that the number of waterborne infauna can be appreciable. Recent experimental work with meiofauna has indicated that water currents can remove meiofaunal invertebrates from sediments into the overlying water column where passive dispersal occurs (Palmer, 1984, 1988a; Eskin and Palmer, 1985; Hicks, 1988). Other work on meiofauna has implicated disturbance by predators (Palmer, 1988b) and active dispersal into the water column (Service and Bell, 1987; Armonies, 1988)

The present research is an outgrowth of research that I performed in 1980 at the same site in False Bay. In the earlier work (Wilson, 1983), I placed small cylinders (28 mm in diameter) of insect screening (1 mm mesh) which had been filled with azoic sediment. In the laboratory, I systematically varied the number of *Pygospio* and *Pseudopolydora* in these cores before planting the cores into the sediment in the field. The densities of both species used were one-third normal density, normal density, three times normal density, and absence. Using all possible combinations of density, sixteen treatments were established. Experiments were run for 1, 2, 4, and 8 weeks. After 8 weeks, the densities of both spionids had converged to natural densities, indicating that emigration from the high-density treatments and immigration into the low-density treatments had taken place over the experimental time period. Furthermore, oligochaetes quickly colonized the experimental cores, achieving densities equal to natural densities within 8 weeks. I inferred from these results, with no direct evidence, that lateral movement of these organisms was occurring and suggested that the organisms could assess the local density around their tube or burrow but could not determine the density in other areas. If an infaunal organism had too many neighbors to allow adequate resource acquisition, that organism should move and then reassess the density of organisms in its new neighborhood.

The present experiments were designed to gain insight into this adult dispersal process. It should be emphasized here that of the six common taxa in this community, only one, *Pseudopolydora*, has the "typical" marine invertebrate life cycle of benthic adults and planktonic larvae. In False Bay, *Pygospio* appears to reproduce entirely by asexual fragmentation (Wilson, 1985). The oligochaetes have direct development. The ostracodes have a nauplius larva in their life cycle, but the adults are highly dispersive themselves. Both *Corophium* and *Cumella* are peracarid crustaceans and hence brood their young to the juvenile stage.

The results of these experiments are surprising both in terms of the scale of dispersal and the mechanism of dispersal. For the A cores, the amount of defaunated sediment encircling the A core had no effect on the rate of colonization of the central area in eleven of twelve cases (Fig. 23.2). These data strongly indicate that the major route of colonization is from the overlying water column. If immigration were lateral, one would expect that A cores with less defaunated sediment around them should have significantly higher abundances of infauna than those with greater areas of surrounding defaunated sediment. It is important to note that the time-frame of the experiment was such that recolonization had not resulted in the attainment of natural densities for most taxa. Therefore, one cannot argue that migration had occurred by lateral movements but had occurred so quickly that the predicted effect of decreasing migrants with increasing total defaunated area was masked.

It was not surprising to find that the actively crawling ostracodes quickly colonized the defaunated areas. However, the rapid colonization by the oligochaetes was unexpected. These animals burrow beneath the sediment surface, and hence one would not predict that they would enter the water column frequently. However, the data clearly indicate that the route of colonization was from the water column. It is not possible to state with the present data whether the oligochaetes actively or passively entered the water column.

The data from the B areas (Fig. 23.3) agree with the data from the A cores. In ten of twelve cases, there was no significant difference among treatments, indicating that the primary route of immigration was from the water column rather than lateral migration from adjacent sediment. In the 1-week experiment, both *Pygospio* and *Pseudopolydora* showed higher densities in the B ring compared to the B-C and B-C-D ring, indicating that a significant amount of migration occurred by lateral movement of these spionids. However, these patterns were not seen in the 2- and 3-week experiments. In agreement with the A core analyses, the results from the B core experiments indicated that oligochaetes and ostracodes colonized at a much faster rate than either of the spionids. The data for oligochaetes and ostracodes strongly indicate that the primary route of colonization was from the water column.

The data from this study cannot distinguish between passive erosion of these invertebrates into the water column and active dispersal. My previous results (Wilson, 1983) showed that emigration by these taxa occurred in a density-dependent fashion, implying that these invertebrates actively choose to disperse. A review of the literature for other infaunal communities indicates a diversity of claims about the importance of active versus passive dispersal. Dobbs and Vozarik (1983) showed that storms can increase the number of waterborne infauna. Ambrose (1986) showed that the amphipod, *Rhepoxynius abronius*, emigrated via the water column in a density-dependent fashion. Floating behavior by the tellinid bivalve, *Macoma balthica*, has been described independently by Sörlin (1988) and Beukema and de Vlas (1989). These bivalves are stimulated to float by temperature and water currents.

Additional studies have demonstrated the importance of adult dispersal although it is not clear if emigration occurs in a density-dependent or density-independent fashion. Highsmith (1985) showed that infaunal invertebrates such as the tanaid crustacean, *Leptochelia dubia*, and the venerid bivalve, *Transennella tantillla*, can disperse by floating or algal rafting. Recolonization of the bottom of Tampa Bay after seasonal defaunation occurs in part by adult immigration (Dauer and Simon, 1976; Santos and Simon, 1980). None of these studies provided data on the relative importance of lateral versus vertical (waterborne) immigration.

In summary, it is apparent that the notion of dispersing larvae and sedentary adults fails to hold true for many infaunal invertebrates. Abundant evidence for the meiofauna indicate that waterborne dispersal of adults is common (e.g., Palmer, 1984, 1988a,b). For soft-sediment macrofauna, there is considerable evidence of adult dispersal (e.g., Levinton, 1979; Wilson, 1983; Ambrose, 1986). The present results, which are consistent with a vertical (waterborne) mode of adult dispersal, are surprising in light of the current dogma that tube-dwelling and burrow-dwelling infauna are sedentary (Woodin, 1974, 1976; Brenchley, 1981). It is clear that models of infaunal community structure will have to incorporate adult dispersal before useful predictions can emerge (Wilson, 1990).

Acknowledgments

I am grateful to Director Dennis Willows for making the facilities of the Friday Harbor Laboratories available to me. The manuscript was improved by the insightful comments of Betsy Brown, Steve Stricker, and two anonymous reviewers. I dedicate this chapter to my friend and colleague, Chris Reed, with whom I shared many pleasant times.

Literature Cited

Ambrose Jr., W. G. 1986. Experimental analysis of density dependent emigration of the amphipod *Rhepoxynius abronius*. Mar. Behav. Physiol. 12: 209–216.

Armonies, W. 1988. Hydrodynamic factors affecting behaviour of intertidal meiobenthos. Ophelia 28: 183–193.

Beukema, J. J., and J. de Vlas. 1989. Tidal current transport of thread-drifting postlarval juveniles of the bivalve *Macoma balthica* from the Wadden Sea to the North Sea. Mar. Ecol. Prog. Ser. 52: 193–200.

Brenchley, G. A. 1981. Disturbance and community structure: An experimental study of bioturbation in marine soft-bottom environments. J. Mar. Res. 39: 767–790.

Costello, D. P. 1946. The swimming of *Leptosynapta*. Biol. Bull. 90: 33–36.

Dauer, D. M., R. M. Ewing, J. W. Sourbeer, W. T. Harlan, and T. L. Stokes, Jr. 1982. Nocturnal movements of the macrobenthos of the Lafayette River, Virginia. In. Revue Ges. Hydrobiol. 67: 761–775.

Dauer, D. M., R. M. Ewing, G. H. Tourtellotte, and H. R. Barker, Jr. 1980. Nocturnal swimming of *Scolecolepides viridis* (Polychaeta: Spionidae). Estuaries 3: 148–149.

Dauer, D. M. and J. L. Simon. 1976. Repopulation of the polychaete fauna of an intertidal habitat following natural defaunation: Species equilibrium. Oecologia 22: 99–117.

Dean, D. 1978a. Migration of the sandworm *Nereis virens* during winter nights. Mar. Biol. 45: 165–173.

———. 1978b. The swimming of bloodworms (*Glycera* spp.) at night, with comments on other species. Mar. Biol. 48: 99–104.

Dobbs, F. C., and J. M. Vozarik. 1983. Immediate effects of a storm on coastal infauna. Mar. Ecol. Prog. Ser. 11: 273–279.

Eskin, R. A., and M. A. Palmer. 1985. Suspension of marine nematodes in a turbulent tidal creek: Species patterns. Biol. Bull. 169: 615–623.

Farke, H., and E. M. Berghuis. 1979. Spawning, larval development and migration of *Arenicola marina* under field conditions in the western Wadden Sea. Nether. J. Sea Res. 13: 529–535.

Hicks, G. R. F. 1988. Sediment rafting: A novel mechanism for the small-scale dispersal of intertidal estuarine meiofauna. Mar. Ecol. Prog. Ser. 48: 69–80.

Highsmith, R. C. 1985. Floating and algal rafting as potential dispersal mechanisms in brooding invertebrates. Mar. Ecol. Prog. Ser. 25: 169–179.

Hulberg, L. W., and J. S. Oliver. 1980. Caging manipulations in marine soft-bottom communities: importance of animal interactions or sedimentary habitat modifications? Can. J. Fish. Aquat. Sci. 37: 1130–1139.

Levinton, J. S. 1979. The effect of density upon deposit-feeding populations: movement, feeding and floating of *Hydrobia ventrosa* Montagu (Gastropoda; Prosobranchia). Oecologia 43: 27–39.

Palmer, M. A. 1984. Invertebrate drift: Behavioral experiments with intertidal meiobenthos. Mar. Behav. Physiol. 10: 235–253.

———. 1988a. Dispersal of marine meiofauna: A review and conceptual model explaining passive transport and active emergence with implications for recruitment. Mar. Ecol. Prog. Ser. 48: 81–91.

———. 1985. Suspension of marine nematodes in a turbulent tidal creek: Species patterns. Biol. Bull. 169: 615–623.

Pamatmat, M. M. 1968. Ecology and metabolism of a benthic community on an intertidal sandflat. In. Revue Ges. Hydrobiol. 53: 211–298.

Santos, S. L., and J. L. Simon. 1980. Marine soft-bottom community establishment following annual defaunation: Larval or adult recruitment? Mar. Ecol. Prog. Ser. 2: 235–241.

Service, S. K., and S. S. Bell. 1987. Density-influenced active dispersal of harpacticoid copepods. J. Exp. Mar. Biol. Ecol. 114: 49–62.

Smith, C. R., and S. J. Brumsickle. 1989. The effects of patch size and substrate isolation on colonization modes and rates in an intertidal sediment. Limnol. Oceanogr. 34: 1263–1277.

Sokal, R. R., and F. J. Rohlf. 1981. Biometry. W. H. Freeman, New York.

Sörlin, T. 1988. Floating behaviour in the tellinid bivalve *Macoma balthica* (L.). Oecologia 77: 273–277.

Thorson, G. 1950. Reproductive and larval ecology of marine bottom invertebrates. Biol. Rev. 25: 1–45.

Wilson, W. H. Jr. 1981. Sediment-mediated interactions in a densely populated infaunal assemblage: The effects of the polychaete *Abarenicola pacifica*. J. Mar. Res. 39: 735–748.

———. 1983. The role of density dependence in a marine infaunal community. Ecology 64: 295–306.

———. 1985. Food limitation of asexual reproduction in a spionid polychaete. Intern. J. Invert. Reprod. Dev. 8: 61–65.

———. 1990. Competition and predation in marine soft-sediment communities. Ann. Rev. Ecol. Syst. 21: 221–241.

Woodin, S. A. 1974. Polychaete abundance patterns in a marine soft-sediment environment: The importance of biological interactions. Ecol. Monogr. 44: 171–187.

———. 1976. Adult-larval interactions in dense infaunal assemblages: Patterns of abundance. J. Mar. Res. 34: 25–41.

24   Seasonal Swimming of Sexually Mature Benthic Opisthobranch Molluscs (*Melibe leonina* and *Gastropteron pacificum*) May Augment Population Dispersal

Claudia E. Mills

ABSTRACT Although adults of the opisthobranch gastropods *Melibe leonina* and *Gastropteron pacificum* are primarily benthic crawlers, they are also capable swimmers. Long-term observations (13–15 years) of plankton in surface waters at Friday Harbor, Washington, reveal that swimming of adult animals in these species is highly seasonal. For *M. leonina*, which appears to have no permanent populations within at least 2 km of the study site, nearly all swimming adults were observed in surface waters between September and March. Most swimming *G. pacificum*, with a resident population on the sloping mud-sand bottom 5–18 m immediately below the observation site, were also seen near the surface between September and February. Swimming individuals from both species were sexually mature, or nearly so, as shown by their ability to reproduce in the laboratory after collection. Although both species have planktonic veliger larvae, commonly considered to be their chief agents for dispersal, these observations suggest the additional importance of seasonally swimming adults in achieving population movements.

Introduction

Dispersal of benthic marine invertebrates can be accomplished by at least three common routes: planktonic larvae or egg capsules, rafting of either egg capsules or juveniles or adults attached to movable substrata, and benthic migration of juveniles or adults. In this chapter, I will discuss a fourth, less common route for population dispersal—intermittent seasonal swimming by reproductively capable adult animals.

Although movement through the water column is a seemingly unlikely mechanism for transport of benthic species, the literature reveals an increasing number of reports that such movement may be an important source of dispersal for postsettlement benthic animals in many phyla (see also ch. 23 of this volume, by W. H. Wilson, Jr.). The list of waterborne benthic adults includes many representatives living in quite different habitats, from infaunal leptosynaptid sea cucumbers (Costello, 1946), bivalves (Highsmith, 1985; Sörlin, 1988), crustaceans (Highsmith, 1985), and polychaetes (Wilson, this volume), to small shelled gastropods and bivalves in the rocky intertidal (Martel and Chia, 1991), to ascidians living attached to eelgrass blades (Worcester, 1992).

In this chapter, I present new data on seasonal swimming by two opisthobranch molluscs

Claudia E. Mills, Friday Harbor Laboratories, University of Washington, 620 University Road, Friday Harbor, WA 98250.

already known to be good periodic swimmers. From many years of observation of the surface plankton, it has become apparent that both *Melibe leonina* (Gould, 1852) and *Gastropteron pacificum* Bergh, 1893, are frequently seen swimming near the surface in the colder months, while only sometimes appearing in the plankton at other times of the year. I have not explored the reason for such seasonality in prolonged swimming, but want to emphasize the strong likelihood that these swimming events are important for dispersing reproductively capable adult animals, which can then spawn after arrival in new locations.

Melibe leonina (Nudibranchia, Dendronotacea) occurs on the west coast of North America from Kodiak Island, Alaska, to the Gulf of California (MacFarland, 1966). In Washington State, *M. leonina* lives primarily on *Zostera marina* eelgrass growing in shallow, protected embayments, where it feeds on small zooplankton closely associated with the eelgrass (Hurst, 1968). Although usually attached to eelgrass blades by its long slender foot, *M. leonina* swims well by side-to-side flexure of the entire body. Animals may swim between grass blades, but swimming in this species is generally considered to be an escape response. *M. leonina* can be easily stimulated to swim by various kinds of physical disturbance (Hurst, 1968) and also swims in response to pinching of the cerata by predatory crabs (Bickell-Page, 1989).

Gastropteron pacificum (Cephalaspidea) has been found from the Aleutian Islands, Alaska, to the Gulf of California (MacFarland, 1966). It is a small infaunal snail with bilateral extensions of its foot (parapodia) that are folded up and over the visceral mass while it creeps in the mud using cilia on the foot. The diet of *G. pacificum* is benthic, especially diatoms and foraminiferans (Reinhart, 1967). *G. pacificum* is also known to be a periodic and active swimmer; the broad, rounded parapodia can be flapped like two large wings for very effective swimming. Swimming by *G. pacificum* is not easily elicited by physical disturbance in the laboratory, but can be predictably stimulated by contact with the predatory cephalaspidean mollusc, *Chelidonura phocae* (Reinhart, 1967).

Materials and Methods

During the course of regular observations and collections of medusae and ctenophores from the floating docks at the Friday Harbor Laboratories (FHL) in the San Juan Archipelago of Washington State, I have observed numerous individuals of *Melibe leonina* and *Gastropteron pacificum* in the surface plankton. These observations have been logged and tabulated from 1979 to 1993 for *M. leonina* and from 1977 to 1993 for *G. pacificum* (Fig. 24.1).

Dock observations of plankton were usually made several hundred times per year during the nearly 15 years reported here; a typical observation period consisted of slowly strolling along the approximately 100 m of the floating docks for 10–30 min at a time. Figure 24.2 represents a rough approximation of the total number of hours spent searching; estimates for the summer months are probably too low by at least one-half, but represent time actually recorded in my notebooks. Observations were fairly evenly spaced throughout each year, except for occasional lapses of 1–5 weeks and from October 1981 to February 1982, January to June 1986 and March to September 1988, all due to my absence. Observations were made in the day as well as night, although night observations were most numerous during the first 5 years of the study.

Animals collected in the surface waters during late autumn and winter were sometimes brought into the laboratory for observations of general healthiness and spawning. These animals were kept, often for months, in a running sea water table at ambient field temperatures (about 9–12°C).

A small number of *Gastropteron pacificum* were also either collected by plankton net in

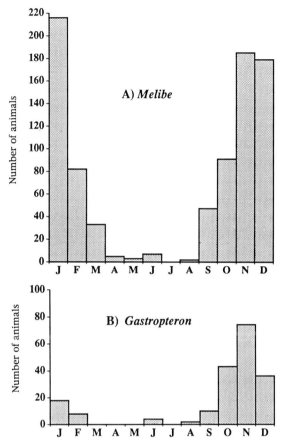

FIGURE 24.1. Total number of individuals seen in the surface plankton in Friday Harbor, Washington, plotted by month of the year. *A. Melibe leonina*, 1979–1993; *B. Gastropteron pacificum*, 1977–1993.

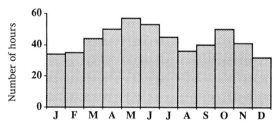

FIGURE 24.2. Approximate total number of hours spent 1977–1993 observing plankton from the floating docks at the Friday Harbor Laboratories to arrive at the numbers presented in Figure 24.1.

nearby Saanich Inlet, British Columbia (1979–80) or seen from the windows of the manned submersible *Pisces IV* in Saanich Inlet or the Strait of Georgia, British Columbia (1980, 1982). A single *Melibe leonina* was seen at the surface in the center of Saanich Inlet, at least 3 km from shore, in November 1980 while I was preparing to make a submersible dive.

Having spoken to numerous marine biologists who work in the vicinity, and made a number of cursory searches myself over the years, I am not aware of any resident populations of *Melibe leonina* existing within Friday Harbor during the course of this study. In late August 1992 I made a thorough 3-day search throughout Friday Harbor for *M. leonina*, using a kayak during a low-tide series. No *M. leonina* were seen, nor were any of its egg masses in evidence on eelgrass. In a bay having resident *M. leonina*, such a search from the surface at low tide should have revealed its presence. It is noteworthy that swimming *M. leonina* began appearing in the plankton in Friday Harbor within 3 weeks of this 1992 search, before any autumn storms had strongly disrupted the region.

Results

Melibe leonina

Although there are no known *Melibe leonina* populations within at least 2 km of the FHL floating docks, 850 animals were observed from the docks either floating or swimming freely in surface waters, or in a few cases attached to drift algae or eelgrass (Fig. 24.1). These "swimmers" were highly seasonal, with 833 (98 percent) of the individuals collected between September and March. Over half of the total number of animals were collected in the autumn-winter 1992–93, which appears to have been an unusually good year for *M. leonina* in this area. Pelagic specimens ranged from approximately 3 to 12 cm in length from expanded hood to tip of the tail; the full range of sizes occurred throughout the swimming season.

Typically, only a few *Melibe leonina* (usually less than 5) were seen swimming or floating near the surface at any given time. Large numbers of *M. leonina* were seen near the docks on ten dates: 17 and 31 December 1987; 30 September 1992; 9, 12, and 24 November 1992; 6 December 1992; and 2, 4, and 5 January 1993. A "large" number of *M. leonina* adrift near the docks would be on the order of fifty to a few hundred (not thousands of) individuals visible over 1–2 hours. To normalize collection times, since invariably I spent more time looking if many were present, only twenty *M. leonina* were recorded on each of these dates.

Most of the pelagic *Melibe leonina* were seen on autumn and winter days when the semidiurnal tide resulted in two high tides in a row, with very little drop between them. On such days, the tide flows in from the south nearly all day, sometimes bringing in oceanic water, as evidenced by the exotic ctenophore fauna it may also carry. When *M. leonina* appeared on such occasions, it was often accompanied by broken pieces of eelgrass or (only during the autumn) the hydromedusa *Polyorchis penicillatus*, which is restricted in habitat to shallow protected bays. It appears that at least some shallow bays may be flushed by such tides.

Melibe leonina is reproductive all year; egg masses have been reported in the field in every month (Hurst, 1967). Swimming and floating animals collected in Friday Harbor surface waters and brought into the laboratory lived for several months and regularly produced viable egg masses in a running sea water table.

Gastropteron pacificum

A resident population of *Gastropteron pacificum* exists 5–18 m below the FHL floats on a sloping mud-sand bottom (Reinhart, 1967). Individuals seen in surface waters at the FHL floats seem likely to have swum up from this population. It is interesting to note, however, that swimming by *G. pacificum* appears to be as seasonal as swimming by *Melibe leonina* in this region and that most swimming *G. pacificum* were also collected on days with double high tides. Of 195 *G. pacificum* seen swimming in surface waters during the collection period, 189 (97 percent) were seen between September and February (Fig. 24.1). On only four dates were more than 4 *G. pacificum* recorded; 6 were seen on 8 October 1991; 5 on 27 November 1991; 6 on 11 November 1992; and 9 on 24 November 1992. On most observation dates, only one swimming individual was recorded.

Autumn–winter 1992–93 was a good year for observations of swimming *G. pacificum*, producing approximately one-fourth of all sightings. Although I report only a few swimmers during the summer (these were juveniles), more juvenile *G. pacificum* can be found swimming in the summer months if the observer devotes hours rather than minutes per day to the collection effort (Reinhart, 1967; R. A. Satterlie, pers. comm.). Autumn and winter swimming *G. pacificum* were usually 9–15 mm long.

About twenty *Gastropteron pacificum* individuals were brought into the laboratory and placed in a silt-covered running sea water table during October–December 1990 and 1991. Two fertilized egg masses were seen in this laboratory water table in February and March 1991; none was seen in spring 1992. In the field, *G. pacificum* lays eggs throughout the year, but they may be most numerous from January through June (Hurst, 1967; Reinhart, 1967).

Gastropteron pacificum was also found high in the water column above a frequently anoxic 180–200 m bottom about 2 km from shore during the winter in Saanich Inlet, indicating that its swimming capabilities are adequate for significant vertical or, more likely, horizontal transport. Single specimens were taken in vertical plankton tows from 25 to 50 m on 13 November 1979, and from 0 to 25 m on 4 November 1980 (replicated in two tows) above a 200-m bottom. *G. pacificum* was not taken in the upper water layers in any other of nearly 600 stratified plankton tows taken bimonthly over a 2-year period in Saanich Inlet. Ten *G. pacificum* were seen from the *Pisces IV* submersible from 16 to 28 m over a 180-m bottom in Saanich Inlet on 6 November 1980, and one was seen at 84 m above a 384-m bottom in the Strait of Georgia, south of Texada Island, British Columbia, again about 2 km from shore.

Discussion

Tidal currents are relatively strong in the San Juan Archipelago. In the area near the study site, current speeds are frequently in the 1–5-km/h range, with occasional currents approaching 8 km/h (U.S. National Ocean Service, 1991). Such currents could well transport swimming opisthobranchs significant distances.

Melibe leonina and *Gastropteron pacificum* exhibit most of their sustained swimming in the upper water column at the same time of the year, but the dispersive consequences of swimming may be different for the two species. *M. leonina* seen from the FHL floating docks have already moved on the scale of kilometers, because there are no known populations nearby. A few of the traveling *M. leonina* were attached to drift algae or eelgrass, but the resulting transport should be about the same as for those swimming solo. It may be speculated that many of the swimming *M. leonina* have been torn from their senescing eelgrass beds in winter storms, but animals are also frequently seen swimming in the plankton during long periods of quiescent autumn and winter weather. Nothing in the literature suggests any seasonal behavior for *M. leonina* merely letting go of their eelgrass in the winter, yet the large number of free individuals suggests that such could be the case (see also the report of Ajeska and Nybakken [1976] below).

Gastropteron pacificum seen swimming off the FHL floating docks are likely to have come from the population immediately below. *G. pacificum* swimming appears to consist mostly of bouts of vertical swimming up from the bottom. Although *G. pacificum* individuals are not very buoyant, and therefore may fall rapidly out of the water column when they stop swimming, they are still likely to be transported at least as far in a single swimming bout in the presence of tidal currents as they might crawl in a day. Reinhart (1967) reports *G. pacificum* swimming continuously in the laboratory for up to $5^{1}/_{2}$ h after collection, so prolonged presence in the plankton with significant transport is certainly possible and perhaps quite common for this species. The correlation of swimming *G. pacificum* with double high tides in the autumn and winter is intriguing, but, so far, unexplained.

Dispersal of both species of opisthobranchs reported here is usually considered to be accomplished by their planktonic larvae. *Melibe leonina* has planktotrophic veliger larvae that are capable of settlement and metamorphosis 30–48 days after hatching (Bickell and Kempf, 1983). *Gastropteron pacificum* also has planktotrophic veliger larvae (Hurst, 1967), and preliminary studies imply that they probably also spend about 4–6 weeks in the plankton (G. D. Gibson, pers. comm.). Displacement of larvae during several weeks in the plankton seems likely to be an important means of dispersal for these species, but one must now recognize the further ability of reproductive adults to be moved during extended swimming bouts in the water column.

Dispersal by swimming or floating adults is thought to be an important source of general population transport for another species of *Melibe*. Large numbers of *Melibe fimbriata*, an Indian Ocean species, were recently reported for the first time in the Mediterranean Sea near Greece, where many swimming and floating adult animals were aggregated in Astakos Inlet of the Ionian Sea (Thompson and Crampton, 1984). It is assumed that *M. fimbriata* entered the Mediterranean Sea through the Suez Canal, but it is noteworthy that the first sighting was of large numbers of adult animals many hundreds of kilometers west of this presumed point of entrance.

Further evidence of mass movement of adults is provided by Ajeska and Nybakken (1976), who studied a large *Melibe leonina* population (an estimated 20,000 individuals) in a kelp bed in California for several months in 1970–71. The entire population, apparently healthy from August onward, disappeared suddenly within 8 days of the last observation in November. Two months later, a smaller population (2000 individuals) was reestablished at this site. According to the authors, *M. leonina* populations in California are typified by periodic rapid appearances and disappearances at given locations.

Populations of *Melibe leonina* in northwest Washington State and southwest British Columbia have also been observed to come and go, although on a less abrupt schedule (G. D. Gibson and L. R. Page, pers, comm.). Agersborg (1923) cites Trevor Kincaid's 1917 observation in Hood Canal, Washington, of a bay filled with "millions" of *M. leonina*. These animals even covered pilings under a dock, an unlikely habitat for this species, further emphasizing, in my view, their recent influx into the area. Near-shore plankton tows taken in late winter in the Strait of Georgia, Washington, included numerous planktonic *M. leonina* whose stomachs were filled with herring larvae (C. M. Eaton, pers. comm.). One might extrapolate that swimming of adults is an important dispersal mechanism of this species in both Washington and California.

As a final note, only one other species of benthic gastropod was seen in the plankton off the FHL floating docks during this study. Eight times between 1990 and 1993, (seven between October and January), *Dendronotus* spp. (probably all *D. iris* Cooper, 1863) nudibranchs were seen swimming at the sea surface (one was rafting on a plastic bag [W. H. Watson, pers. comm.]). On five of these occasions, *Dendronotus* was collected together with *M. leonina. D. iris* feeds on cerianthid anemones on muddy bottoms, but it is also known to be a strong and capable swimmer (Agersborg, 1922; Robilliard, 1970). No pattern should be deduced from the small number of *Dendronotus* observed, although individuals were sometimes seen at the surface for more than 30 min., again offering the opportunity for reproductive adults to travel substantial distances by means of tidal currents.

Altogether, these data appear to support the hypothesis that transport of reproductive adults by swimming up into the water column provides a second mechanism for population dispersal of some opisthobranch, in addition to the more familiar migrations by planktonic veliger larvae. I have not attempted to evaluate the contribution to recruitment afforded by this mechanism. A forthcoming study by S. E. Worcester (pers. comm.) makes a much more thorough attempt to quantify the importance of such adult transport in an ascidian population in California, and in fact, preliminary results (Worcester, 1992) indicate that in this case the seemingly unlikely rafting by adult tunicates may be more important than larval dispersal in colonizing new habitats.

Acknowledgments

I am grateful to the staff and scientists at the Friday Harbor Laboratories for use of the facilities and intellectual companionship over many years. Thanks to Tom Schroeder, Louise Page, Glenys Gibson, Alan Kohn, Richard Strathmann, Jon Havenhand, Richard Palmer, and Herb Wilson for comments on this manuscript and discussions about the animals involved.

Literature Cited

Agersborg, H. P. 1922. Notes on the locomotion of the nudibranchiate mollusk *Dendronotus giganteus* O'Donoghue. Biol. Bull. 47: 257–266.

———. 1923. A critique on Professor Harold Heath's *Chioraera dalli*, with special reference to the use of the foot in the nudibranchiate mollusk, *Melibe leonina* Gould. Nautilus 36: 86–96.

Ajeska, R. A., and J. Nybakken. 1976. Contributions to the biology of *Melibe leonina* (Gould, 1852) (Mollusca: Opisthobranchia). Veliger 19: 19–26.

Bickell, L. R., and S. C. Kempf. 1983. Larval and metamorphic morphogenesis in the nudibranch *Melibe leonina* (Mollusca: Opisthobranchia). Biol. Bull. 165: 119–138.

Bickell-Page, L. R. 1989. Autotomy of cerata by the nudibranch *Melibe leonina* (Mollusca): ultrastructure of the autotomy plane and neural correlate of the behaviour. Philos. Trans. Roy. Soc. Lond. B. 324: 149–172.

Costello, D. P. 1946. The swimming of *Leptosynapta*. Biol. Bull. 90: 93–96.

Highsmith, R. C. 1985. Floating and algal rafting as potential dispersal mechanisms in brooding invertebrates. Mar. Ecol. Prog. Ser. 25: 169–179.

Hurst, A. 1967. The egg masses and veligers of thirty Northeast Pacific opisthobranchs. Veliger 9: 255–288.

———. 1968. The feeding mechanism and behaviour of the opisthobranch *Melibe leonina*. Symp. Zool. Soc. Lond. 22: 151–166.

MacFarland, F. M. 1966. Studies of opisthobranchiate mollusks of the Pacific coast of North America. Mem. Cal. Acad. Sci. 6: 1–546.

Martel, A., and F.-S. Chia. 1991. Drifting and dispersal of small bivalves and gastropods with direct development. J. Exp. Mar. Biol. Ecol. 150: 131–147.

Reinhart, J. M. 1967. The swimming behavior of *Gastropteron pacificum*. Master's thesis. University of Washington, Seattle.

Robilliard, G. A. 1970. The systematics and some aspects of the ecology of the genus *Dendronotus*. Veliger 12: 433–479.

Sörlin, T. 1988. Floating behavior in the tellinid bivalve *Macoma balthica* (L.). Oecologia 77: 273–277.

Thompson, T. E., and D. M. Crampton. 1984. Biology of *Melibe fimbriata*, a conspicuous opisthobranch mollusc of the Indian Ocean, which has now invaded the Mediterranean Sea. J. Moll. Stud. 50: 113–121.

U.S. National Ocean Service. 1991. Tidal Current Tables 1992: Pacific Coast of North America and Asia North. U. S. Department of Commerce, Coast and Geodetic Survey, Washington, D. C.

Wilson, W. H. This volume. Dispersal of soft-bottom benthos: Migration through the water column or through the sediment? Ch. 23.

Worcester, S. E. 1992. Adult rafting and larval swimming: Does dispersal mode affect the recruitment of a colonial ascidian into new habitats? Am. Zool. 32: 122A.

Index